Climate Change Modeling Methodology

This volume collects selected topical entries from the *Encyclopedia of Sustainability Science and Technology* (ESST). ESST addresses the grand challenges for science and engineering today. It provides unprecedented, peer-reviewed coverage of sustainability science and technology with contributions from nearly 1,000 of the world's leading scientists and engineers, who write on more than 600 separate topics in 38 sections. ESST establishes a foundation for the research, engineering, and economics supporting the many sustainability and policy evaluations being performed in institutions worldwide.

Philip J. Rasch
Editor

Climate Change Modeling Methodology

Selected Entries from the Encyclopedia of Sustainability Science and Technology

Springer

Editor
Philip J. Rasch
Pacific Northwest National Laboratory
Richland, WA, USA

This book consists of selections from the Encyclopedia of Sustainability Science and Technology edited by Robert A. Meyers, originally published by Springer Science+Business Media New York in 2012.

ISBN 978-1-4614-5766-4 ISBN 978-1-4614-5767-1 (eBook)
DOI 10.1007/978-1-4614-5767-1
Springer New York Heidelberg Dordrecht London

Library of Congress Control Number: 2012954280

Printed on acid-free paper

Springer is part of Springer Science+Business Media (www.springer.com)

Contents

Chapter 1
Climate Change Modeling Methodology, Introduction

Philip J. Rasch

In the early 1980s, Eugene F. Stoermer began informally using the term "anthropocene" to refer to a time in the history of the Earth in which humans introduce impacts that at times rival natural processes in the global functioning of the planet [7]. While there is some uncertainty about when the anthropocene began (with some dating it as far back as 8,000 years ago), there is little doubt that humans are affecting the planet today, and one of the ways is through our impact on the Earth's climate.

Concern about the possible impact of mankind on climate dates back at least as far as 1896, when Svante Arrhenius [1] suggested that mankind might inadvertently warm the planet by increasing the concentration of carbon dioxide, and the idea that man might have an impact on the Earth's climate has been studied ever since. Concern about the matter has been increasing since the 1970s, and in 1995, a very large group of climate scientists, in an assessment produced by the Intergovernmental Panel on Climate Change [2], wrote that "*the balance of evidence suggests a discernible human influence on global climate*." That statement, based on a careful evaluation of observational evidence, theory, and model studies, ignited a firestorm of comment and controversy. There is little doubt that the vast majority of climate scientists were concerned then, and study since that time has made the arguments that man is affecting climate even more compelling. In February of 2007, the IPCC released its fourth assessment [3], which found that "*human actions are 'very likely' the cause of global warming,*" with "very likely" specifically defined to mean a 90% or greater probability of explaining an increase in average global temperatures over the last 100 years, with many other attendant changes. Some debate continues about these conclusions to this day, with a few scientists and more nonscientists

This chapter was originally published as part of the Encyclopedia of Sustainability Science and Technology edited by Robert A. Meyers. DOI:10.1007/978-1-4419-0851-3

P.J. Rasch (✉)
Pacific Northwest National Laboratory, P.O. Box 999, Richland, WA 99352, USA
e-mail: philip.rasch@pnl.gov

P.J. Rasch (ed.), *Climate Change Modeling Methodology: Selected Entries from the Encyclopedia of Sustainability Science and Technology*,
DOI 10.1007/978-1-4614-5767-1_1,

indicating doubt that mankind is responsible for the changes, but almost no doubt that the warming is real.

There are many reasons for skepticism about climate change: some scientific, some not. It is not my intention that this section provides much information about climate change itself. The assessment report, the "synthesis document," and the "summaries" intended for policymakers, laypersons, and scientists from other fields [2–5] provide much more detail than could possibly be included here.

Instead, these entries are designed to introduce readers to the "methodologies" used in climate science. What are the tools used for? How is confidence gained in the use of those tools? It is impossible to provide a thorough review with the few entries, and few pages of each entry in this section, but it is hoped that they provide readers with an introduction to the science, and methodology, and pointers to places where more detail can be found.

Climate science is a complex field. Climate is governed by processes that interact and operate on a vast array of time and space scales. The processes involving radiative transfer, chemistry, and phase changes of water are most easily described at atomic and molecular scales; the influence of ice sheets, continents, and planetary scale circulations controlling the basic energy balance of the planet operate at continental scales; even planetary orbital and solar variations operating at millennial time scales cannot be ignored. Many of those processes are well understood; other less so.

Models are the means that scientists use to express their knowledge about a process, or a set of processes using equations that can be solved on a computer. It is impossible to represent everything known about all the processes that govern climate in detail on today's computers because it is simply beyond the computer power currently available to us, and approximations must be made. There are also many things scientists recognize that are "not known" that may be important in understanding climate change. It is hoped that these entries will provide hints on how climate models are constructed, how they are assessed and tested, and how they are used to provide insight into the changing climate of the past, and the possible changes that will take place in the future.

Here is a brief description of the entries of this section, and the rationale for their inclusion.

A broad overview of coupled climate system modeling is provided in Coupled Climate and Earth System Models by Gent. This entry introduces the components of a coupled model, how they work together, and the kind of scientific problems that they are designed to address.

The individual component models are described in more detail in the next set of entries. Each of these component models can be used on its own to study a piece of the climate system by essentially "prescribing" the behavior of the other components, or they can be used as a part of the bigger model to explore interactions. Bitz and Marshall describe treatments of the cryosphere (ice and snow in the Earth System) in the entry Cryosphere, Modeling of; Yoon and Ma cover ocean processes in Oceanic General Circulation Models. Bacmeister discusses global atmospheric models from a weather modeling point of view in

Weather Prediction Models, and Rasch discusses these models from a climate perspective in the entry Atmospheric General Circulation Modeling. Niu and Zeng provide an introduction to land surface models in Earth System Model, Modeling the Land Component of. These models provide scientists with a picture of how physical components of the climate system respond to "forcing" from other components.

It is also interesting to explore the interaction between the physical system and society. Integrated Assessment Modeling by Edmonds et al. describes "Integrated Assessment" modeling and climate change that begin to assess how humans affect climate and how climate affects humans.

Some aspects of climate change are most easily explored at a very high resolution with more detail than can be afforded on today's computer systems. This motivates the entry on Regional Climate Models by Leung. Methods used to assess uncertainties in climate change studies are outlined in Climate Change Projections: Characterizing Uncertainty Using Climate Models by Sanderson and Knutti.

Climate varies on many time scales. The entry by Goddard on Climate Predictions, Seasonal-to-Decadal provides a very nice introduction to the issues of prediction of climate processes on multi-decadal time scales. The climate section concludes with an example by Wang and Lau of one way that the mankind may be influencing a very important feature of the Earth system that influences the lives of billions of people, the Indo-Asian monsoon (Monsoon Systems, Modeling of).

It is hoped that scientists and engineers find the entries interesting, and that they stimulate additional study in this important topic in Sustainability Science.

Bibliography

1. Arrhennius S (1896) On the influence of carbonic acid in the air upon the temperature of the ground. Philos Mag 41:237–276
2. IPCC (1996) Climate change 1995: the science of climate change. In: Houghton JT, Meira Filho LG, Callander BA, Harris N, Kattenberg A, Maskell K (eds) Contribution of working group I to the second assessment report of the intergovernmental panel on climate change. Cambridge University Press, Cambridge. ISBN 0-521-56433-6 (pb: 0-521-56436-0)
3. IPCC (2007-SRM) Climate change 2007: synthesis report. In: Core Writing Team, Pachauri RK, Reisinger A (eds) Contribution of working groups I, II and III to the fourth assessment report of the intergovernmental panel on climate change. IPCC, Geneva, 104 pp
4. IPCC, 2007-WG1: Climate change 2007: The physical science basis. In: Solomon S, Qin D, Manning M, Chen Z, Marquis M, Averyt KB, Tignor M, Miller HL (eds) Contribution of working group I to the fourth assessment report of the intergovernmental panel on climate change. Cambridge University Press, Cambridge/New York
5. IPCC 2007-WGII, Barker T, Bashmakov I, Bernstein L, Bogner JE, Bosch PR, Dave R, Davidson OR, Fisher BS, Gupta S, Halsnæs K, Heij GJ, Kahn Ribeiro S, Kobayashi S, Levine MD, Martino DL, Masera O, Metz B, Meyer LA, Nabuurs G-J, Najam A, Nakicenovic N, Rogner H-H, Roy J, Sathaye J, Schock R, Shukla P, Sims REH, Smith P, Tirpak DA, Urge-Vorsatz D, Zhou D (2007) Technical summary. IPCC-WGIII, climate change 2007: mitigation. In: Metz B, Davidson OR, Bosch PR, Dave R, Meyer LR (eds) Contribution of working group

III to the fourth assessment report of the intergovernmental panel on climate change. Cambridge University Press, Cambridge/New York
6. Parry ML, Canziani OF, Palutikof JP, Co-authors (2007) Climate change 2007: impacts, adaptation and vulnerability. In: Parry ML, Canziani OF, Palutikof JP, van der Linden PJ, Hanson CE (eds) Contribution of working group II to the fourth assessment report of the intergovernmental panel on climate change. Cambridge University Press, Cambridge, pp 23–78
7. Steffen W, Grinevald J, Crutzen P, McNeil J (2011) The anthropocene: conceptual and historical perspectives. Philos Trans R Soc 369:842–867. doi:10.1098/rsta.2010.0327

Chapter 2
Coupled Climate and Earth System Models

Peter R. Gent

Glossary

Climate model	A numerical model consisting of four components: atmosphere, ocean, land, and sea ice.
Earth system model	A climate model with additional components, which must include a carbon cycle in the land, atmosphere, and ocean components.
Troposphere	The lower part of the atmosphere where most of the weather occurs.
Stratosphere	The region of the atmosphere above the troposphere, and is the location of the ozone layer.
Carbon cycle	The processes by which carbon in all its forms interacts and moves around in the land, atmosphere, and ocean components of the climate system.
Positive feedback	A set of processes whereby a small perturbation in the climate system amplifies and increases in size.
Negative feedback	A set of processes whereby a small perturbation in the climate system decays and reduces in size.
Control simulation	A run of a climate model or earth system model where the forcing is kept constant in time.
Ensemble simulations	A set of runs which have the identical forcing, but start from slightly different initial conditions.

This chapter was originally published as part of the Encyclopedia of Sustainability Science and Technology edited by Robert A. Meyers. DOI:10.1007/978-1-4419-0851-3

P.R. Gent (✉)
Oceanography Section, National Center for Atmospheric Research,
P. O. Box 3000, Boulder, CO 80307-3000, USA
e-mail: gent@ucar.edu

P.J. Rasch (ed.), *Climate Change Modeling Methodology: Selected Entries from the Encyclopedia of Sustainability Science and Technology*,
DOI 10.1007/978-1-4614-5767-1_2, © Springer Science+Business Media New York 2012

Chaotic system	A system of equations with the property that two runs starting from slightly different initial conditions diverge from each other, often quite quickly.
Climate projection	A simulation of the climate system into the future with prescribed forcing, where the model has not been initialized to the observed climate.
Climate forecast	A simulation of the climate system into the future with prescribed forcing, where the model has been initialized to the observed climate.
Equilibrium climate sensitivity	The increase in the globally averaged surface temperature in a model when the atmosphere concentration of carbon dioxide is doubled.
Atmosphere Model Intercomparison Project	A standard simulation of the atmosphere component of a climate or earth system model, which allows different models to be compared to each other.
El Nino-Southern Oscillation	The largest interannual signal in the climate system, which occurs primarily in the tropical region of the Pacific Ocean.
Thermohaline circulation	The overturning circulation in the global oceans where water sinks at very high latitudes, spreads very slowly horizontally to all the ocean basins, and then slowly returns toward the surface.
Conveyor belt	Another popular name for the thermohaline circulation.
Deep water formation	The process by which very dense water near the surface sinks to near the ocean bottom at high latitudes, which forms the sinking part of the thermohaline circulation.

Definition of the Subject

We are all familiar with weather forecasts that predict the local weather for the next few days. These are made using a high-resolution numerical model of the atmosphere, and sometimes extend out as far as 10 days. Most meteorological centers also produce seasonal outlooks, which give probabilities of the average temperature and precipitation being above, near, or below normal. These outlooks do not forecast the weather for a particular day, but give predictions of the seasonal averages. Seasonal outlooks are also made with an atmosphere model, but use climatological observed values for the evolving state of the surface ocean, land, and sea ice conditions. However, if forecasts are to be made more than a season ahead, then using just an atmosphere model is not sufficient, and the evolution of the ocean, land, and sea ice states must also be made using numerical models for these

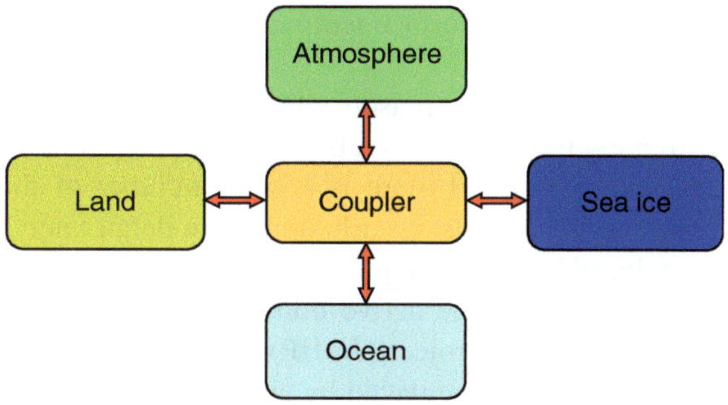

Fig. 2.1 Configuration of a climate model

components of the climate system. The reason is that the surface ocean, land, and sea ice states interact strongly with the atmosphere and influence its future evolution because they change on a much slower timescale than the atmosphere.

A climate model is used to understand how the climate system works, and how the various components interact with each other. It is used to simulate the present day climate, the recent past climate, and the climates of different paleoclimatic epochs. It can also be used to simulate the future statistical state of the atmosphere a decade or a century into the future, but does not predict the local weather on particular days. The atmosphere resolution of a climate model is much reduced compared to that used in a weather forecast, so that climate information is given on regional to global scales, and not on local scales. The climate state a long time ahead depends on the future levels of quantities that force the climate system, such as the concentrations of carbon dioxide and other greenhouse gases, several different atmospheric aerosols, and the levels of solar and volcanic activity. Therefore, these climate projections depend on many future choices to be made by mankind, which will determine the concentrations of greenhouse gases and aerosols over the next century. Each climate projection needs a scenario for the future concentrations of greenhouse gases and aerosols before it can be carried out.

Thus, a physical climate model consists of four components; atmosphere, ocean, land, and sea ice. These components are used to calculate the future state of the component given an initial state and the various quantities that force the component. These four basic components have to interact with each other, so that most climate models have a fifth component, often called the coupler, see Fig. 2.1, which has two main functions. The first function is to start, oversee the time evolution, and finish each model simulation. The second is to receive all the information from each component that is required by the other components and to send back to each component all the information that it requires to continue its simulation forward in time. For example, the ocean component needs the atmosphere-ocean wind stress that drives the ocean currents, the net heat flux and net fresh water flux (precipitation plus river runoff and sea ice melt minus evaporation) going from the atmosphere, ice, and land into the ocean. These are most often calculated in the coupler, and depend on the atmosphere surface wind, temperature, and humidity, etc., and the ocean sea surface temperature and currents, which are fields that are sent to the coupler.

There is another reason why the set up using a coupler shown in Fig. 2.1 is extremely useful. Only a relatively small fraction of climate model runs are in fully coupled mode, and there is a large number of different ways to run the model components. In runs described in more detail later, one or more of the components is replaced by its data equivalent, which provides the observed data required by the coupler to force the active components. For example, in an Atmosphere Model Intercomparison Project (AMIP) run, the numerical ocean and sea ice components are turned off and replaced by simple data components that provide observed time series of surface ocean and sea ice temperatures to the coupler. The coupler framework shown in Fig. 2.1 then ensures that the fluxes exchanged between various components are always calculated consistently, whether using observations or predicted model fields.

There is no universally accepted definition of an Earth System Model (ESM), but it must have more components than the four in a climate model. The usual additional components are a model for the distribution of carbon on the land surface, and an ocean ecosystem component, which are required if the ESM is to simulate the earth's carbon cycle. However, an ESM will often have additional components as well. The commonest of these is an atmospheric chemistry component, but some ESMs have an atmosphere component that simulates the upper levels of the atmosphere, including the stratosphere, not just the troposphere, which is the lowest layer of the atmosphere where most of the weather takes place. Finally, several ESMs will soon include a component that simulates the Greenland and Antarctic ice sheets, in order to estimate the future rate of ice loss that will raise the level of the earth's oceans.

Introduction

Numerical model simulation of the atmosphere has a long history that goes back over 60 years. The first integrations were done on the ENIAC machine at the Advanced Study Institute in Princeton by 1950 [1]. It took another 10 years for this to develop into weather forecasts that used models that had vertical structure and could be initialized using atmospheric observations. The first numerical ocean models were developed in the mid 1960s by Kirk Bryan at the Geophysical Fluid Dynamics Laboratory (GFDL) in Princeton [2], which used simplified sector geometry for the ocean basins. The first coupled atmosphere/ocean model was developed at GFDL when the global atmosphere model of Syukuro Manabe was coupled to Bryan's ocean model, and the results were published in 1969 [3]. However, the first real coupled climate model that had realistic geometry for the ocean basins and very elementary components for the land and sea ice was developed over the first half of the 1970s. The first results were published in two landmark papers by Manabe, Bryan, and coworkers in 1975 [4, 5]. The horizontal grid-spacing of this model was $5° \times 5°$, and there were nine vertical levels in the atmosphere component, and five levels in the ocean component. Even this coarse resolution was sufficient that the climate model ran slowly on GFDL's supercomputer of the early 1970s. Other meteorological and weather centers in several countries followed the GFDL lead and produced similar climate models of their own over the 1980s. As supercomputers became faster and larger, so the four components became more sophisticated, and the resolution of climate models improved.

However, there was a serious problem with all climate models when trying to obtain a control run for the present day climate. All the components would be initialized using the best set of observations available. It is most important to initialize the ocean component because it has by far the largest heat capacity, and its evolution is governed by much longer time scales than the other components. The problem was that, as the control run continued in time, the ocean and sea ice solutions would drift away from the realistic initial conditions. The drift was fast enough that rather quickly the model climate became significantly different than that of the present day earth.

The cause of this problem was diagnosed as follows. When the atmosphere and ocean components were run in standalone mode with the other component replaced by a data component that provides observations, the fluxes of heat and fresh water at the air–sea interface can be calculated. These fluxes from the atmosphere and ocean were very different, so that they were incompatible when coupled together. The problem was overcome by a very unphysical fix called flux correction [6]. The diagnosed heat and fresh water fluxes from atmosphere and ocean stand alone runs were differenced, and this difference was added to the fluxes exchanged between the atmosphere and ocean every time step of the coupled run. This enabled a climate model to maintain a non-drifting solution in a present day control run. However, it disguised the fact that the climate model components needed further development work to improve the simulations and make their fluxes of heat and fresh water across the air–sea interface compatible with each other. Use of flux corrections in climate models remained the standard method of running until the late 1990s.

The first model that could maintain the present day climate in a control run without the use of flux corrections was the first version of the Community Climate System Model (CCSM) developed at the National Center for Atmospheric Research (NCAR). A 300 year present day control simulation that showed virtually zero drift was run during the second half of 1996 and documented in 1998 [7]. The reason for this success was further refinement of the atmosphere, and especially the ocean [8] components, so that the surface heat and fresh water fluxes produced by the two components were compatible. Quite quickly, the climate centers in Australia and the UK implemented two of the new ocean parameterization improvements from the CCSM and were also able to run their models without flux corrections [9, 10]. Now, a large majority of climate models run without flux corrections, although some of the coarser resolution models still use this technique. Coarse horizontal resolution now means a grid-spacing of about $3° \times 3°$, whereas many climate models currently use about 1° × 1° grid-spacing, or slightly higher, for their standard runs.

The number of climate models maintained around the world has steadily increased over the last decade, so that results from 18 different models were submitted to the 4th Assessment Report of the Intergovernmental Panel on Climate Change (IPCC), which was published in February 2007. This 4th Assessment Report [11] was the joint recipient of the 2007 Nobel Peace Prize.

Earth System Models

All ESMs contain components that enable the carbon cycle in the land, ocean, and atmosphere to be predicted, rather than being passive in simulations of the earth's climate, for the following reason. Only about half the carbon dioxide (CO_2) emitted into the atmosphere over the past 150 years has stayed in the atmosphere; the other half has been taken up by the land and oceans in about equal measure. Climate models need past and future concentrations of CO_2 and other greenhouse gases in order to simulate the past and future climates. For future climate projections, it is currently assumed that the land and oceans sinks will continue to be as effective as in the past in taking up CO_2, so that future atmosphere concentrations will be based on about half of the future emissions staying in the atmosphere. However, there are real concerns that in the future, the ocean especially will not be able to take up the same fraction of CO_2 emissions because it is becoming warmer and more saturated with CO_2 [12]. Whether the land will continue to take up the same fraction of CO_2 is also not obvious and strongly depends on future land use practices. Over the last 30 years, deforestation of tropical forests has rapidly increased, which results in less CO_2 taken up by the land and more emitted into the atmosphere if the wood is burnt. This is now the cause of a substantial fraction of the recent increase in atmospheric CO_2 concentration. In contrast, there has been reforestation at some locations in the northern hemisphere mid-latitudes, such as the eastern part of the USA. Rather than assuming how much of the emitted CO_2 stays in the atmosphere, this fraction is predicted by an ESM with a carbon cycle. Thus, if the model predicts that the ocean will take up less CO_2 in the future, then a larger fraction will stay in the atmosphere to act as a greenhouse gas. This is a positive feedback in the climate system that is in ESMs, but not in climate models. Interactive carbon cycles have been put into a number of climate models around the world, and there has been an intercomparison project that compares their results [13]. The strength of the positive feedback from the carbon cycle is quite different in these various models, so the strength of this positive feedback is presently quite uncertain and needs to be constrained better.

There is some evidence that the stratospheric circulation can affect phenomena such as the Arctic and Antarctic Oscillations [14, 15] and will be important in how quickly the observed "ozone hole" in the southern hemisphere stratosphere will recover over the first half of the twenty-first century. If these processes are to be included in an ESM, then the atmosphere component needs to include all of the stratosphere, which is located above the troposphere. The region usually modeled by the atmosphere component of a climate model is the troposphere and just the lower part of the stratosphere. How important these processes are to the future trajectory of climate change has not been fully evaluated at present. In addition, an atmospheric chemistry component may be important to model the future levels of atmospheric aerosols. These are important in reflecting incoming solar radiation and in the formation of clouds, which are extremely important in the radiation budget of the atmosphere. A chemistry component is also needed if an ESM is to evaluate future levels of natural and man-made pollution in the very large cities of the future.

Another component of the earth system that has recently taken on more importance is the role of the Greenland and Antarctic ice sheets. There is growing evidence that the Greenland ice sheet has lost mass more quickly in the first decade of the twenty-first century than previously [16, 17], and there are changes in how quickly it is moving [18]. There are also observations of accelerations in Antarctic glaciers, especially after small ice shelves have collapsed [19, 20]. This has two important effects. The first is that the fresh water input to the ocean from these ice sheets increases the mean sea level [21], although it is important to note that this increase is not uniform over the ocean. The second is that fresh water input from the Greenland ice sheet can possibly cause a future weakening of the so-called thermohaline circulation in the North Atlantic Ocean [22, 23]. This circulation carries a lot of heat northward and certainly affects the climate of Western Europe, and is discussed in more detail in the next section. These possible future effects are not included in climate models at present. A new ice sheet component to evaluate these future climate change possibilities will be a vital component of ESMs over the next few years.

Climate Model Simulations

One or Two Active Components

As a climate model is being built and assembled, the first type of simulation that is performed and analyzed is runs using either one or two of the components in active mode, with the other components being replaced by simple data components that provide observed time series of the required fields. The best known of this type of run is when the atmosphere and land components are active, and the ocean and sea ice are replaced by observations of sea and sea ice surface temperature. When the observations are over the period 1960–2005, this is called an AMIP run, which is named after the Atmosphere Model Intercomparison Project, which first formalized this type of run. Results from AMIP runs made with the atmosphere and land components of many different climate models have been compared in this type of intercomparison for 20 years or more [24]. These comparisons have given, and continue to give, insight into the validity of the parameterizations used to simulate the many important processes in the atmosphere component of different climate models.

Scientists developing the land component of a climate model use these AMIP runs to validate their component. However, in order to isolate parameterizations in the land component, they frequently make simulations with just the land component active. In this type of run, the land is forced by a time series of observations from 1960 to 2005 of all the surface atmosphere variables that are required to force the land component.

This same time series of surface atmosphere variables, but over the oceans, is very frequently used to force the ocean component of climate models run in standalone mode. This type of run is done to validate the ocean component because the ocean observations available for comparison are mainly from the period 1960–2005. One of the difficulties in setting up this type of run is how to force the ocean under sea ice. The interaction between the ocean and sea ice is very important, especially when ice is being formed. Sea ice is formed with a salinity of about 5 parts per 1000 from sea water with a salinity of about 35 parts per 1000. Therefore, this process rejects brine into the surface water, and at cold temperatures, the ocean salinity is more important than temperature in determining its density. Thus, sea ice formation produces very dense surface water, and when this is denser than the water below, the water column overturns down to a depth of 2 km or more, resulting in what is called "deep water formation." This only occurs in winter in a very few locations in the world oceans. Off Antarctica, it occurs in the Weddell and Ross Seas, producing Antarctic Bottom Water, which is the densest water mass in the oceans. It also occurs in the North Atlantic Ocean in the Greenland–Iceland–Norwegian Seas north of Iceland and in the Labrador Sea between Canada and western Greenland. This forms North Atlantic Deep Water, which flows south at 2–3 km depth, and is the return flow of the North Atlantic thermohaline circulation. This overturning circulation is often called the "Conveyor Belt," following Broecker [25], and a schematic is shown in Fig. 2.2. The deep water formation regions in the North Atlantic and off Antarctica are the sinking branches of the Conveyor Belt. The dense water near the bottom of the ocean very slowly makes its way into the Indian and Pacific Oceans, and then slowly rises toward the surface in all the oceans. It has been estimated from ocean observations and models that deep water formed near Antarctica, which then goes into the Pacific Ocean, will take between 800 and 1,000 years before it returns to the ocean surface. It is also interesting to note that deep water formation does not occur in the North Pacific Ocean. The main reason is that the salinity there is much less than in the North Atlantic, and the surface water is never dense enough to overturn.

In order to overcome the difficulty of how to force the ocean under ice, the ocean and sea ice components are sometimes run together in active mode, forced by the time series of atmospheric surface observations. Often the scientists developing the sea ice component wish to isolate that component, and make stand alone sea ice runs forced by atmospheric observations, and allowing the sea ice to interact with a much simpler ocean component called a slab ocean. A slab ocean component only models the upper mixed layer near the ocean surface. This is needed because there are no observations of the surface ocean under ice, so that a slab ocean component is used

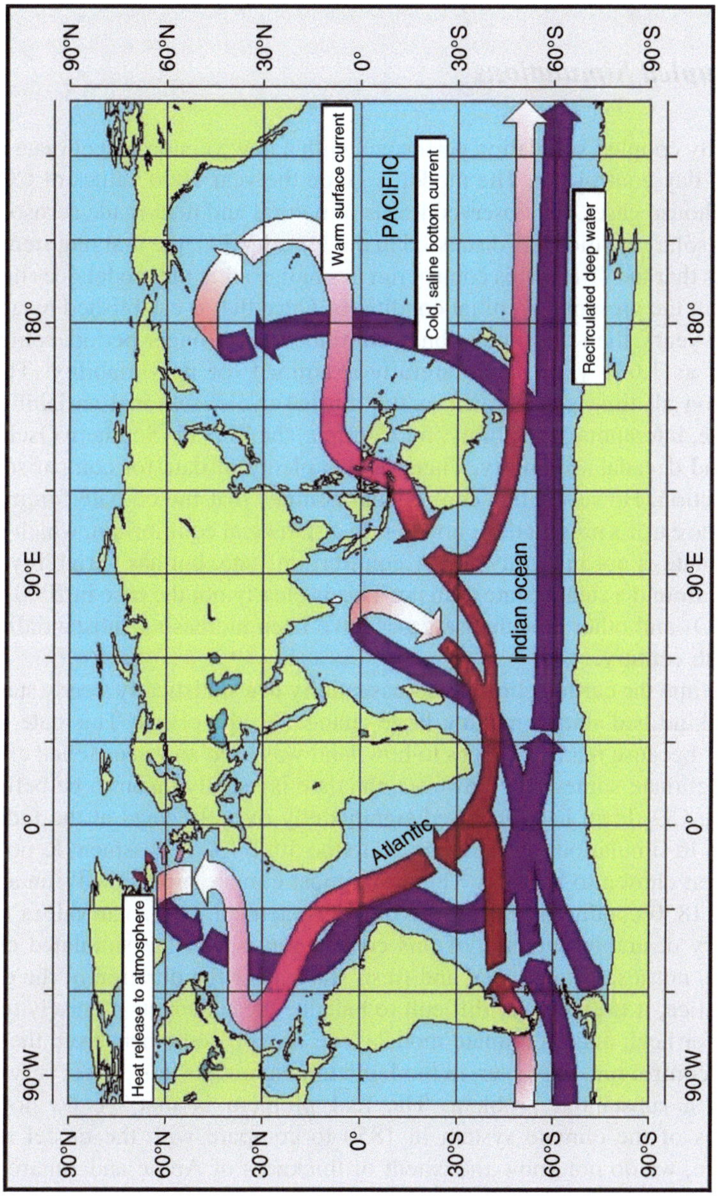

Fig. 2.2 Schematic of the thermohaline circulation, or the “Conveyor Belt,” after Broecker [25]

which exchanges heat and salt with the sea ice above. As stated earlier, given this very large variety of ways required to run the climate model components, it becomes obvious why the setup using a coupler shown in Fig. 2.1 is extremely useful.

Fully Coupled Simulations

The first fully coupled simulation performed with a new version of a climate model is a present day control run. The model is given the year 2000 values of CO_2 and other greenhouse gases, the observed levels of natural and man-made aerosols, and the level of solar radiation. As discussed in the Introduction, the first requirement of the model is that the drift in this control run is small, so that the model does not drift very far from the present day initial conditions. Once that is established by a run of at least 100 years, then the simulation is continued for a longer period, sometimes for as long as 1,000 years, and carefully examined for its variability. There is variability on all time scales, such as the diurnal cycle, seasonal variability, the annual cycle, interannual variability, for example, the El Nino-Southern Oscillation (ENSO), and decadal variability. There is also plenty of data for comparison, see the next section. However, this control run assumes that the climate forcings are fixed, and the earth's present day climate is in a statistical equilibrium, which means that the climate is not in a truly steady equilibrium state, but has variability on all time scales around a steady state climate. This is clearly not the case in 2000, as the levels of CO_2 and other greenhouse gases have been increasing substantially over the twentieth century.

The last time the earth's climate was essentially in a statistically steady state was before mankind had started making large changes to the planet. This date can be argued over because man's changes to how land was used and trees felled changed the earth's climate somewhat. However, the date is usually taken to be before the atmospheric CO_2 level had increased significantly over the level at the Industrial Revolution. In simulations to be submitted to the 5th IPCC Assessment Report, this date has been chosen to be 1850. Therefore, most climate models will run another control for 1850 conditions, forced by the CO_2, aerosol, and solar values of that year. A very desirable outcome of this control run is that the simulated climate system does not lose or gain heat and fresh water over the duration of the control run. In practice, it is extremely difficult to balance these budgets precisely to zero, especially for heat, and all climate models lose or gain some heat from the ocean during any control run. However, in modern climate models this drift is very small, and is not a substantial problem. The real problem is that we do not have observations of the climate system in 1850 to compare with the model results. For example, we do not know the extent or thickness of Arctic and Antarctic sea ice in 1850.

The real purpose of an 1850 control run is to provide initial conditions for runs that simulate the period from 1850 to 2005, which are often called twentieth century runs. Time series over this period of four quantities are needed to force this type of run. They are the atmospheric concentrations of CO_2 and other greenhouse gases, the levels of natural and man-made aerosols, the level of solar output, and the level of aerosols in the atmosphere from volcanic eruptions. The last quantity is determined from the observed levels of aerosols from recent eruptions, such as El Chichon in 1982 and Pinatubo in 1991, and then scaled by the size of significant eruptions earlier in the 1850–2005 period. Very often an ensemble of these twentieth century simulations is run, where the initial conditions are taken from different times in the 1850 control run. If a climate model is to be useful, then its twentieth century runs must reproduce well many of the observed changes in the earth's climate over the last 150 years. Most of the comparisons with observations will use the last 50 years of these runs, which is when virtually all of the observations were made.

Note that for ESMs, which have an active carbon cycle, the twentieth century runs will be forced by CO_2 emissions, rather than atmospheric concentrations. A severe test for twentieth century ESM simulations will be to reproduce the time history of atmospheric CO_2 concentration over the time period 1850–2005. The reason is that to accomplish this, the ocean and land components of the ESM will have to take up the correct fraction of CO_2 emitted into the atmosphere. This nicely illustrates the fact that as a climate model or ESM becomes more complicated with more components, then it is required to perform at a higher level. The reason is that very important quantities, such as the atmospheric CO_2 concentration, are now being predicted by the model, instead of being prescribed from observations.

The ensemble of twentieth century runs will then be continued to make projections of future climate changes over the rest of the twenty-first century. In order to make a future climate projection, time series of two quantities are required: the atmospheric CO_2 concentrations (for climate models) or emissions (for ESMs) and other greenhouse gases, and the levels of natural and man-made aerosols. In these projections, the solar output is kept constant at its 2005 level, and only a background level of volcanic aerosols is used to account for future small volcanic eruptions. In all climate models, the magnitude of future climate change depends crucially on the concentrations of CO_2 and other greenhouse gases in the future, and to a smaller extent on the future levels of man-made aerosols, which are expected to keep reducing, as they have done over the last 30 years. For the 4th IPCC Assessment Report, three scenarios for the future concentrations of CO_2 were used, which all had CO_2 levels strongly increasing until 2100. For the 5th Assessment Report, scenarios will be used where the CO_2 concentrations increase at a much slower rate during the second half of the twenty-first century because it has been assumed that emissions will be much reduced over that period.

The second crucial factor that determines the magnitude of a model's future climate change over the twenty-first century is its climate sensitivity. Equilibrium climate sensitivity (ECS) is defined as the increase in the globally averaged surface temperature that results from a doubling of CO_2 in the atmosphere component when it is coupled to a slab ocean model. This setup of a climate model only takes about 30 years to come into equilibrium, whereas the full depth ocean component takes about 3,000 years. However, it has recently been shown [26] that the ECS using a full depth ocean is not very different than that obtained using a slab ocean model. Transient climate sensitivity (TCS) is defined as the increase in globally averaged surface temperature that occurs when CO_2 has doubled after 70 years of a transient simulation where CO_2 concentration increases at the rate of 1% per year. In general, a model with a small (large) ECS will also have a small (large) TCS, but the relationship is not one-to-one because models differ in the rate of heat uptake into the ocean and the timescales of other feedbacks. It is interesting to note that the ECS of every climate model ever developed has been positive, which is a very strong indication that the equilibrium climate is warmer when there is an increased concentration of atmospheric CO_2. Almost all models used in the IPCC 4th Assessment Report have an ECS in the range of 2°C–4.5°C. Despite dramatic improvements in climate models over the last 20 years, this range of ECS is the same as in the IPCC 1st Assessment Report [27]. It can be viewed as a disappointment that the range of ECS in climate models has not been reduced over this time period, but it reflects the fact that climate models still have to parameterize several important processes that affect climate sensitivity, the most important of which is clouds. The earth's climate sensitivity has also been estimated using observations [28], but this estimate has also not reduced the possible spread in its value. This brings up the subject of how climate models are validated.

Model Validation

The atmosphere component is the easiest to validate because there is a whole host of observations to compare its results against. These include observations taken by instruments, including satellites, and the so-called atmospheric reanalyses, which use a numerical model to assimilate many different observations to provide a time history of the state of the global atmosphere. These observations and reanalyses are compared with the results from AMIP simulations, which are described in the previous section. AMIP runs use a time history of observed sea surface temperature (SST), which is a relatively accurately observed quantity, especially since the start of the satellite era. There are a large number of variables that can be compared, which include temperature, winds, pressure, cloud amount, precipitation, shortwave solar radiation, and long-wave radiation emitted by the earth. These quantities can also be compared on many timescales from diurnal, seasonal, annual to interannual variability. In general, most of these comparisons are quite good, with cloud amount and precipitation being two of the more difficult variables for the atmosphere component to simulate well. As an example, Fig. 2.3a shows the mean annual cycle of precipitation from an AMIP simulation using the CCSM4 atmosphere component compared to long-term observations made at the Southern Great Plains site in Oklahoma. The difference between the model and observations is plotted in Fig. 2.3b, and shows that the model has too little precipitation during the fall and winter, but has too much precipitation in the late summer. Overall, the

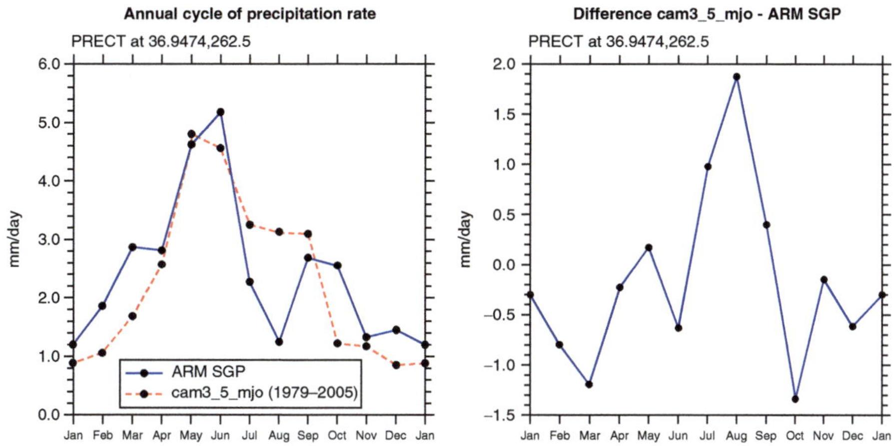

Fig. 2.3 Mean annual cycle of precipitation from an AMIP run of the CCSM4 atmosphere component and observations in the Southern Great Plains of the USA

comparison is reasonable because the annual mean values from the model and observations are quite close. Literally hundreds of such comparisons can be made, but what is a lot more difficult is how to synthesize and interpret the comparison results in order to produce better parameterizations for the clouds and precipitation in the atmosphere component.

It is a different story for the ocean component because there are far fewer observations to compare to ocean alone simulations. There is a compendium of temperature and salinity observations at prescribed depths [29] that can be used to compare to average conditions in the late twentieth century. In the best observed oceans, these observations can be split into the four seasons, so that the annual cycle in the upper ocean can be verified. It should be pointed out that satellites can only measure surface ocean quantities, so that their observations do not give information about the ocean vertical structure, unlike the atmosphere. However, there are direct observations in a few regions of the ocean, such as the upper, tropical Pacific Ocean, which can be used to make comparisons. Figure 2.4 shows the zonal current along the equator in the upper 400 m of the Pacific Ocean from an ocean alone simulation of the CCSM4 and observations [30]. It shows that the component does quite a good job in reproducing the westward surface current, and the very strong eastward equatorial undercurrent, which is one of the fastest ocean currents with a maximum speed of about 100 cm/s. The model simulation depends on the atmosphere winds used to force it, as well as some of the model parameterizations, and it is frequently difficult to decide whether a poor comparison with observations is the result of poor forcing fields or a problem with the model parameterizations.

The situation is worse for sea ice because there are even fewer observations. Sea ice extent and concentration were not well known until they began to be observed from satellites in 1979. Sea ice thickness is still not well observed, although the general spatial patterns are known from accumulating point measurements over the

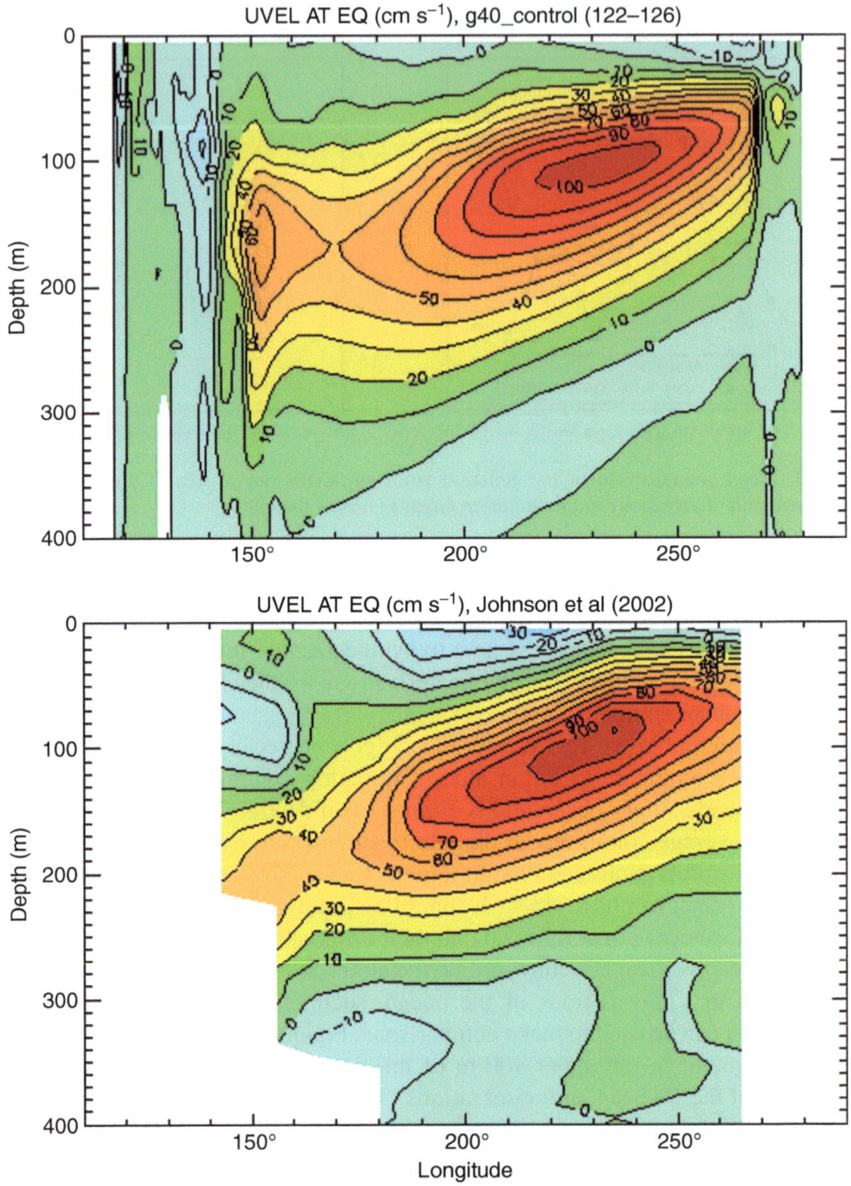

Fig. 2.4 Comparison of zonal velocity along the equator in the upper Pacific Ocean between the ocean component of the CCSM4 and the observations in [30]

years. However, there are many processes that affect sea ice, such as ridging, the formation of polynas, and melt ponds, and how snow aging affects the albedo that have to be parameterized, although there are few observations of them. There are also not too many measurements to compare with the variables in the land component, although more than for sea ice. The measurements of quantities such as soil temperature and moisture, albedo, and the leaf area amount have to be taken in areas with natural vegetation, as well as in man-made areas such as croplands. Again, there has been a large increase in observations over the past 20 years or so during the satellite era, and from land based observations at several specific sites.

As mentioned in the previous section, there are difficulties comparing both present day and 1850 control simulations with observations because the present climate is not in equilibrium, and there are not many observations from 1850. The only quantity from that time that can be estimated directly from observations is the globally averaged surface temperature. Also, the global SST pattern from 1850 to the present has been estimated in the HadISST dataset [31] by determining the principle variation patterns from the period when SST has been well measured and using these patterns to produce global data in the early part of the period when there were only a few measurements. However, the best simulations to compare with observations are the ensemble of twentieth century runs from 1850 to 2005.

There are a very large number of variables that can be compared to observations from the second half of the twentieth century, but some of the most important are large-scale patterns of interannual variability, such as ENSO and the North Atlantic Oscillation [32]. ENSO is the largest interannual signal in the earth's climate and much about it has been learned from observations over the last 25 years. The variable that is most often used to characterize ENSO is called the nino3 SST, which is the SST averaged over the area 90°W–150°W, 5°S–5°N in the central Pacific Ocean. Figure 2.5 shows the nino3 monthly SST anomalies, and a wavelet analysis, which is a method to plot the time variation of the amplitude of the anomalies as a function of the frequency content. The three smaller boxes show the power spectrum (variance against period in years), the autocorrelation against lag time in months, and the annual cycle of the variance amplitude.

Figure 2.5 shows that the amplitude of nino3 SST anomalies in the CCSM4 is a little smaller than in the HadISST observations, especially the warm events which have a maximum amplitude of just over 3°C in the data, but are only 2.5°C in the model. This means that the wavelet and power spectrum are also a little weak in the CCSM4. However, the amplitude of nino3 SST anomalies from earlier periods of the twentieth century run is larger than the HadISST data, which shows there is strong decadal variability in the CCSM4 ENSO amplitude. The CCSM4 power spectrum peaks at a period of 3–4 years compared to 4–5 years in the data, the autocorrelation compares quite well, and the annual cycle of variance is quite good with a minimum in May compared to April in the HadISST data. This good comparison is independent of the period of the run examined and is a very important improvement over all the previous versions of the CCSM, which had ENSO spectra that had a dominant peak at 2 years. This improvement was due to two changes made to the convection parameterization scheme in the atmosphere component [33]. The CCSM was one of many climate models that had a poor ENSO simulation for a long time [34], which was not a good situation given that ENSO is the largest interannual signal in the earth's climate.

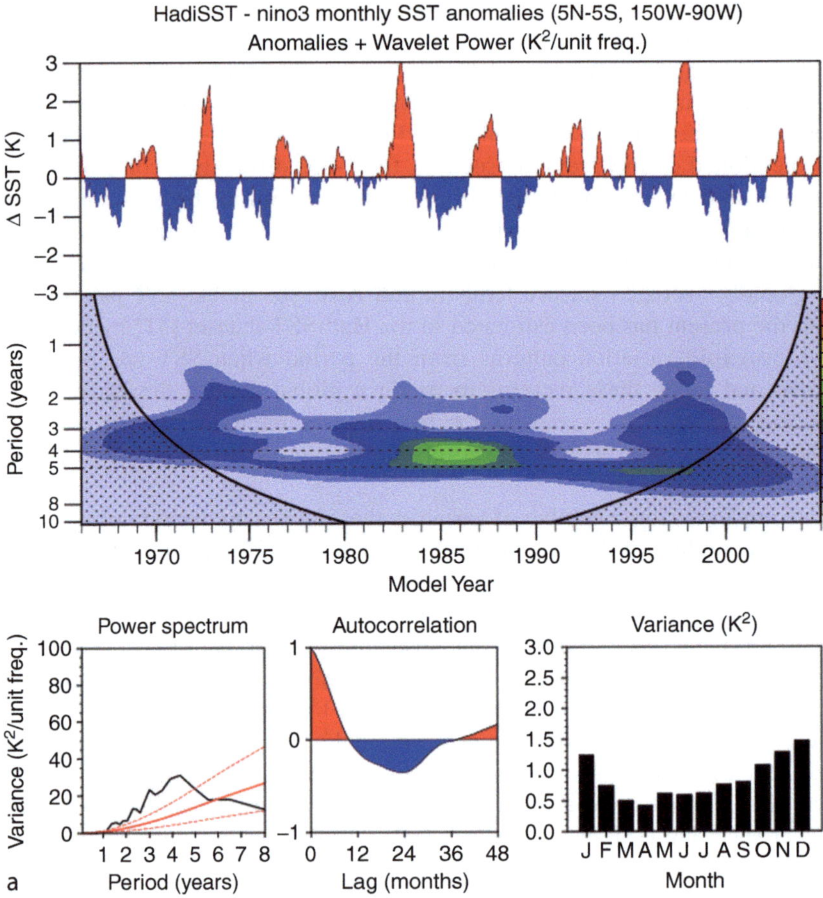

Fig. 2.5 Nino3 monthly SST anomalies between 1966 and 2005, wavelet power, power spectrum, autocorrelation, and variance from (**a**) HadISST observations, and (**b**) a twentieth century simulation using the CCSM4

Probably the only well-measured variable that can be compared to the model over the whole period of a twentieth century run is the globally averaged surface temperature. Figure 2.6 shows this comparison over 1890–2000 between the HadISST data and an ensemble of twentieth century runs using the CCSM version 3. The red line is the mean value from the ensemble, and the shading indicates the standard deviation across the eight member ensemble. This comparison is not perfect, but the data is not too often outside the shading. The model was then integrated forward to make an ensemble of projections for the twenty-first century [35] that were submitted to the IPCC 4th Assessment Report.

Another quantity that has been given a lot more attention in recent years is the sea ice extent in the Arctic Ocean. In order to give realistic projections of the future state of Arctic sea ice, a climate model must simulate it well at the end of the

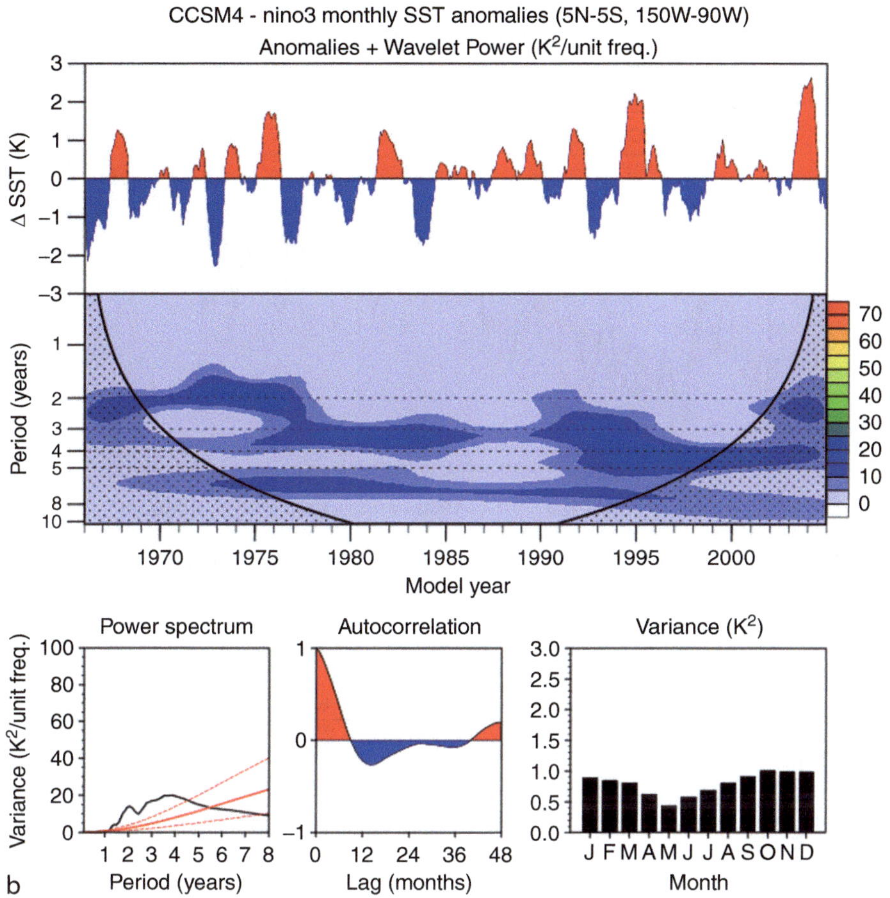

Fig. 2.5 (continued)

twentieth century. Arctic sea ice has a minimum extent in September, and this has been well measured by satellite since 1979. Figure 2.7 shows the observed September Arctic sea ice extent from observations and the latest two versions of the CCSM. For the CCSM3, the twentieth century run forced by observed concentrations of CO_2 ends in 2000, and the model then used a scenario for the future levels of CO_2. The projected decline in the CCSM3 ice extent between 2000 and 2009 is not quite as large as has been observed. However, the actual rise in CO_2 concentration in the earth's atmosphere in the decade since 2000 has been somewhat larger than in the forcing scenario used in the CCSM3 projection shown in Fig. 2.7. It is important to remember that results from future projections strongly depend on the forcing scenario used. This same projection suggests that the Arctic Ocean will become virtually ice free in September by 2040 [36], but again the actual year when this might occur will depend on the concentrations of CO_2 and

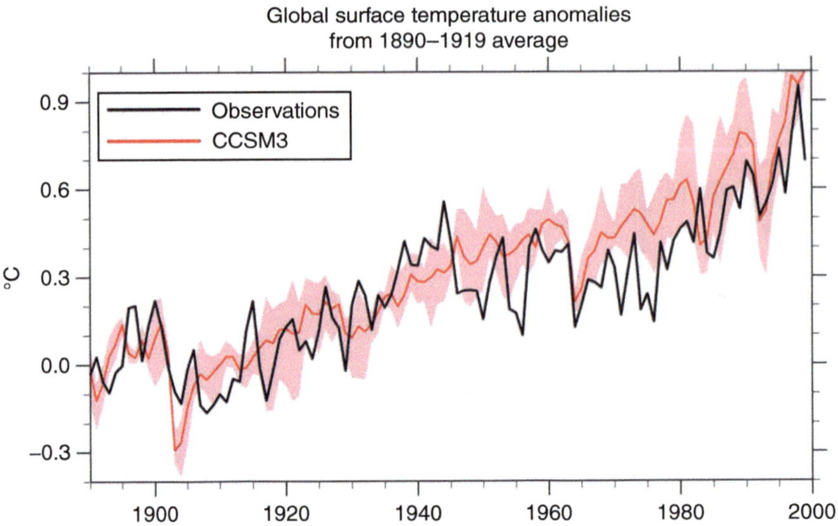

Fig. 2.6 Globally averaged surface temperature anomalies from HadISST data plotted against the results of an ensemble of twentieth century runs using the CCSM3

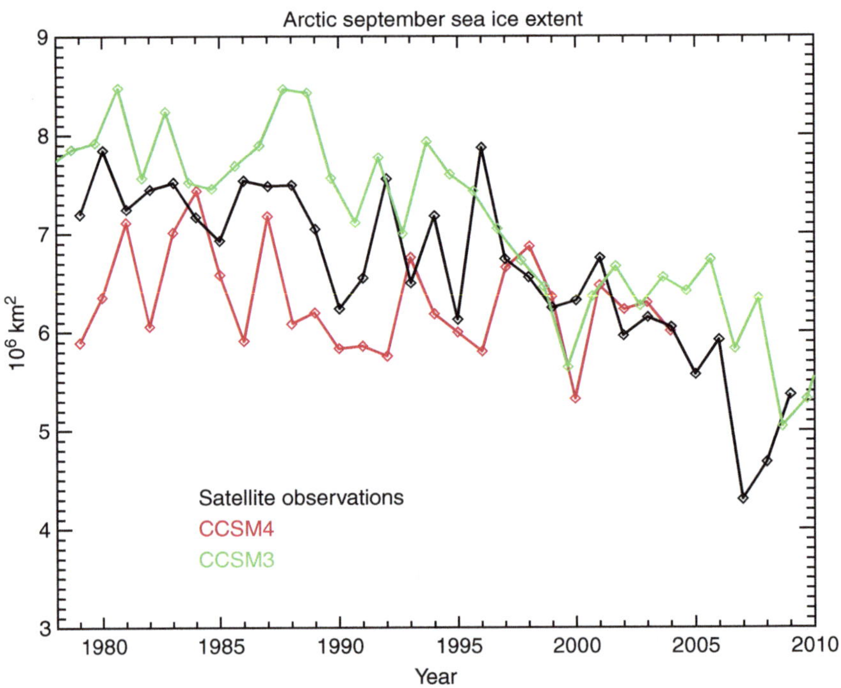

Fig. 2.7 Arctic September sea ice extent from observations, CCSM3, and CCSM4

other greenhouse gases over the next 30 years. The result from the CCSM4 shown in Fig. 2.7 is just from a twentieth century run, which goes to the beginning of 2005. Again the comparison is good, and the CCSM4 will be used to make future projections for the IPCC 5th Assessment Report.

Other comparisons to validate the ability of a climate model to simulate the historical evolution of the earth's climate can be made, but the observations are probably not as accurate as for surface temperature and sea ice extent. Two examples are ocean heat content and the distribution of chlorofluorocarbon-11 (CFC-11) in the ocean. The time history of CFC-11 concentration in the atmosphere is well known, so this can be used as an input to the ocean component during a twentieth century simulation. Observations of both ocean heat content and CFC-11 are sparse in both time and space, but estimates of their changes can be made and compared to model results [37]. This comparison helps to determine whether the ocean component is taking up quantities at the correct rate. It is very important in an ESM that the ocean takes up the correct fraction of the CO_2 that is emitted into the atmosphere. It is more difficult to make comparisons of changes in the land component because the largest changes in land use over the twentieth century are man-made and not changes in the natural vegetation. Changes in how land has been, and might be, used are often imposed in the land component during twentieth and twenty-first century simulations, which allows an assessment of how these changes have affected the past climate and might affect future climate changes [38]. On the global scale, these changes are much smaller than changes due to increases in CO_2 and other greenhouse gases, but they can be important in affecting the climate locally.

Climate Forecasts

First, the difference between a forecast and a projection needs to be explained. When a weather forecast is made, there are two separate factors that determine the quality of the forecast. The first is the quality of the atmosphere model used; all models are not perfect, but some are better than others. However, just as important is the quality of the analysis of the current state of the atmosphere that is used as the initial condition for the forecast. Even if the model were perfect, if the initial condition is slightly incorrect, then the model forecast and the real evolution of the atmosphere will diverge. The reason is that both the real atmosphere and the forecast model are examples of a chaotic system. What defines a chaotic system of equations is that, if they are integrated forward from two very slightly different initial conditions, then the two future solutions will diverge from each other, often quite quickly. Chaos theory was founded by a famous meteorologist, Edward Lorenz, who published a classic paper in 1963 [39]. He made drastic approximations to the equations that represent the atmosphere to produce a set of three, quite simple ordinary differential equations. When he integrated them forward in time from two slightly different initial states, the solutions diverged, which is the characteristic of what came to be called a chaotic system. In practice, it is very difficult to separate these two sources of error in a weather forecast because the model is also used to create the initial conditions on the model grid using all the latest meteorological observations from around the world. In order to reduce the likelihood of a bad weather forecast, an ensemble of forecasts is made using a set of slightly different initial conditions. If all the ensemble forecasts predict that something will occur, then it is forecast with a high probability; whereas if the ensemble forecasts differ markedly, then it is forecast with a low probability.

Virtually all long climate model simulations of the future done so far are projections, not forecasts. The reason is that most twenty-first century runs are just a continuation of a twentieth century run of the model, and no attempt is made to initialize the climate model to the observed climate in 2005, or whatever year the twenty-first century run starts. All results submitted to the IPCC 4th Assessment Report were from future climate projections. If something, such as a large ENSO event or a sudden reduction in the September extent of Arctic sea ice for example, occurs in 2015 in a projection, then this is not a forecast that it will actually happen in 2015, but a strong indication that this type of event might very well occur in the years around 2015. The case of ENSO events is interesting and instructive because ENSO forecasts up to a year in advance are now regularly made by a number of centers around the world using climate models [40].

For a weather forecast using an atmosphere model, it is important to start with the correct initial state of the atmosphere. However, for a seasonal or ENSO forecast, a full climate model must be used because the land, ocean, and sea ice evolve on these time scales. For these forecasts, therefore, initial conditions for the climate model are needed, and the most important component to initialize correctly is the ocean because it has the slowest time scales and by far the largest heat capacity. For an ENSO forecast, what is needed is the correct thermal state of the upper 300–400 m of the tropical Pacific Ocean between about 15° north and south. ENSO forecasts could not become a reality until there was an observing system in the tropical Pacific to continuously measure and report upper ocean temperatures [41]. An analysis is performed on these observations to produce a temperature field on the model grid, and this is used as the ocean component initial condition. As the forward integration starts, the tropical atmospheric circulation comes into balance with the sea surface temperature field in about a week, which is why it is not necessary to initialize the atmosphere component. As always, an ensemble of ENSO forecasts is made by slightly changing the initial conditions used in the ocean component. For an ENSO forecast, it is important to initialize correctly the upper tropical Pacific Ocean, but for a climate forecast over a decade, there are many more aspects of the climate model that need to be initialized correctly: the ocean deeper than the upper 400 m, especially in the North Atlantic Ocean where the thermohaline circulation occurs, the sea ice distribution in both the Arctic and Antarctic, and some aspects of the land component, such as where the soil moisture content is above or below normal. We do not know precisely all the quantities that need to be initialized correctly, but we are absolutely certain that there are not adequate observations of all these quantities.

Forecasts of climate changes over the next decade on a regional basis are what would be most helpful in planning for the future. A few preliminary decadal forecasts have been made by centers in the UK and Germany [42, 43], and many centers will submit a suite of decadal forecasts to the IPCC 5th Assessment Report. However, the science of decadal climate forecasts is in its infancy [44], and there is a very large amount of research to be done before they will become reliable. Decadal forecasts are now where weather forecasts were 50 years ago, but they have another disadvantage. Weather forecasts are made and verified every day, so that there is a very large number of realizations that can be used to make improvements. By their very nature, decadal forecasts are only verified after 10 years, so that the number of opportunities to compare model predictions to observations is reduced enormously. However, the outlook for decadal forecasts has improved over the last few years. First, there is now an ocean observing system called ARGO floats (named after the mythical Greek ship used by Jason and the Argonauts to seek the Golden Fleece), that since about 2003 has been giving nearly global coverage of temperature and salinity down to a depth of 2 km, which has improved enormously our ability to correctly initialize the ocean component [45]. However, no decadal forecast initialized using ARGO data, which can start at the beginning of 2005 at the earliest, has yet had enough time to be verified. There are also satellite observations of Arctic and Antarctic sea ice extent and the soil moisture content over the continents, which could potentially be used in the initialization. Second, as the computing capacity continues to increase, then the resolution of climate models used for predictions will continue to improve, which will enable decadal forecasts to be more accurate on the regional scales that are required for future planning.

Future Directions

The computational power available to climate modelers will continue to increase in the future, so how should it be used? Should it be used to increase the resolution of present day climate models or used to increase the range of components in ESMs? This is an extremely difficult question to answer definitively. Increased resolution will undoubtedly improve some aspects of climate model simulations, but omitting an additional component may well leave out feedbacks that are potentially important. The answer will almost certainly be to push forward in both directions because scientists with different interests will lead the work in the two different directions. Another possibility is to increase the ensemble size used in future projections and predictions, which will give more reliability to simulated changes in extreme events [46], for example, which is a very important factor in planning for the future.

As mentioned above, clouds have to be parameterized in the atmosphere, and the way this is done can change the ECS of a climate model. Clouds also have to be parameterized in weather forecast models, but are often done so in a different way because the weather forecast model is run at a much higher resolution than the climate model. Over the last few years, there has been the suggestion, called Seamless Prediction, that the same atmosphere component should be used in both weather and climate prediction. In this situation, the cloud parameterization used would have to work well across all the scales involved in both weather and climate predictions. This is not as easy as it sounds because both groups have developed their own parameterizations over past years, which make rather different assumptions. There have even been suggestions that both models should use extremely fine resolution on the order of 1 km, so that clouds can be resolved rather than parameterized, but running climate models at this resolution is still many years away. Seamless Prediction is a long-term goal, but it will probably not be realized over the next few years.

Another example of a phenomenon that is not resolved in present day climate models is mesoscale eddies in the ocean. These are the equivalent of atmospheric highs and lows, but occur at a range of scales from 200 to 300 km near the equator to 20–30 km in the very high latitude oceans. Only the equatorial eddies are partially resolved if the ocean component has a grid-spacing of about 1°. So, the effect of these energetic eddies on the large-scale mean flow has to be parameterized in present day climate models. However, it has been shown that a majority of these eddies can be resolved when using a grid-spacing of 1/10° in the ocean component [47, 48]. Diagnosis of these simulations has shown that the eddy parameterization used in the 1° simulations works quite well, but still the question remains: will future climate change projections in models that resolve the meso-scale eddies give very similar answers to future projections where they are parameterized? The answer to this important question should be found in the next few years because some climate change runs with resolved eddies are now possible with the available computer time.

Examples of new components that are currently being incorporated into ESMs have been discussed earlier, and include chemistry-air quality, hydrological, dynamic vegetation, and crop model components. The new component to simulate the Greenland and Antarctic ice caps is a very nice example of an important new component. However, there is a long list of possible feedbacks that have not been included in any ESM so far. Good examples are the increased release of methane, which is a very potent greenhouse gas, from Arctic tundra as the Arctic region warms [49], the possible release of methane from ocean clathrates [50, 51], and the possible fast breakup of the West Antarctic Ice Sheet [52]. All these are examples of possible abrupt climate changes that could result in large future changes that would have very far reaching consequences. However, all are very difficult to simulate accurately in an ESM and to assess quantitatively the possibilities that they will occur.

The science of decadal forecasts will also be advanced in the near future, both by new ideas and experience in how they should be initialized, and by increasing the resolution of the components used that will give more regional information. As explained in the previous section, there is a lot to learn and much experience needs to be gained before decadal forecasts become reliable. However, they will produce the most useful kind of information that is required by people planning for the future.

Finally, what motivates the scientists working to develop climate models and ESMs? First, it is a very stimulating intellectual challenge to understand what controls the earth's past, present and future climates, and to build an ESM that gives a faithful representation of this. This requires the expertise of many scientists across a large and diverse set of sciences ranging from several earth sciences to computer science. It is a real challenge to make these models run correctly and efficiently on several of today's massively parallel supercomputers. I also know from experience, that managing an ESM project is very challenging because it is such a diverse scientific enterprise. A second motivating factor is also very important to many scientists working on ESMs. It is that they believe these models are the best means we have available to anticipate possible future changes to the earth's climate, and that their results should be made freely and widely available to anyone who wants to see them.

Bibliography

Primary Literature

1. Charney JG, Fjortoft R, von Neumann J (1950) Numerical integration of the barotropic vorticity equation. Tellus 2:237–254
2. Bryan K, Cox MD (1967) A numerical investigation of the oceanic general circulation. Tellus 19:54–80
3. Manabe S, Bryan K (1969) Climate calculations with a combined ocean- atmosphere model. J Atmos Sci 26:786–789
4. Manabe S, Bryan K, Spelman MJ (1975) A global ocean-atmosphere climate model. Part I. The atmospheric circulation. J Phys Oceanogr 5:3–29
5. Bryan K, Manabe S, Pacanowski RC (1975) A global ocean-atmosphere climate model. Part II. The oceanic circulation. J Phys Oceanogr 5:30–46
6. Sausen R, Barthels RK, Hasselmann K (1988) Coupled ocean-atmosphere models with flux correction. Clim Dyn 2:154–163
7. Boville BA, Gent PR (1998) The NCAR climate system model, version one. J Clim 11:1115–1130
8. Gent PR, Bryan FO, Danabasoglu G, Doney SC, Holland WR, Large WG, McWilliams JC (1998) The NCAR climate system model global ocean component. J Clim 11:1287–1306
9. Hirst AC, McDougall TJ (1996) Deep-water properties and surface buoyancy flux as simulated by a z-coordinate model including eddy-induced advection. J Phys Oceanogr 26:1320–1343
10. Gordon C, Cooper C, Senior CA, Banks H, Gregory JM, Johns TC, Mitchell JF, Wood RA (2000) The simulation of SST, sea ice extents and ocean heat transports in a version of the Hadley Centre coupled model without flux adjustments. Clim Dyn 16:147–168
11. IPCC (2007) The physical science basis. In: Solomon S, Qin D, Manning M, Chen Z, Marquis M, Averyt K, Tignor M, Miller H (eds) Contribution of working group 1 to the 4th assessment report of the Intergovernmental Panel on Climate Change, Cambridge University Press, Cambridge

12. Fung IY, Doney SC, Lindsay K, John J (2005) Evolution of carbon sinks in a changing climate. Proc Natl Acad Sci 102:11201–11206
13. Friedlingstein P, Cox P, Betts R, Bopp L, von Bloh W, Brovkin V, Cadule P, Doney S, Eby M, Fung I, Bala G, John J, Jones C, Joos F, Kato T, Kawamiya M, Knorr W, Lindsay K, Matthews H, Raddatz T, Rayner P, Reick C, Roeckner E, Schnitzler K, Schnur R, Strassmann K, Weaver A, Yoshikawa C, Zeng N (2006) Climate-carbon cycle feedback analysis: Results from the C^4MIP model inter-comparison. J Clim 19:3337–3353
14. Thompson DW, Baldwin MP, Solomon S (2005) Stratosphere-troposphere coupling in the southern hemisphere. J Atmos Sci 62:708–715
15. Son SW, Polvani LM, Waugh DW, Akiyoshi H, Garcia RR, Kinnison D, Pawson S, Rozanov E, Shepherd TG, Shibata K (2008) The impact of stratospheric ozone recovery on the southern hemisphere westerly jet. Science 320:1486–1489
16. Steffen K, Nghiem SV, Huff R, Neumann G (2004) The melt anomaly of 2002 on the Greenland Ice Sheet from active and passive microwave satellite observations. Geophys Res Lett 31:L20402. doi:10.1029/2004GL02044
17. Hanna E, Huybrechts P, Steffen K, Cappelen J, Huff R, Shuman C, Irvine-Fynn T, Wise S, Griffiths M (2008) Increased runoff from melt from the Greenland Ice Sheet: A response to global warming. J Clim 21:331–341
18. Rignot E, Kanagaratnam P (2006) Changes in the velocity structure of the Greenland Ice Sheet. Science 311:986–990
19. Joughin I, Rignot E, Rosanova CE, Lucchitta BK, Bohlander J (2003) Timing of recent accelerations of Pine Island Glacier, Antarctica. Geophys Res Lett 30:1706. doi:110.1029/2003GL017609
20. Scambos TA, Bohlander JA, Shuman CA, Skvarca P (2004) Glacier acceleration and thinning after ice shelf collapse in the Larsen B embayment, Antarctica. Geophys Res Lett 31:L18402. doi:10.1029/2004GL020670
21. Shepherd A, Wingham D (2007) Recent sea-level contributions of the Antarctic and Greenland Ice Sheets. Science 315:1529–1532
22. Gerdes R, Hurlin W, Griffies SM (2006) Sensitivity of a global ocean model to increased run-off from Greenland. Ocean Model 12:416–435
23. Hu A, Meehl GA, Han W, Yin J (2009) Transient response of the MOC and climate to potential melting of the Greenland ice sheet in the 21st century. Geophys Res Lett 36:L10707. doi:10.1029/2009GL037998
24. Gates WL, Rowntree PR, Zeng QC (1990) Validation of climate models. In: Houghton JT, Jenkins GJ, Ephraums JJ (eds) Climate change: the IPCC scientific assessment, Cambridge University Press, Cambridge, pp 93–130
25. Broecker WS (1991) The great ocean 'conveyor'. Oceanography 4:79–89
26. Danabasoglu G, Gent PR (2009) Equilibrium climate sensitivity: Is it accurate to use a slab ocean model? J Clim 22:2494–2499
27. Houghton JT, Jenkins GJ, Ephramus JJ (eds) (1990) Scientific assessment of climate change. Report of working group 1 to the 1st assessment report of the Intergovernmental Panel on Climate Change, Cambridge University Press, Cambridge
28. Gregory JM, Stouffer RJ, Raper SC, Stott PA, Rayner NA (2002) An observationally based estimate of the climate sensitivity. J Clim 15:3117–3121
29. Locarnini RA, Mishonov AV, Antonov JI, Boyer TP, Garcia HE (2006) In: Levitus S (ed) World ocean atlas 2005, volume 1: Temperature, NOAA Atlas NESDIS 61. U.S. Government Printing Office, Washington, DC
30. Johnson GC, Sloyan BM, Kessler WS, McTaggart KE (2002) Direct measurements of upper ocean currents and water properties across the tropical Pacific Ocean during the 1990's. Prog Ocean 52:31–61
31. Rayner NA, Parker DE, Horton EB, Folland CK, Alexander LV, Rowell DP, Kent EC, Kaplan A (2003) Global analyses of sea surface temperature, sea ice, and night marine air temperature since the late nineteenth century. J Geophys Res 108:D14, 4407. doi:10.1029/2002JD002670

32. Hurrell JW (1995) Decadal trends in the North Atlantic Oscillation: Regional temperature and precipitation. Science 269:676–679
33. Neale RB, Richter JH, Jochum M (2008) The impact of convection on ENSO: From a delayed oscillator to a series of events. J Clim 21:5904–5924
34. Guilyardi E, Wittenberg A, Fedorov A, Collins M, Wang C, Capotondi A, van Oldenborgh GJ, Stockdale T (2009) Understanding El Nino in ocean-atmosphere general circulation models: progress and challenges. Bull Am Met Soc 90:325–340
35. Meehl GA, Washington WM, Santer BD, Collins WD, Arblaster JM, Hu A, Lawrence DM, Teng H, Buja LE, Strand WG (2006) Climate change projections for the twenty-first century and climate change commitment in the CCSM3. J Clim 19:2597–2616
36. Holland MM, Bitz CM, Tremblay B (2006) Future abrupt reductions in the summer Arctic sea ice. Geophys Res Lett 33:L23503. doi:10.1029/2006GL028024
37. Gent PR, Bryan FO, Danabasoglu G, Lindsay K, Tsumune D, Hecht MW, Doney SC (2006) Ocean chlorofluorocarbon and heat uptake during the twentieth century in the CCSM3. J Clim 19:2366–2381
38. Feddema JJ, Oleson KW, Bonan GB, Mearns LO, Buja LE, Meehl GA, Washington WM (2005) The importance of land-cover change in simulating future climates. Science 310:1674–1678
39. Lorenz EN (1963) Deterministic nonperiodic flow. J Atmos Sci 20:130–141
40. See the web site: iri.columbia.edu/climate/ENSO/currentinfo/SST_table.html
41. McPhaden MJ, Busalacchi AJ, Cheney R, Donguy JR, Gage KS, Halpern D, Ji M, Julian P, Meyers G, Mitchum GT, Niiler PP, Picaut J, Reynolds RW, Smith N, Takeuchi K (1998) The Tropical Ocean-Global Atmosphere observing system: A decade of progress. J Geophys Res 103:14169–14240
42. Smith DM, Cusack S, Colman AW, Folland CK, Harris GR, Murphy JM (2007) Improved surface temperature prediction for the coming decade from a global climate model. Science 317:796–799
43. Keenlyside N, Latif M, Junclaus J, Kornblueh L, Roeckner E (2008) Advancing decadal climate scale prediction in the North Atlantic. Nature 453:84–88
44. Meehl GA, Goddard L, Murphy J, Stouffer RJ, Boer G, Danabasoglu G, Dixon K, Giorgetta MA, Greene A, Hawkins E, Hegerl G, Karoly D, Keenlyside N, Kimoto M, Navarra A, Pulwarty R, Smith D, Stammer D, Stockdale T (2009) Decadal prediction: Can it be skillful? Bull Am Meteorol Soc 90:1467–1485
45. Balmaseda M, Anderson D, Vidard A (2007) Impact of Argo on analyses of the global ocean. Geophys Res Lett 34:L16605. doi:10.1029/2007/GL030452
46. Meehl GA, Tebaldi C, Teng H, Peterson TC (2007) Current and future U.S. weather extremes and El Niño. Geophys Res Lett 34:L20704. doi:10.1029/2007GL031027
47. Maltrud ME, McClean JL (2005) An eddy resolving global $1/10^0$ ocean simulation. Ocean Model 8:31–54
48. Bryan FO, Hecht MW, Smith RD (2007) Resolution convergence and sensitivity studies with North Atlantic circulation models. Part 1: The western boundary current system. Ocean Model 16:141–159
49. Brook E, Archer D, Dlugokencky E, Frolking S, Lawrence D (2008) Potential for abrupt changes in atmospheric methane. In: Abrupt climate change. US Climate Change Science Program. US Geological Survey, Reston, pp 163–201
50. Dickens GR (2003) A methane trigger for rapid warming. Science 299:1017
51. Archer D (2007) Methane hydrate stability and anthropogenic climate change. J Geophys Res Biogeo 4:521–544
52. Bamber GL, Riva RE, Vermeersen BL, LeBrocq AM (2009) Reassessment of the potential sea-level rise from a collapse of the West Antarctic ice sheet. Science 324:901–903

Books and Reviews

Griffies SM (2004) Fundamentals of ocean climate models. Princeton University Press, Princeton
Lorenz EN (1993) The essence of chaos. University of Washington Press, Seattle
Philander SG (1990) El Nino, La Nina, and the southern oscillation. Academic, San Diego
Randall DA (ed) (2000) General circulation model development: Past, present and future. Academic, San Diego
Sarachik ES, Cane MA (2010) The El-Nino-southern oscillation phenomenon. Cambridge University Press, Cambridge
Siedler G, Church J, Gould J (eds) (2001) Ocean circulation and climate: observing and modeling the global ocean. Academic, San Diego
Trenberth KE (ed) (1992) Climate system modeling. Cambridge University Press, Cambridge
Washington WM, Parkinson CL (2005) An introduction to three-dimensional climate modeling, 2nd edn. University Science Books, Sausalito
Wigley TM, Schimel DS (eds) (2000) The carbon cycle. Cambridge University Press, Cambridge

Chapter 3
Cryosphere, Modeling of

Cecilia M. Bitz and Shawn J. Marshall

Glossary

Ablation	Snow and ice removed from an ice mass via meltwater runoff, sublimation, wind scour, or glacial calving (mechanical fracturing and separation).
Accretion	Increase in ice mass by basal growth in the case of floating ice, the compression of snow into ice, or freezing of water that has pooled on the ice or percolated into snow from rain, meltwater, or flooding of sea/lake/river water.
Accumulation	Snow and ice added to an ice mass via snowfall, frost deposition, rainfall that freezes on/in the ice mass, refrozen meltwater, wind-blown snow deposition, and avalanching.
Glacier	A perennial terrestrial ice mass that shows evidence of motion/deformation under gravity.
Grounding line	The transition zone between grounded and floating ice.
Ice sheet	A large (i.e., continental-scale) dome of glacier ice that overwhelms the local bedrock topography, with the ice flow direction governed by the shape of the ice cap itself.

This chapter was originally published as part of the Encyclopedia of Sustainability Science and Technology edited by Robert A. Meyers. DOI:10.1007/978-1-4419-0851-3

C.M. Bitz (✉)
Atmospheric Sciences Department, University of Washington, Atmospheric Sciences, Box 351640, Seattle, WA 98195-1640, USA
e-mail: bitz@atmos.washington.edu

S.J. Marshall
Department of Geography, Earth Sciences, University of Calgary, 2500 University Dr NW, Calgary, AB T2N 1N4, Canada
e-mail: shawn.marshall@ucalgary.ca

P.J. Rasch (ed.), *Climate Change Modeling Methodology: Selected Entries from the Encyclopedia of Sustainability Science and Technology*, DOI 10.1007/978-1-4614-5767-1_3, © Springer Science+Business Media New York 2012

Ice shelf	Glacier ice that has flowed into an ocean or lake and is floating, no longer supported by the bed.
Icefield	A sheet of glacier ice in an alpine environment in which the ice is not thick enough to overwhelm the local bedrock topography, but is draped over and around it; glacier flow directions in an icefield are dictated by the bed topography.
Lake/river ice	Floating ice on rivers or lakes, usually freshwater ice.
Mass balance	The overall gain or loss of mass for a component of the cryosphere over a specified time interval, typically 1 year. This can be expressed as a rate of change of mass (kg $year^{-1}$), ice volume (m^3 $year^{-1}$), or water-equivalent volume (m^3 w.eq. $year^{-1}$). It is also common to express this as the area-averaged rate of change or the specific mass balance rate, with units of kg m^{-2} $year^{-1}$ or m w.eq. $year^{-1}$.
Permafrost	Perennially frozen ground, technically defined as ground that is at or below 0°C for at least 2 years.
Sea ice	Floating ice from frozen seawater.
Snow	Ice-crystal precipitation that accumulates on the surface.
Soil ice	Ice in permafrost.

Definition of Subject

The global cryosphere encompasses snow and ice in all its forms in the natural environment, including glaciers and ice sheets, sea ice, lake and river ice, permafrost, seasonal snow, and ice crystals in the atmosphere. Cryospheric models are mathematical and numerical descriptions of these components of the Earth system, designed to simulate snow and ice processes and feedbacks in the context of the global climate system. This entry covers all forms of ice except ice crystals in the atmosphere, which are more appropriately combined with clouds and appear in entry Solar Radiation Management, Cloud Albedo Enhancement. We focus on models that are appropriate for global climate modeling. There are regional climate modeling applications that include some of the cryospheric components discussed here as well.

The cryosphere is critical to understanding global climate change owing to its control on Earth's surface reflectivity. In addition, storage of freshwater by the cryosphere and exchange of freshwater between the cryosphere and ocean are fundamental to ocean circulation and sea level rise. Sea ice and snow insulate the underlying surface and usually allow lower atmospheric temperatures than snow and ice environments. Biogeochemical cycles in sea ice and permafrost influence carbon dioxide and methane concentrations in the atmosphere, and sea ice hosts organisms and nutrients important to marine ecosystems. Modeling cryosphere-climate interactions generally requires modeling the mass balance of ice on Earth and the key features of ice that interact with the climate system.

Perennial ice covers 10.8% of Earth's land, with most of this ice area in the great polar ice sheets in Greenland and Antarctica. Smaller glaciers and icefields are numerous – the global population is estimated at 160,000 – but these ice masses cover less than 1% of the landscape. An additional 15.4% of Earth's land surface is covered by permafrost: frozen ground that ranges from a few meters to 100 s of meters deep.

In contrast to this permanent land ice, seasonal snow and ice fluctuate dramatically. Snow cover is the largest-varying element of the cryosphere, with complete summer loss of this snow everywhere on Earth except over Antarctica, the interior of Greenland, and in the accumulation areas of other high-altitude and polar ice caps. In winter, the northern hemisphere snow cover reaches an average maximum extent of 45.2×10^6 km [1], based on data from 1966 to 2004). This amounts to 49% of the Northern Hemisphere land mass. Because the southern hemisphere continents are situated at lower latitudes (excepting Antarctica), southern snow cover is less extensive, with the seasonal snow cover estimated at 1.2×10^6 km. This combines with the permanent blanket of snow over Antarctica to give a peak southern hemisphere terrestrial snow cover of 15.0×10^6 km.

Relative to the snowpack, seasonal sea ice cycles are more hemispherically symmetric, although there are interesting north–south contrasts. Based on passive microwave remote sensing for the period 1979–2010, the average minimum Northern Hemisphere sea ice area is 4.4×10^6 km, typically reached in September [2]. Maximum ice cover is usually attained in March, with an average area of 13.1×10^6 km. The annual average Northern Hemisphere sea ice area is 9.4×10^6 km. The Southern Hemisphere sea ice has a larger seasonal cycle, with relatively little multiyear ice. Annual mean sea ice cover in the south is 8.7×10^6 km, varying from 1.9×10^6 km (February) to 14.5×10^6 km (September).

Introduction

The early energy balance modelers of the 1960s who investigated global climate recognized the importance of the high albedo of ice in the climate system [3, 4]. Their models parameterized snow and ice albedo by varying the land or ocean surface albedo with surface temperature, but no other physical characteristics of ice, such as its ability to insulate or store energy and freshwater, was simulated. When subject to climate forcing, such as a perturbation to the solar constants, the models nonetheless exhibited an amplification of temperature change in the high latitudes – a phenomenon widely known as polar amplification. The albedo difference between ice-covered and ice-free regions determined the strength of the polar amplification. Today we know ice-albedo feedback is only one of many important ways that ice and snow contribute to the climate system.

The next progression in ice modeling among global climate models was to include the major elements of the mass balance of snow on land [5] and sea ice [6], both implemented in the Geophysical Fluid Dynamics Laboratory (GFDL) in 1969. Yet the GFDL model and all other climate models for several more decades had no ice dynamics, continued to treat ice sheets as shallow snow fields with prescribed representative topography, and ignored soil ice altogether. The more advanced aspects of cryosphere modeling evolved in parallel with, but independent of global climate models.

In the 1950s, John Nye [7, 8] laid the theoretical foundations for glacier models, through the elucidation of the essential physics as well as several well-judged simplifications that permit analytical solutions. Theoretical and laboratory analyses by a contemporary materials scientist, John Glen, led to the understanding that glacier ice deforms as a nonlinear viscous fluid. Glen established a constitutive relation for the rheology of glacier ice that endures to this day [9, 10]. Combined with the conservation equations for mass, momentum, and energy, this provides the basis for modeling glaciers and ice sheets. By the late 1960s and 1970s, emerging computer power presented the opportunity to develop numerical models of glaciers [11–13]. Model development through the 1980s largely focused on regional simulations [14] and ice flow in different regimes, such as ice shelves and ice streams [15, 16].

In the late 1980s, Philippe Huybrechts developed the first pragmatic, operational 3D thermomechanical ice sheet model [17, 18]. This model has been applied extensively to the Greenland and Antarctic Ice Sheets, helping to understand their past and present evolution, e.g., [19–22]. The Huybrechts model also underpins the projections of ice sheet response to climate change in the IPCC reports. Similar models have now been developed in several research groups, and intercomparison exercises have been carried out to evaluate model strengths and weaknesses [23, 24].

A few global climate models have incorporated a Huybrechts-type ice sheet model such that changes to the global climate can interact with the shape and extent of ice sheets in Greenland and Antarctica. The interaction occurs through changes in surface albedo as the ice sheet retreats or advances over bare soil, elevation-temperature feedbacks, and through changes in the atmospheric and oceanic circulation [21, 25–27].

The low-order glacier models implemented in global climate models to date are based on a simplified representation for ice sheet flow (see section "Land Ice Dynamics"), which is not well-suited to ice shelves, ice streams, and ice sheet margins. Because these features of ice sheets are known to be changing rapidly, current development efforts are focusing on higher-order solutions of ice dynamics [28–32]. Models of valley glacier dynamics have followed a similar evolution from simplified representations of ice dynamics, e.g., [33], to recent simulations that include a more complete representation of the glacier stress and strain regime [34–38]. Parallel to this recent effort are formal, community-scale programs to couple more sophisticated, high-resolution ice sheet models into global climate models [39], although fully coupled efforts with global climate models are still in their infancy.

The development of sea ice models took a rather different path than glacial models. Model development began later, but the methods, albeit often simplified, migrated more quickly into global climate models. Today the lag between developing new sea ice model physics and implementing them in global climate models is often only a few years.

The first sea ice rheology was proposed in the 1970s, roughly two decades after the glacier rheology proposed by Nye and Glen. Initially a plastic rheology was put forward as a way to produce deformation (ridging and rafting) for the Arctic Ice Dynamics Joint Experiment (AIDJEX) model spearheaded by Max Coon [40]. Models attempting to treat ice as a plastic were only appropriate for local-scale problems of a few weeks to a month duration and could not be used to investigate ice-climate interactions owing to their inherent numerical complexity [41]. In the late 1970s, William Hibler proposed a nonlinear viscous-plastic (VP) rheology – a simplification motivated on physical grounds [42]. In a 1979 landmark paper, Hibler applied his sea ice model to the whole Arctic Ocean basin and ran an 8-year simulation [43]. The AIDJEX and Hibler sea ice dynamics remains the foundation of modern sea ice models.

A subgridscale parameterization of the variety of sea ice thicknesses that are found in a typical model grid box, known as an ice thickness distribution (ITD), was developed by Alan Thorndike and colleagues [44] and implemented by Hibler at the basin-scale in 1980 [43]. Modeling the intricacies of sea ice thermodynamics to account for the thermal inertia of brine-pocket physics was developed slightly earlier by Norbert Untersteiner and Gary Maykut [45, 46], but it was only implemented in 1D until the present decade.

The first global climate models treated sea ice as a uniform slab without leads (openings among floes), melt ponds, or brine pockets, based on the simplifications proposed by Albert Semtner [47]. If the sea ice moved at all, it was advected with the surface currents – in what is known as "free drift." Once the sea ice thickness reached some threshold (4 m was common), it was then held motionless to prevent the sea ice from building to excess in regions of convergence [48–50].

It was not until Flato and Hibler [51] simplified the VP model by treating sea ice as a cavitating fluid (CF) that global climate modelers attempted to implement sea ice dynamics with a constitutive law. However, the lack of shear strength in the CF model degraded the accuracy of the simulation compared to the VP model. Soon after Hunke and Ducowicz [52] developed a technique of treating sea ice as an elastic-viscous-plastic (EVP) material – a numerical approximation to the VP model that asymptotes to the full VP solution and yet is efficient, highly parallelizable, and offers flexible grid choices. Zhang and Hibler [53] followed suit by making the VP numerics more efficient and parallelizable. These new dynamical schemes ushered in a time of rapid improvement in the sea ice dynamics in climate models, and now EVP and VP dynamics are in wide use among climate models. Thermodynamic advances in global climate models have been slower. The use of an ITD and brine-pocket physics has only been recent in climate models [54, 55]. Melt pond parameterizations and radiative transfer that includes scattering have only been developed in one dimension and are in the early stages of production runs in the next generation of models.

Snow and permafrost models used for one- and two-dimensional applications also progressed significantly before their developments were brought into climate models, e.g., [56–58]. Permafrost models are typically based on one-dimensional thermal diffusion in the upper 2 km of the Earth surface, consisting of bedrock, sediments, and soils. Where temperatures are below freezing, ground ice occupies pore space and fractures in the rock. A freezing front propagates to a depth that is limited by the geothermal heat flux. Permafrost models simulate the aggradation or degradation of permafrost, based on mean annual surface (ground) temperatures and subject to geothermal heat flux from below. Detailed near-surface models can be added to simulate the seasonal melting/freezing of the surface active layer. In reality, more complex thermodynamic and hydrological processes are involved in permafrost dynamics. For instance, water flow can advect heat and there are 3D thermal effects associated with horizontal gradients in surface temperature, arising from variable surface vegetation, micro-topography, snow cover, and surface lakes. In addition, freezing of subsurface water is not always limited to the pore space; ice can accrue as "massive ice" deposits (e.g., ice lenses) through migration of water to the interstices of soils and sediments, driven by low capillary pressures. Simple permafrost models, as used in climate models, typically neglect these processes and focus on large-scale predictions of permafrost depth and 1D ground temperature structure.

For snow models, the chief physical processes that prove challenging to model are snow aging and grain size evolution influences on the albedo and density; liquid water infiltration and storage; snow blowing, redistribution, and collection by vegetation; and active layer depth dynamics. Many global climate models have developed terrestrial snow schemes to treat all but snow redistribution in the last few decades because these processes have a major impact on surface albedo, surface hydrology, and soil carbon storage [59–64]. Subgridscale snow distribution models exist [65] but have yet to be implemented in any global climate model that we are aware of. Models of snow on sea ice tend to be more primitive because transporting every variable that describes the snow is necessary and expensive.

In subsequent sections we describe the equations and methods used to model the cryosphere, followed by a brief outlook of future directions and priorities in cryospheric modeling. Components of the cryosphere are grouped into four types based on common characteristics and physics: Land ice refers to ice above the soil/bedrock (e.g., glaciers and ice sheets) and ice shelves; floating ice refers to river, lake, or sea ice; frozen soil includes permafrost and seasonally frozen soil; and finally snow, which may overly land ice, floating ice, or soil. The underlying equations for the dynamics of land and floating ice are rooted in a blend of continuum mechanics and fluid dynamics, but there are significant differences in their implementation, justifying separate descriptions. In contrast, thermodynamics cuts across all aspects of the cryosphere, and much about it can be described more generally.

Thermodynamics of the Cryosphere

Conservation of Energy

The governing equation for the thermodynamics of the cryosphere is conservation of energy:

$$\frac{Dq}{Dt} + \nabla \cdot (\mathbf{u}_v \phi q_v) = -\nabla \cdot (k\nabla T) + Q_{sw} + \Phi \tag{3.1}$$

which is written in terms of an enthalpy (q) for ice, snow, or mixed soil and ice, where q is expressed in units of Jm^{-3}, ϕ is the void (or pore) fraction within the solid where liquid water/vapor may exist, $\mathbf{u}_v$ is the velocity of liquid water/vapor moving in the voids, T is temperature in the solid and voids, q_v is the enthalpy of liquid water/vapor in the voids, k is the conductivity in solids and voids, Q_{sw} is the absorption of shortwave radiation over a finite thickness of the ice (W m^{-3}), which is assumed to be significant only in ice, and Φ is the strain heat production due to deformation work (significant for glaciers). The first term is a Lagrangian derivative of the heat required to raise the temperature and change the phase between solid and liquid. It is Lagrangian to account for the horizontal advection of heat in moving land and floating ice (see sections "Land Ice Dynamics" and "Floating Ice Dynamics"). The second term represents heat transport due to liquid water or vapor transport in the voids, and the third term is diffusion of heat in both solid and voids. In models of practical use today, vertical gradients in the enthalpy are far greater than those in the horizontal; therefore, the horizontal diffusion and horizontal liquid/vapor transport in the pore space are generally neglected.

The enthalpy can be written to a good approximation

$$q = c_i(T - T_o) + L_o\,(1 - \phi) \tag{3.2}$$

where c_i is the volumetric specific heat of ice or snow (or the ice and soil combination for permafrost), L_o is the volumetric latent heat of fusion for ice or snow, and ϕ is the fraction of the volume that is not composed of ice. This includes air- and liquid-filled pore space in the volume, and in frozen ground, it also includes the soil or rock matrix. Modern models of snow and frozen soil may allow liquid water to infiltrate and possibly supercool, and therefore, additional equations are needed to describe liquid infiltration and the conversion of liquid to ice and vice versa. Such equations may be devised such that the second term in Eq. 3.2 is not needed. The reader may refer to the Community Land Model, CLM version 4 for a description of one possible model of snow and frozen soil that is designed for a global climate model but is relatively complete [66].

The boundary conditions of Eq. 3.1 depend on the cryosphere component and possibly the climatic conditions. First consider the special case of floating ice where the basal temperature is always assumed to be at T_f. In these materials, the bottom boundary of Eq. 3.1 has a temperature boundary condition, and the net flux into the surface excluding the conductive flux in the ice is equal to the heat flux from the water below: $F|_{bottom} = F_W$ (fluxes are taken as positive toward the surface). In contrast, the base of land ice and the lowest point considered in a soil or snowpack model are generally not at the freezing point, in which case Eq. 3.1 has a flux boundary condition at the base equal to the geothermal heat flux: $F|_{bottom} = F_G$. At the top surface of ice or snow, one must test if the net flux into the top surface can balance the conductive flux (taking z as positive down):

$$F(T)|_{top} = k\frac{\partial T}{\partial z}\bigg|_{top},$$

for a surface temperature below the freezing point ($T < T_f$). If so, then a flux boundary condition is used at that surface in Eq. 3.1. If the surface temperature is at or above the freezing point, then a temperature boundary condition is used at that surface in Eq. 3.1 with $T = T_f$.

Ablation, Accretion, and Accumulation

Models of land ice, floating ice, and snow usually employ a fixed number of layers and therefore layer thicknesses vary in time. This so-called moving boundary method requires a Stefan condition to describe ablation, accretion, and accumulation. The rate of change of the top surface position (z_o, positive down) for snow or ice from ablation is

$$\begin{array}{l} q\dfrac{dz_o}{dt}\bigg|_{\text{ablation}} = F(T)|_{top} - k\dfrac{\partial T}{\partial z}\bigg|_{top} \\ \text{if } F(T)|_{top} > k\dfrac{\partial T}{\partial z}\bigg|_{top} \end{array} \quad . \tag{3.3}$$

An additional equation is needed to account for snow processes that contribute to increasing the mass of ice and/or snow at the top surface:

$$\frac{dz_o}{dt}\bigg|_{\text{accumulation}} = -S_{fall} + \delta, \tag{3.4}$$

where S_{fall} is the rate of falling snow, and δ is from snow densification, snow to ice conversion, snow redistribution, etc. The rate of change of the bottom surface position (z_b, positive down) for ice from accretion (if liquid water is available) or ablation is

$$q\frac{dz_b}{dt} = -F(T)\Big|_{bottom} - k\frac{\partial T}{\partial z}\Big|_{bottom}. \tag{3.5}$$

Floating ice may experience lateral accretion or ablation as well. Land and floating ice must also take into account the horizontal transport of ice, which is described in sections "Land Ice Dynamics" and "Floating Ice Dynamics".

Surface Energy Balance, Radiation, Surface Albedo, and Melt Ponds

The net flux entering the top surface of land and floating ice, soil, or snow is a sum of radiative and turbulent heat fluxes:

$$F(T)|_{top} = F_r(1-\alpha) - I_0 + F_L - \varepsilon\sigma T^4 + F_s + F_e, \tag{3.6}$$

where $F_r(1-\alpha)$ is the net downward solar irradiance above the top surface, α is the surface albedo, I_o is the solar irradiance that penetrates the top surface, F_L is the downward longwave irradiance, $\varepsilon\sigma T^4$ is the upward longwave irradiance (for T in Kelvin), ε is the emissivity, σ is the Stefan-Boltzmann constant, F_s and F_e are the downward sensible and latent heat fluxes, respectively.

In many ice and snow models, the surface albedo is a function of various quantities such as temperature, snow grain size, snow age, impurities, snow depth, ice thickness, and melt pond coverage. Often shortwave radiation is absorbed in the ice interior based on Beer's law, although Beer's law is inappropriate in materials with depth-dependent surface albedo parameterizations. Usually the temperature dependence of the surface albedo is a proxy for modeling melt pond, grain size, and/or surface scattering characteristics. These relatively crude methods are being revamped considerably in models at this time.

A better way is to design a highly interdependent set of physics for radiative transfer, ponding, and liquid infiltration. Ideally one would have radiative transfer account for multiple-scattering and be based on intrinsic optical properties that vary with impurity concentrations, snow grain size, ice bubbles, and brine pockets. Ponds would accumulate water above sea level when there is insufficient hydraulic connectivity to drain meltwater, and they would accumulate below sea level when there is hydraulic connectivity that is high enough to allow liquid water to rise up from below and flood the surface.

Influence of Salts and Other Impurities

Dissolved impurities such as salts in the pore water depress the freezing point. For ice to remain in local thermodynamic equilibrium, the pore water is always at the freezing point and freezing or melting must occur at the pore-ice interface to dilute or concentrate the solute. If we assume for simplicity that the voids are completely filled with liquid, then the void fraction is a function of temperature and the bulk solute concentration. The solute concentration in the voids, assuming freezing expels the solute completely into the voids, is a function of temperature according to the liquidus relation from the phase diagram of the binary material (ice plus solute). A few global climate models today have sea ice models that allow for such pore-ice interchange (also known as brine-pocket physics in sea ice) [54, 55]. However, these models assume the bulk solute concentration is fixed in time, and hence, they neglect the heat and solute transport in the voids. These simplifications made it possible for the first step, but they cannot capture the important structural evolution of young sea ice or rapidly changing permafrost. These processes should not be neglected in models that aim to model biogeochemistry in the polar regions.

Land Ice Dynamics

Glaciers are perennial ice masses that are large enough to experience gravitational deformation: the flow of ice under its own weight. Glaciers and ice sheets nucleate where snow accumulation exceeds snow and ice ablation over a period of many years or decades. With time, the accumulated snow is buried and compressed, metamorphosing into firn and then glacier ice. Ice behaves as a nonlinear, visco-plastic fluid; once the ice thickness is sufficient, internal gravitational stresses cause the ice to deform.

Snow accumulation is primarily meteoric (derived from atmospheric precipitation), but snow can also accumulate at a site through wind deposition or avalanching. Ablation refers to the loss of snow and ice through melting, sublimation, wind erosion, or calving, a process where slabs of ice at the glacier margin mechanically fracture and detach from the main ice mass. Iceberg calving is a very effective ablation mechanism for glacier and ice sheets that are in contact with the ocean. Melting occurs at the glacier surface – the ice-atmosphere interface – but there is also melting internally (englacially), at the glacier bed (subglacially), and on vertical ice cliffs that are common at the ice margin, particularly where glaciers reach the sea and a large area of ice can be in contact with water. Melting only leads to ablation in the case where meltwater runs off and is removed from the system; some surface meltwater in the upper regions of glaciers and ice sheets percolates into the snowpack or ponds at the surface, where it can refreeze.

A glacier's mass balance is determined by the net accumulation minus ablation over a specific time period, typically 1 year. This can be expressed as a total balance (kg year^{-1} or m^3 year^{-1} water-equivalent for the entire glacier) or a specific balance, per unit area of the glacier (kgm^{-2} year^{-1} or m year^{-1} water-equivalent). There is no simple "threshold temperature" for a glacier to be viable; a mean annual temperature below 0°C is not a necessary or sufficient condition for glacier ice to

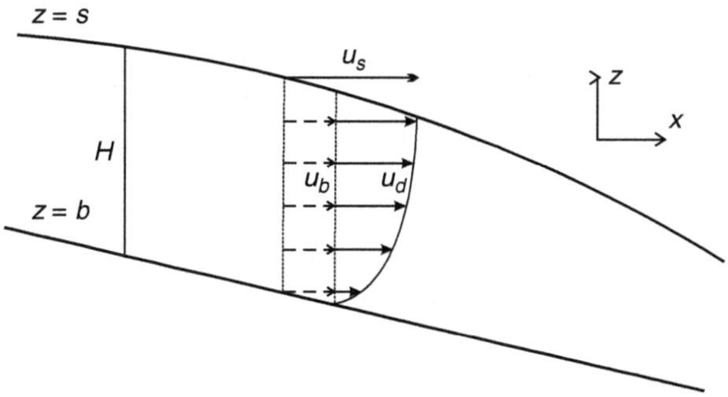

Fig. 3.1 Schematic of glacier flow mechanisms. The surface velocity $u_s = u_b + u_d(s)$. Basal velocity is the sum of deformation of underlying sediments and decoupled sliding at the ice-bed interface. Where ice is moving at the bed, these two processes can operate together, or only one of them may be active

exist. Tidewater glaciers are vivid examples of this. Because ice flow delivers large fluxes of ice to low elevations, glaciers can extend to sea level environments where mean annual temperatures are several degrees above 0°C. While most glaciers do not reach the ocean, this feature is intrinsic to all glaciers; glacier ice in the ablation area does not grow in situ, but is a consequence of ice transport from the accumulation area to the ablation area.

Glacier flow occurs through three different mechanisms: internal "creep" deformation, decoupled sliding at the ice-bed interface, and deformation of subglacial sediments (Fig. 3.1). The former is a function of ice rheology and the stress regime in the ice, while the latter two mechanisms are governed by conditions at the base of the glacier [67]. These processes are described in more detail below.

Governing Equations for Glacier Dynamics

Similar to models of atmosphere or ocean dynamics, the flow of glaciers and ice sheets is mathematically described from the equations for the conservation of mass, momentum, and energy. For a point on the glacier with ice thickness H, the vertically integrated form of the conservation of mass is most commonly employed in glaciological models:

$$\frac{\partial H}{\partial t} = -\nabla \cdot (\bar{\mathbf{u}}H) + b. \tag{3.7}$$

Here $\bar{\mathbf{u}}$ is the average horizontal velocity in the vertically integrated ice column and b is the mass balance rate. The mass balance is computed generally from Eqs. 3.3–3.5 by

$$b = -\left.\frac{dz_o}{dt}\right|_{\text{ablation}} - \left.\frac{dz_o}{dt}\right|_{\text{accumulation}} + \frac{dz_b}{dt}. \tag{3.8}$$

The first term on the right-hand-side describes the horizontal divergence of ice flux, while the second term describes the net local source or sink of mass associated with accumulation and ablation. The vertically averaged velocity includes ice flow due to both internal deformation and basal flow: $\overline{\mathbf{u}} = \overline{\mathbf{u_d}} + \mathbf{u_b}$. Glacial ice moves slowly, so a year is typically adopted as the most convenient unit of time; hence, ice velocities are reported in m year^{-1} and b is expressed as m year^{-1} of ice-equivalent gain or loss of mass.

The main challenge in modeling glaciers and ice sheet is evaluation of the velocity field. Acceleration and inertial terms are negligible in glacier flow, so the Navier–Stokes equations that describe conservation of momentum reduce to a case of Stokes flow, where gravitational stress is balanced by internal deformation in the ice:

$$\nabla \cdot \sigma = -\rho \mathbf{g}, \tag{3.9}$$

where σ is the ice stress tensor, ρ is ice density, and $\mathbf{g}$ is the gravitational acceleration. Expanding this into the three directional components,

$$\begin{aligned}
\frac{\partial \sigma_{xx}}{\partial x} + \frac{\partial \sigma_{xy}}{\partial y} + \frac{\partial \sigma_{xz}}{\partial z} &= 0, \\
\frac{\partial \sigma_{yz}}{\partial x} + \frac{\partial \sigma_{yy}}{\partial y} + \frac{\partial \sigma_{yz}}{\partial z} &= 0, \\
\frac{\partial \sigma_{xz}}{\partial x} + \frac{\partial \sigma_{yz}}{\partial y} + \frac{\partial \sigma_{zz}}{\partial z} &= -\rho_i g.
\end{aligned} \tag{3.10}$$

Ice deformation is independent of confining (i.e., hydrostatic) pressure, so ice rheology is usually couched as a function of the deviatoric stress tensor, σ'. Under the assumption that ice is incompressible, the momentum equations can be analyzed to give the horizontal balances

$$\begin{aligned}
2\frac{\partial \sigma'_{xx}}{\partial x} + \frac{\partial \sigma'_{yy}}{\partial x} + \frac{\partial \sigma'_{xy}}{\partial y} + \frac{\partial \sigma'_{xz}}{\partial z} &= -\rho g \frac{\partial h}{\partial x}, \\
\frac{\partial \sigma'_{yz}}{\partial x} + 2\frac{\partial \sigma'_{yy}}{\partial y} + \frac{\partial \sigma'_{xx}}{\partial y} + \frac{\partial \sigma'_{yz}}{\partial z} &= -\rho g \frac{\partial h}{\partial y}.
\end{aligned} \tag{3.11}$$

The terms on the right-hand-side represent the surface slope, for the glacier surface $h(x, y)$. Greve and Blatter [129] present a detailed derivation of this full system of equations and their solution. A constitutive relation is needed to express internal stresses in terms of strain rates in the ice: $\dot{\varepsilon}_{ij} = 0.5(\partial u_i/\partial x_j + \partial u_j/\partial x_i)$. Equation 3.11 can then be rewritten as a function of the 3D ice velocity field, providing a framework to solve for ū and integrate Eq. 3.7 to model the evolution of glacier geometry in response to variations in ice dynamics or climate. The next section describes the constitutive relation that is most commonly used in glacier modeling.

Because ice rheology is strongly temperature-dependent, an additional equation is needed to solve for the 3D temperature distribution. As discussed in section "Thermodynamics of the Cryosphere," the local energy balance gives the governing equation for temperature evolution in the ice sheet in Eq. 3.1.

Given a 3D temperature distribution through the ice sheet, the effective rheology of the ice can be evaluated and the velocity field can be numerically determined. Knowledge of the temperature field is also essential to assessing whether the base of the ice sheet is at the pressure melting point or not; if so, liquid water can be present at the bed and the glacier or ice sheet is subject to basal flow.

Internal Deformation: Ice Rheology

In order to solve the Stokes flow diagnostic equation, ice sheet stresses need to be expressed as velocity fields, via a constitutive relation for ice. The rheology of polycrystalline glacier ice is well-studied in laboratory and field environments [68]. Lab studies of tertiary ice deformation reveal that ice deforms as a nonlinear viscous fluid [9, 10, 69, 70]. The original form of the flow law proposed by Glen [9, 10] is broadly supported by field studies of tunnel and borehole deformation [7, 71], as well as observations and modeling of large-scale ice motion [72, 73].

This constitutive relation is known as Glen's flow law and is written as a function of the second invariant of the deviatoric stress tensor, $\sum_2^{'} = \left(\sigma_{ij}^{'}\sigma_{ji}^{'}\right)/2$,

$$\dot{\varepsilon}_{ij} = B(T) \sum\nolimits_2^{'(n-1)/2} \sigma_{ij}^{'}. \tag{3.12}$$

$B(T)$ is an "ice softness" term that follows an Arrhenius temperature dependence,

$$B(T) = B_0 \exp\left(\frac{-Q}{RT}\right). \tag{3.13}$$

B_0 is called the Glen flow law parameter, R is a constant, and Q is the creep activation energy. Lab and field studies of ice deformation suggest that B_0 can vary by a factor of about 10 [68]. Including the effects of strain-softening and temperature, the effective viscosity of ice varies by several orders of magnitude.

This formulation is an isotropic flow law that allows the first-order effects of ice temperature and deviatoric stress regime to be incorporated in estimates of ice deformation. Ice deformation is typically modeled as an $n = 3$ process. Where shear stress and shear deformation are dominant, as is often the case, this is well-approximated by

$$\dot{\varepsilon}_{xz} = B(T)\, \sigma'^{3}_{xz}. \tag{3.14}$$

Glen's flow law is for pure, isotropic ice. There are numerous other complicating factors for ice deformation, such as anisotropic ice fabric [74–76], the potential impact of grain size [77, 78], and the presence of impurities and intergranular liquid water content [79]. These effects are not explicitly resolved in ice sheet models, but the flow rate parameter, B_0, is typically tuned to approximate the bulk effects of crystal fabric, grain size, and impurity content.

The strain rates in Eq. 3.12 or 3.14 can be expressed as velocity gradients, and then vertically integrated or inverted and substituted into the momentum balance, Eq. 3.11, to give a set of equations for the horizontal ice velocity. Various numerical solutions to these equations have been adopted in glacier and ice sheet modeling, outlined in more detail in the next section.

Basal Flow

In addition to internal deformation, glacier ice can flow at the base where the bed is at the pressure melting point, through some combination of subglacial sediment deformation and decoupled sliding over the bed. Large-scale basal flow generally requires pressurized subglacial meltwater, which can lubricate the bed, float the ice, or weaken subglacial sediments. Subglacial hydrology therefore plays a pivotal role in fast-flowing glaciers and ice streams [80, 81]. High subglacial water pressures can decouple the ice from the bed by reducing or eliminating basal friction. On local scales this may not entice a significant ice-dynamical response, as resistive stresses can be taken up at adjacent well-coupled regions of the bed, by side drag from valley walls or adjacent ice, or by longitudinal stress bridging (upstream and downstream resistance to flow). However, numerous observational studies report occasions where inputs of surface meltwater to the bed overwhelm these resistive stresses and produce localized speedups in valley glaciers, e.g., [82, 83] and in polar icefields [84–86].

For large-scale ice stream flow or surging of outlet glaciers, subglacial water must occupy a significant portion of the glacier bed, at pressures which are sufficient to drown geologic and topographic pinning points [83, 87]. In this situation, widespread ice-bed decoupling can permit high rates of basal flow (100–1,000 s of m year^{-1}) and a regime in which ice fluxes are dominated by basal flow.

In ice sheets that exhibit high rates of basal flow, there is ongoing uncertainty as to the relative importance of sliding flow along the ice-bed interface versus deformation of the underlying glacial sediment. High subglacial water pressures are conducive to both processes. Fast flow in West Antarctica's Siple Coast ice streams is associated with plastic failure of a thin layer of saturated marine sediment [88, 89], and similar processes are expected to be important wherever subglacial sediments and topographic features offer a relatively smooth, low-friction substrate [90].

Models make some allowance for basal flow, usually through a local sliding "law" relating basal flow, u_b, to gravitational shear stress at the bed, raised to some power m, $u_b \propto \tau_d^m$. In some applications, the effects of subglacial water pressure on basal flow are introduced, typically through the effective pressure p_e. This is the difference between glaciostatic (ice) pressure and subglacial water pressure: $p_e = pi - p_w$. An effective pressure of 0 indicates that ice is floating, so low or negative effective pressures promote ice-bed decoupling and enhanced basal flow. While it is safe to assume that $u_b \propto p_e^{-k}$, for some unknown power k, there is likely no generalized local relationship between u_b and p_e; actual basal flow is affected by regional-scale ice dynamics, not just local conditions. A prescription of the form $u_b = A\tau_d^m/p_e^k$ is unstable as this blows up as $p_e \to 0$. Local flotation is commonly observed in nature, so $p_e = 0$ is a physically acceptable possibility. The mathematical instability is simply a failure of the local form of the basal flow law. An alternative is to introduce a parameterization in terms of the flotation fraction p_w/p_i, with $u_b = 0$ when $p_w = 0$ and basal flow increasing with p_w/p_i. The local expression $u_b = A\tau_d^m f(p_w/p_i)$ can represent this, e.g., [91]. Basal flow observations are notoriously difficult to make so there is no clear recommendation as to the functional form of $f(p_w/p_i)$. Hydrological enabling of basal flow is expected to be a nonlinear, threshold process [92].

Modeling Glaciers and Ice Sheets

This section describes the physical approximations and numerical techniques in place for simulations of glacier and ice sheet dynamics, based on solution to Eqs. 3.7, 3.11, and 3.1.

One common reduction in ice sheet modeling, known as the shallow-ice approximation, involves a number of scaling assumptions that are valid when horizontal gradients in ice thickness and velocity are small and ice flows dominantly by vertical shear deformation. Under these conditions, Eq. 3.11 simplifies to

$$\frac{\partial \sigma'_{iz}}{\partial z} = -\rho g \frac{\partial h}{\partial x_i}. \tag{3.15}$$

where subscript i refers to the horizontal direction of interest (x, y). Equation 3.15 can be vertically integrated to give

$$\sigma'_{iz}(z) = -\rho g (h - z) \frac{\partial h}{\partial x_i}. \tag{3.16}$$

This vertical shear stress is commonly known as the gravitational driving stress, τ_d. It vanishes at the glacier surface ($z = h$) and $\tau_d = -\rho g H \nabla h$ at the glacier bed. Substituting for the strain rates based on Glen's flow law and representing strain rates in terms of velocity gradients gives

$$u_i(z) = u_{bi} - 2(\rho g)^n |\nabla h|^{n-1} \frac{\partial h}{\partial x_i} \int_b^z B(T)(h-z)^n dz, \tag{3.17}$$

for basal velocity $\mathbf{u_b}$.

For the shallow-ice approximation represented in Eq. 3.17, the vertically averaged ice velocity due to deformation is nonlinearly proportional to local surface slope and ice thickness,

$$\mathbf{u_d} \propto (\nabla h)^n H^{n+1}. \tag{3.18}$$

This treatment of ice dynamics has been adopted in most ice sheet modeling studies to date, e.g., [19, 20, 93]. The nonlinearity of Glen's flow law gives ice deformation rates that are exceptionally sensitive to the glacier thickness and surface slope. Under this approximation, ice velocity is solely a function of local ice geometry. This neglects "farfield" effects and other complicating influences on ice flow, such as longitudinal stretching/compression of the ice and horizontal shearing due to the friction of valley walls. The influence of longitudinal stress coupling on ice dynamics is significant in complex terrain such as valley glaciers and mesoscale icefields [34, 94]. It is also important in settings such as tidewater glaciers, ice shelves, ice streams, transition regions from inland to floating ice dynamics, and ice sheet divides [14, 28]. Many of the most interesting questions in ice sheet behavior involve these parts of the system, so more complete representations of ice dynamics are of great interest.

Doug MacAyeal [16, 130] introduced an approximation for ice shelf and ice stream flow that is essentially the complement of the shallow-ice approximation in Eq. 3.15. Because flow in ice streams is predominantly at the base, a plug-flow approximation, assuming no vertical shear, is reasonable. The gravitational driving stress is then taken up by longitudinal stress, horizontal shear stress, and basal friction. In ice shelves, the basal traction also vanishes. Under this approximation, Eq. 3.11 can be written

$$\begin{aligned} \frac{\partial}{\partial x}(2\sigma'_{xx} + \sigma'_{yy}) + \frac{\partial}{\partial y}(\sigma'_{xy}) &= -\rho g \frac{\partial h}{\partial x}, \\ \frac{\partial}{\partial x}(\sigma'_{xy}) + \frac{\partial}{\partial y}(\sigma'_{xx} + 2\sigma'_{yy}) &= -\rho g \frac{\partial h}{\partial y}. \end{aligned} \tag{3.19}$$

Substituting for strain rates and defining an effective viscosity μ_e from the inverted form of Glen's flow law [16],

$$\begin{aligned} \frac{\partial}{\partial x}\left[\mu_e\left(2\frac{\partial u}{\partial x}+\frac{\partial v}{\partial y}\right)\right] + \frac{\partial}{\partial y}\left[\frac{\mu_e}{2}\left(\frac{\partial u}{\partial y}+\frac{\partial v}{\partial x}\right)\right] &= -\rho g\frac{\partial h}{\partial x}, \\ \frac{\partial}{\partial x}\left[\frac{\mu_e}{2}\left(\frac{\partial u}{\partial y}+\frac{\partial v}{\partial x}\right)\right] + \frac{\partial}{\partial y}\left[\mu_e\left(\frac{\partial u}{\partial x}+2\frac{\partial v}{\partial y}\right)\right] &= -\rho g\frac{\partial h}{\partial y}. \end{aligned} \tag{3.20}$$

A vertically integrated form of Eq. 3.20 is readily derived under the assumption that there is no vertical variation in strain rates. These equations can be numerically solved to give the horizontal velocity fields, subject to prescription of a basal shear stress as a boundary condition at the bed.

This set of equations provides a good representation of ice dynamics where vertical shear deformation is negligible. MacAyeal and colleagues have had good success in simulating Antarctic ice stream and ice shelf dynamics with this method. This approach to modeling ice dynamics has also been applied to former ice streams in the Laurentide Ice Sheet [95] and in studies of the inland propagation of ice-marginal thinning in the Amundsen Sea sector of West Antarctica [31]. Using control theory (inverse methods), these equations also provide an opportunity to construct the basal friction from known surface velocity fields [96, 97].

The assumptions that underlie both shallow-ice models and the "plug-flow" equations are limiting in mixed flow regimes where vertical and horizontal shear stresses and longitudinal stresses are all important. This is the case where ice flow goes through a transition from grounded to floating conditions (the grounding line), where there are large spatial gradients in basal flow, at ice divides, and in valley glaciers, where the shallow-ice approximation is not valid because the ice thickness is of similar magnitude to horizontal variations in ice thickness and velocity. To address this and provide a modeling framework that is generic and self-consistent for all environments, recent efforts have explored solutions to the full Stokes system, Eq. 3.10, as well as intermediate stages of complexity between the shallow-ice approximation and the full Stokes solution. The development of theoretical and numerical solutions of full stress-field solutions is promising, and has become tractable on regional scales [28, 29, 32]. Full solutions are still computationally unwieldy on continental scales, and have yet to be applied to whole ice sheet simulations in Greenland or Antarctica.

Progress has been slower in numerical simulations of the subglacial geological and hydrological processes that give rise to basal flow. This remains absent or oversimplified in models, where basal sliding is often specified as a function of gravitational driving stress, subject to thermal regulation. Warm-based ice can slide, while ice frozen to the bed is subject to a no-slip boundary condition. Some recent efforts include explicit models of subglacial water flow and storage, permitting sliding-law formulations that model this control on basal flow [98–100], but the physical understanding and the numerical representation of these processes remain limited.

Computational Considerations

For whole ice sheet simulations, the horizontal grid spacing is typically 20–50 km, with some 30 layers in the vertical. Recent simulations have adopted grid sizes of 5–10 km. Such resolutions are tractable but the computational demand increases by an order of magnitude for each factor of two reduction in the resolution, due to a fourfold increase in grid cells and the necessary reduction in the solution time step.

Part of the computational challenge for full-stress solutions arises from the need for long spin-up simulations for the polar ice sheets. Both Greenland and Antarctica contain ice that is more than 200 k year old, and it is essential to model the temperature and thickness evolution of the ice sheets through the last one or two glacial cycles in order to provide a reasonable internal temperature field for present-day studies or future projections of the ice sheets. Present-day dynamical adjustments are ongoing from the long-term evolution of the ice sheets, and this provides a background signal that must be understood in order to evaluate the response of the ice sheets to recent climate change. Regions of the ice sheets are thickening and thinning as a result of the ice sheet and climate history, largely associated with the long timescales of thermal advection and diffusion. The ice sheets are also adjusting to the geometric changes that attended the last deglaciation, e.g., [101]. These "secular" effects are further compounded by the slow (10^3–10^4 year) timescale of isostatic adjustment to the last glacial maximum and the subsequent deglaciation, which also needs to be simulated for the ice sheet system.

Finite difference and finite element approaches have both been applied to glacier and ice sheet models. Finite elements are more versatile and applicable to the complex geometry of mountain glaciers and fixed, limited spatial domains such as ice shelves, and these have advantages for solution of the diagnostic equations for ice velocity. Time-adaptive, moving grids are needed to simulate the evolution of glacier geometry, however, and this combines with the simplicity of finite-difference methods to make these more popular for simulation of large-scale ice sheet dynamics.

Priorities and Challenges

Fundamental glaciological data are still sparse in large sectors of the polar ice sheets and mountain glaciers, including knowledge of ice thickness, thermal regime, and subglacial conditions. It is difficult to apply hydrological and basal flow models to much of Greenland and Antarctica, where groundwater drainage, sediment properties, and other details of the subglacial environment are poorly known. Better understanding of the essential subglacial processes physics is also needed [67], as well as methods to parameterize these subgrid-scale processes in large-scale models. Similar physical and numerical challenges are involved in simulation of iceberg calving; there is no established "calving law," as the underlying physics and environmental controls are not fully understood.

There is a recent emphasis on glacial systems modeling, including subglacial processes and hydrological evolution, e.g., [92], but these efforts are still in early stages. With these limitations in mind, there is potential for major progress on at least two fronts based on what is currently well-understood about glaciers and ice sheets: (1) through high-resolution, full Stokes solutions, which are on the horizon, and (2) through improved coupling of ice sheet and climate models. Improved coupling with climate models is necessary and many research groups have initiated this effort in recent years, for both individual glaciers [102] and continental ice sheets [25, 103–105].

Improved representation of stress–strain regimes at ice sheet margins, within floating ice and grounding zones, and in drainage basins that feed fast-moving outlets will provide a better representation of interannual and decadal variability of the ice sheets. This will also improve model capability in simulating glacier and ice sheet sensitivity to climate change. Resolutions (and input bedrock and climate-forcing datasets) of order 1 km are needed for this advance in the polar ice sheets, at least in the areas of complex flow and steep gradients at the ice sheet margin. For valley glaciers, input fields and ice dynamics need to be simulated at resolutions of order 100 m, and closer to 10 m if one wishes to simulate interannual glacier terminus response to climate change. It is also important to recognize that even with full Stokes solutions, ice streams, surging behavior, summer speedups, ice shelf instabilities, and areas of fast flow within the ice sheets will not spontaneously arise in the correct places; these are associated with processes and forcings that are absent in most models. In particular, there is limited two-way interaction between oceans and ice sheets in modeling studies to date, despite recent evidence that ice sheets are highly responsive to ocean warming [106].

Floating Ice Dynamics

State-of-the-art climate models today treat the jumble of floating ice floes as a continuum. At present, in global models, the only floating ice with dynamics is sea ice, and lake and river ice models are generally thermodynamics only. There are specialized models of lakes and river ice with dynamics, and they are based on the same principles as in sea ice models, so we discuss their dynamics together here.

The physics of floating ice is strongly dependent on the thickness. Therefore, floating ice is generally described in terms of a distribution of ice thicknesses at the subgrid-scale. The ice motion is also considered for a continuum, rather than for individual floes. With this brief overview, a global-scale floating ice model can be developed from the description of thermodynamics given in section "Thermodynamics of the Cryosphere" and the additional governing equations for the dynamics and ice thickness distribution given in this section.

Ice-Thickness Distribution

The formulation of a floating ice model begins with the ice thickness distribution equation. The ITD is a probability density function (pdf), usually written $g(h)$, that describes the probability that the ice cover in a particular region has thickness h. A cruder alternative is to model the mean thickness of the pdf and the total ice concentration.

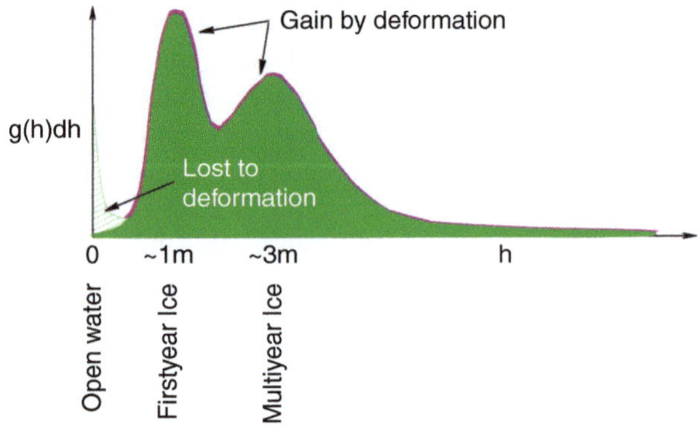

Fig. 3.2 Illustration of deformation. The portion of the distribution labeled "lost to deformation" is also known as the ice that participates in redistribution. It subsequently is redistributed to thicker parts of $g(h)$, where it is labeled "gain by deformation"

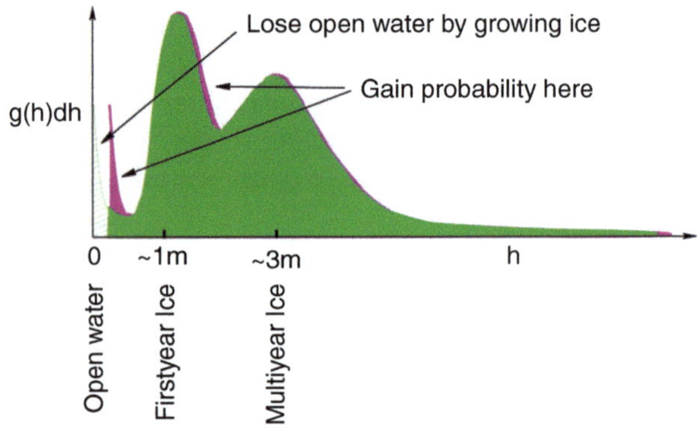

Fig. 3.3 Illustration of advection in thickness space

In a floating ice model, the ITD describes the pdf of a grid cell and thus it is sometimes called a subgrid-scale parameterization. A parameterization typically represents processes that are too small-scale or complex to be represented explicitly. For example, deformation is parameterized with a set of rules in a continuum model because there is no known differential equation that describes deformation in this type of model. The rules must select the portion of the ITD that will deform and then redistribute it within the ITD. In contrast, ice growth and melt alter the ITD in a way that is computed from first principles. Hence the ITD actually includes both parameterized and explicit physics.

The ITD equation is

$$\frac{Dg}{Dt} = -g\nabla \cdot \mathbf{u} + \Psi - \frac{\partial}{\partial h}(fg) + \mathcal{L} \tag{3.21}$$

where the left-hand-side is the Lagrangian derivative of g following an ice "parcel" and terms on the right-hand-side are the rate of change of g from parcel convergence, mechanical redistribution (Ψ, see Fig. 3.2), advection of g in ice thickness space from growth/melt (see Fig. 3.3), and the reduction rate of g from lateral melt ($\mathcal{L}$). Here, **u** is the horizontal ice velocity (vertical ice velocity is ignored) and *f* is the net growth rate. We use the variable *f* as is traditional in the ITD literature, though it is essentially equivalent to the mass balance *b* used to describe glaciers in Eqs. 3.7 and 3.8. The ITD equation was introduced by [44]. Models that specify the ITD, e.g., [107], or only permit a single ice thickness in the ice-covered fraction of a model grid cell, e.g., [108], would have an equation for the gridcell mean ice thickness instead.

There are two parts to deformation: a rate of opening (creating open water) and closing (closing open water and/or deforming and redistributing the ice), which depend on **u** and $g(h)$. The opening and closing rates depend on the convergence and/or shear in the ice motion field. It may not be obvious that shear would cause deformation. Imagine that the ice pack is composed of pieces with jagged edges. When shearing, the jagged edges can catch on one another and cause deformation, which converts kinetic energy into potential energy from piling up ice, or shearing can cause frictional loss of energy and no deformation. Thus the closing rates also depend on assumptions made about frictional losses, see e.g., [51, 54].

A mathematical representation of lead opening and mechanical redistribution is most conveniently written in terms of the divergence and shear of **u**, $\dot{\epsilon}_I$ and $\dot{\epsilon}_{II}$, respectively. The parameterization follows from [44]:

$$\psi = |\dot{\epsilon}|[\alpha_0(\theta)\delta(h) + \alpha_r(\theta)w_r(h,g)], \tag{3.22}$$

where $|\dot{\epsilon}| = (\dot{\epsilon}_I^2 + \dot{\epsilon}_{II}^2)^{1/2}$, $\theta = \tan^{-1}(\dot{\epsilon}_{II}/\dot{\epsilon}_I)$, $\delta(h)$ (the delta function) is the opening mode and $w_r(h,g)$ is the ridging mode. The coefficients $|\dot{\epsilon}|\alpha_0(\theta)$ and $|\dot{\epsilon}|$ $\alpha_r(\theta)$ are known as the lead opening and closing rates, respectively, and they are related such that their difference equals the divergence, $\dot{\epsilon}|\alpha_0(\theta) - |\dot{\epsilon}|\alpha_r(\theta) = \dot{\epsilon}_I$.

For the redistribution process, some portion of $g(h)$ is identified as potentially able to “participate” in redistribution (see Fig. 3.2). This is usually the thinnest 15% of $g(h)$. If the open water fraction exceeds 15%, then no redistribution takes place, and instead the open water closes under convergence and nothing happens under shear. The so-called participation function is weighted according to its thickness, so that the thinnest ice is most likely to deform. Then a rule is needed to redistribute the ice that ridges. Originally, Thorndike et al. [44] proposed that ridged ice would end-up five times thicker than its starting thickness. Other more complex redistribution schemes have been used since then [43, 109]. The interested reader can refer to these references, for examples, of parameterizations for $w_r(h,g)$.

The other primary mechanism that changes $g(h)$ is ice growth or melt, which cause $g(h)$ to shift in thickness space. This process is illustrated in Fig. 3.3. The growth/melt rate depends on thickness, so $g(h)$ becomes distorted in the process.

An explicit conservation equation for ice volume (or mass) is not given in this section because conservation of volume is contained in the equation for $g(h)$ (or at least it depends on the way $g(h)$ is discretized).

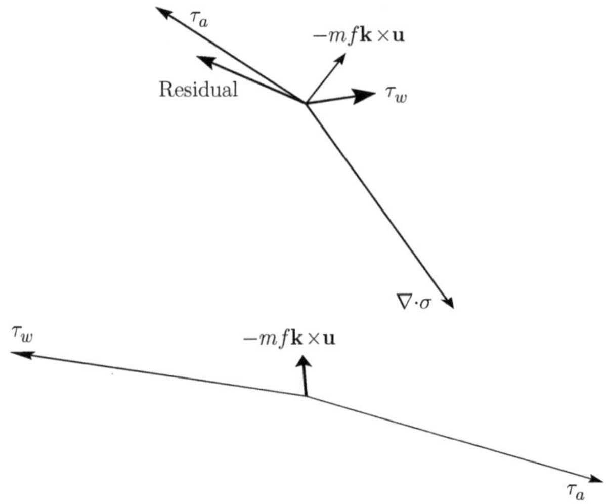

Fig. 3.4 Illustration of force balance from CCSM4 in October at 85N 65W (*upper*) and 64S 164W (*lower*). The residual force is the force needed to balance the other forces that are shown, and it is equal to the acceleration minus the force due to ocean tilt, the latter normally is very small. The forces from internal stress and the residual are negligible at the location of the *lower panel*

Momentum Equation

The second governing equation is conservation of momentum:

$$m\frac{D\mathbf{u}}{Dt} = -mf\mathbf{k} \times \mathbf{u} + \boldsymbol{\tau}_a + \boldsymbol{\tau}_{\mathrm{w}} - \mathrm{mg_r}\nabla\mathrm{Y} + \nabla \cdot \sigma, \tag{3.23}$$

where the left-hand-side is the Lagrangian derivative of **u** following an ice parcel, the right-hand-side begins with a term representing Coriolis force, air and water stresses, the force due to ocean surface tilt, and the ice internal force. In this equation m is mass per unit area, f is the Coriolis parameter, g_r is gravity, Y is the sea surface height, and σ is the ice stress.

The force balance at two locations, one just north of Greenland and the other in the Weddell Gyre, is given in Fig. 3.4. The internal stress constitutes a very large internal force in the sea ice near Greenland, where the sea ice is converging against the coast. In contrast, at the Weddell location, the ice is nearly in free drift and the main force balance is between air and water drag. In both cases, the ice motion is perpendicular to the Coriolis Force, to the left in the upper panel and to the right in the lower.

Ice Rheology and the Constitutive Law

A constitutive law characterizes the relationship between the ice stress and strain rate and defines the nature of the ice dynamics. A simplistic picture of a converging ice pack with uniform thickness under an imposed compressive wind force is given in

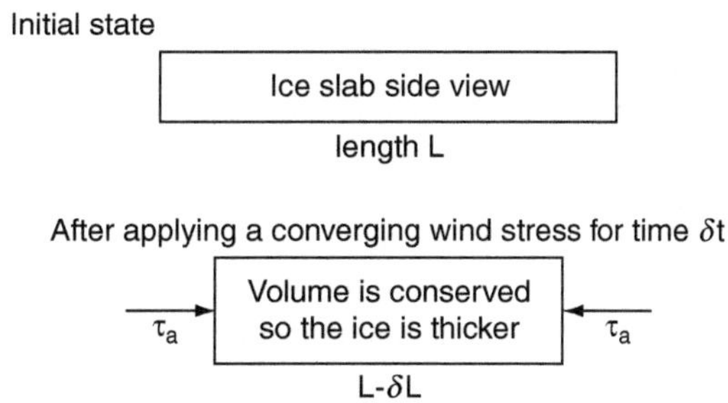

Fig. 3.5 Illustration of ice slab that deforms under compressive force

Fig. 3.5. Floating ice generally repels the compressive force somewhat even if it is deforming. The resulting internal force is associated with a nonzero stress state. In Fig. 3.5 the ice pack is converging such that its length L on one side decreases by δL in some time δt, so the ice experiences a strain $\dot{\varepsilon} = \delta L/L$ and a strain rate $\dot{\varepsilon} = \delta L/L\delta t$. A modeler chooses the constitutive law to relate σ and $\dot{\varepsilon}$, which are actually two-dimensional tensors, not scalars as shown in Fig. 3.3 for illustrative purposes only. In two dimensions, the divergence and shear of the ice velocity can be conveniently used to describe the strain rate tensor as they are invariant across grid transformations.

The momentum equation depends on the ice internal force, which in turn depends on the ice stress tensor. The most common rheology used today is the viscous-plastic rheology from [108] (or a close derivative thereof) where the ice behavior is plastic at normal strain rates and viscous at very small strain rates.

The constitutive law for the viscous-plastic model of floating ice can be cast in terms of the invariants of stress (σ_I and σ_{II}) and strain rate ($\dot{\varepsilon}_I$ and $\dot{\varepsilon}_{II}$) with the pair of equations

$$\begin{aligned} \sigma_I &= \zeta \dot{\varepsilon}_I - P/2 \\ \sigma_{II} &= \eta \dot{\varepsilon}_{II}, \end{aligned} \tag{3.24}$$

where P is the ice strength and ζ and η are bulk and shear viscosities, respectively. The relationship between the viscosities and strain rate invariants is chosen so that the stress state lies on an elliptical yield curve,

$$\frac{(\sigma_I + P/2)^2}{(P/2)^2} + \frac{\sigma_{II}^2}{(P/2e)^2} = 1, \tag{3.25}$$

where e is the ratio of the principal axes of the ellipse (the original Hibler model [108] used $e = 2$, which is still common today). Thus requiring

$$\begin{aligned} \zeta &= \frac{P}{2\Delta} \\ \eta &= \frac{\zeta}{e^2} \\ \Delta &= (\dot{\varepsilon}_I^2 + \dot{\varepsilon}_{II}^2 e^{-2})^{1/2} \end{aligned} \tag{3.26}$$

for plastic behavior. To avoid infinite viscosities as $\Delta \rightarrow 0$, ζ and η are assigned large constant values. In this case, the stress state lies inside the elliptical yield curve and the ice behaves like a viscous fluid, exhibiting creep.

The ice strength must be parameterized somehow. For a model with ridging, P can be based on the energetics argument proposed by Rothrock [110] for plastic deformation, where the compressive strength is equated to the potential energy increase per unit strain in pure convergence. Following [43], the potential energy is multiplied by a constant to account for the dissipation of kinetic energy by frictional energy loss during ridging,

$$P = ZC_p \int_0^\infty h^2 w_r(h) dh \tag{3.27}$$

where $C_p = 0.5(\rho_i/\rho_w)\hat{g}\,(\rho_w - \rho_i)$ and Z is the ratio of total energy dissipated to potential energy gain (ρ_i and ρ_w are the densities of ice and water and $\hat{g}$ is the acceleration due to gravity). [111] and [112] recommend $Z = 17$.

Computational Considerations

Computation time given to the sea ice component of a typical global climate models is roughly 10–20% for a ~100 km resolution ice grid, although it could easily be more or less depending on the relative resolution, sophistication, and optimization among components. Lake and river ice in global climate models take a trivial portion of the total computation time because their area is so small compared to the rest of the globe.

A key consideration for climate modelers is to design a model that can run the scenarios of interest in a time to meet publication deadlines. Choices must be made to balance model resolution, physics, parameterizations, and numerics. High-performance computers today have tens of thousands of cores and many have been constructed for the purpose of running climate models. Codes must parallelize well to take advantage of these machines, and sea ice codes generally do because there are relatively many processes that operate at the grid-scale (e.g., vertical heat equation) and subgrid-scale (e.g., deformation). The Los Alamos Sea Ice model CICE has already been successfully scaled beyond 10,000 cores on Cray XT equipment [113].

Future Directions

Most of the progress to date in modeling glaciers, ice sheets, permafrost, and snow science has stemmed from independent efforts by individual researchers. Advances have paralleled the much larger, coordinated research programs dedicated to the development of climate models, with little crossover or integration. This contrasts with modeling developments in sea ice studies, which are now fully integrated in most ocean–atmosphere general circulation models.

To date there have been few efforts to couple climate and ice sheet models. Most ice sheet simulations use "offline" forcing where climate model–derived fields are used to drive simulations of ice sheet evolution, through one-way forcing. The long timescale of ice sheet evolution means that this may be a reasonable approach, but evolution of ice sheet albedo and topography feeds back on atmospheric conditions on timescales of decades to millennia. Further, there are direct seasonal-timescale links between climate and the dynamics of glaciers and ice sheets, including the effects of surface meltwater and a possible link between coastal ocean temperature and sea ice conditions and outlet-glacier dynamics. These processes all require improved coupling between climate and ice sheet models. This is particularly true at the ice-ocean interface, where mass and energy exchanges are not physically modeled in current simulations.

In general, the mass balance fields (accumulation minus melt) simulated by climate models are not accurate enough for fully coupled ice sheet-climate modeling, nor can sophisticated atmospheric models be integrated over the millennial time scales of interest for ice sheet evolution. However, the development of improved regional-scale meteorological and glacier mass balance models, e.g., [114–116], permit direct estimates of surface mass balance from meteorological models, and offer a good physically based method to simulate how these fields can be expected to change with ice sheet geometry.

The challenge is greater for the mass balance fields and surface climatological forcing of mountain glaciers and permafrost. Mountain glaciers reside in complex terrain where temperature and precipitation gradients are steep. The topography and relevant meteorological processes are not faithfully resolved in even regional climate models, so some form of climatic downscaling is needed to prescribe mass balance fields for glacier modeling [102]. These methods generally do not conserve energy or mass, and improved treatments are needed. There are different considerations for permafrost modeling, but they also relate to the resolution of the landscape and surface climate. The mean annual surface temperature that governs permafrost aggradation and degradation depends on local-scale vegetation, snowpack depth, hydrology, and soil properties, which commonly vary over a spatial scale of meters. Hence, global-scale permafrost and snow models should be interpreted "statistically" for a region (i.e., via a distribution of permafrost thickness and snow depth in a given climate model grid cell, based on the range of ground cover and snow conditions in the region).

In contrast to the terrestrial components of the cryosphere, the needs of global climate models often drive sea ice model developments. For example, models that use eddy-resolving ocean components are approaching the large floe-scale, where sea ice (assumed to be on the same grid) can no longer be considered a continuum. Rheologies that attempt to treat ice as a granular material hold promise as a means to extend the utility of continuum models to smaller length scales. Such rheologies often use Mohr-Coulomb or decohesive type yield curves, which depart from the standard elliptical yield curve particularly where the stress states on the ellipse resist breaking under tension. The floe models to date do well at modeling deformation but the process of joining floes at freeze-up has not been solved [111, 117].

Recent observations of sea ice reveal very long, narrow openings, or leads, in the ice that suggest oriented weaknesses occur. These regions have been dubbed linear kinematic features [118]. Such features occur in sea ice models with isotropic rheologies, such as the viscous-plastic type, with some level of realism at fine resolution [119]. However, models that account for such anisotropy explicitly by keeping track of lead orientation and computing resistance to opening depending on the orientation with respect to leads, e.g., [120, 121], may be able to match observations better [122].

The drive to model the carbon cycle and hence ecosystem dynamics in global climate models has spawned an effort to model sea ice algae and nutrient cycling. When seawater freezes, algae stick to the ice particles, and as the ice particles combine they trap brine, nutrients, and organisms in brine pockets. Biomass concentrations can be hundreds of times greater in sea ice than in seawater, and carbon and key nutrients in sea ice are a substantial fraction of the total in ice-covered regions. When sea ice melts it deposits these materials into the ocean precisely when solar input and meltwater runoff is highest, creating prime conditions for an ocean bloom. Models of these processes have been developed offline [123–127] and will soon be implemented in sea ice models of global climate models. Models of ecosystem dynamics in sea ice also require a treatment of seawater infiltration, brine drainage, and meltwater flushing, which necessarily involves modeling sea ice salinity [127].

Modeling the cryosphere today is concerned with far more than describing Earth's surface albedo. Global climate models need to be coupled to components of the cryosphere with adequate sophistication to investigate modern scientific problems involving sea level rise, Arctic sea ice retreat, permafrost thawing, and more. There is also evidence that models with better physics are among the models that agree best with observations [128].

Bibliography

1. Lemke P, Ren J, Alley RB, Allison I, Carrasco J, Flato G, Fujii Y, Kaser G, Mote P, Thomas RH, Zhang T (2007) Observations: changes in snow, ice and frozen ground. In: Solomon S, Qin D, Manning M, Chen Z, Marquis M, Avery K, Tignor M, Miller H (eds) Climate change 2007: the physical science basis. Contribution of working group I to the fourth assessment report of the intergovenmental panel on climate change. Cambridge University Press, Cambridge, UK, pp 337–384
2. Fetterer F, Knowles K, Meier W, Savoie M (2002) Sea ice index. National Snow and Ice Data Center, Boulder, CO, USA. Digital Media, updated 2009
3. Budyko MI (1969) The effect of solar radiation variations on the climate of the earth. Tellus 21:611–619

4. Sellers WD (1969) A global climate model based on the energy balance of the earth-atmosphere system. J Appl Meteorol 8:392–400
5. Manabe S (1969) Climate and the ocean circulation I. The atmospheric circulation and the hydrology of the Earth's surface. Mon Weather Rev 97:739–774
6. Bryan K (1969) Climate and the ocean circulation III. The ocean model. Mon Weather Rev 97:806–827
7. Nye JF (1953) The flow law of ice from measurements in glacier tunnels, laboratory experiments, and the Jungfraufirn borehole experiment. Proc R Soc Lond Ser A 219:477–489
8. Nye JF (1957) The distribution of stress and velocity in glaciers and ice sheets. Proc R Soc Lond Ser A 275:87–112
9. Glen JW (1955) The creep of polycrystalline ice. Proc R Soc Lond Ser A 228:519–538
10. Glen JW (1958) The flow law of ice. A discussion of the assumptions made in glacier theory, their experimental foundations and consequences. Int Assoc Hydrol Sci Publ 47:171–183
11. Budd WF (1970) The longitudinal stress and strain-rate gradients in ice masses. J Glaciol 9:29–48
12. Mahaffy MW (1976) A three-dimensional numerical model of ice sheets: tests on the Barnes Ice Cap, Northwest Territories. J Geophys Res 81:1059–1066
13. Jenssen D (1977) A three-dimensional polar ice sheet model. J Glaciol 18:373–389
14. Raymond CF (1983) Deformation in the vicinity of ice divides. J Glaciol 29:357–373
15. Thomas RA, MacAyeal DR (1982) Derived characteristics of the Ross ice shelf, Antarctica. J Glaciol 28(100):397–412
16. MacAyeal DR (1989) Large-scale flow over a viscous basal sediment: theory and application to Ice stream B, Antarctica. J Geophys Res 94(B4):4071–4088
17. Huybrechts P, Oerlemans J (1988) Evolution of the east Antarctic ice sheet: a numerical study of thermo-mechanical response patterns with changing climate. Ann Glaciol 11:52–59
18. Huybrechts P (1990) A 3-D model for the Antarctic ice sheet: a sensitivity study on the glacial-interglacial contrast. Clim Dyn 5:79–82
19. Huybrechts P, Letréguilly A, Reeh N (1991) The Greenland ice sheet and greenhouse warming. Palaeogeogr Palaeoclimatol Palaeoecol 89(4):399–412
20. Huybrechts P, de Wolde J (1999) The dynamic response of the Greenland and Antarctic ice sheets to multiple-century climatic warming. J Clim 12:2169–2188
21. Huybrechts P, Janssens I, Poncin C, Fichefet T (2002) The response of the Greenland ice sheet to climate changes in the 21st century by interactive coupling of an AOGCM with a thermomechanical ice-sheet model. Ann Glaciol 35:409–415
22. Huybrechts P, Gregory J, Janssens I, Wild M (2004) Modelling Antarctic and Greenland volume changes during the 20th and 21st centuries forced by GCM time slice integrations. Glob Planet Chang 42:83–105
23. MacAyeal DR, Rommelaere V, Huybrechts P, Hulbe CL, Determann J, Ritz C (1996) An ice-shelf model test based on the Ross ice shelf. Ann Glaciol 23:46–51
24. Payne AJ, Huybrechts P, Abe-Ouchi A, Calov R, Fastook JL, Greve R, Marshall SJ, Marsiat I, Ritz C, Tarasov L, Thomassen MPA (2000) Results from the EISMINT model intercomparison: the effects of thermomechanical coupling. J Glaciol 46(153):227–238
25. Ridley JK, Huybrechts P, Gregory J, Lowe J (2005) Elimination of the Greenland ice sheet in a high co_2 climate. J Clim 18:3409–3427
26. Driesschaert E, Fichefet T, Goosse H, Huybrechts P, Janssens I, Mouchet A, Munhoven G, Brovkin V, Weber SL (2007) Modeling the influence of Greenland ice sheet melting on the Atlantic meridional overturning circulation during the next millennia. Geophys Res Lett 34: L10707
27. Mikolajewicz U, Vizcaíno M, Jungclaus J, Schurgers G (2007) Effect of ice sheet interactions in anthropogenic climate change simulations. Geophys Res Lett 34(L18706). doi:10.1029/2007GL031173
28. Pattyn F (2003) A new three-dimensional higher-order thermomechanical ice sheet model: basic sensitivity, ice stream development and ice flow across subglacial lakes. J Geophys Res 108(B8):2382. doi:10.1029/2002JB002329

29. Pattyn F, Huyghe A, Brabander SD, Smedt BD (2006) The role of transition zones in marine ice sheet dynamics. J Geophys Res 111(F02004). doi:10.1029/2005JF000394
30. Pattyn F et al (2008) Benchmark experiments for higher-order and full stokes ice sheet models (ISMIP-HOM). The Cryosphere 2:95–108
31. Payne AJ, Vieli A, Shepherd A, Wingham DJ, Rignot E (2004) Recent dramatic thinning of largest west-Antarctic ice stream triggered by oceans. Geophys Res Lett 31:L23401
32. Price SF, Conway H, Waddington ED, Bindschadler RA (2008) Model investigations of inland migration of fast-flowing outlet glaciers and ice streams. J Glaciol 54:49–60
33. Oerlemans J, Anderson B, Hubbard A, Huybrechts P, Jóhannesson T, Knap WH, Schmeits M, Stroeven AP, van de Wal RSW, Wallinga J, Zuo Z (1998) Modelling the response of glaciers to climate warming. Clim Dyn 14:267–274
34. Blatter H (1995) Velocity and stress fields in grounded glaciers: a simple algorithm for including deviatoric stress gradients. J Glaciol 41:333–344
35. Albrecht O, Jansson P, Blatter H (2000) Modelling glacier response to measured mass balance forcing. Ann Glaciol 31:91–96
36. Schneeberger C, Albrecht O, Blatter H, Wild M, Hock R (2001) Modelling the response of glaciers to a doubling in atmospheric co2: a case study of storglaciaren, northern Sweden. Clim Dyn 17(11):825–834
37. Pattyn F (2002) Transient glacier response with a higher-order numerical ice-flow model. J Glaciol 48(162):467–477
38. Jarosch AH (2008) Icetools: a full stokes finite element model for glaciers. Comput Geosci 34(8):1005–1014
39. Rutt IC, Hagdorn M, Hulton NRJ, Payne AJ (2009) The glimmer community ice sheet model. J Geophys Res 114(F02004). doi:10.1029/2008JF001015
40. Coon MD, Maykut GA, Pritchard RS, Rothrock DA, Thorndike AS (1974) Modeling the pack ice as an elastic–plastic material. AIDJEX Bull 24:1–105
41. Coon MD (1980) A review of AIDJEX modeling. In: Pritchard RS (ed) Sea ice processes and models. University of Washington Press, Seattle, pp 12–27
42. Hibler WD (1980) Modeling pack ice as a viscous-plastic continuum: some preliminary results. In: Pritchard RS (ed) Sea ice processes and models. University of Washington Press, Seattle, pp 163–176
43. Hibler WD (1980) Modeling a variable thickness ice cover. Mon Weather Rev 108:1943–1973
44. Thorndike AS, Rothrock DS, Maykut GA, Colony R (1975) The thickness distribution of sea ice. J Geophys Res 80:4501–4513
45. Untersteiner N (1961) On the mass and heat budget of arctic sea ice. Arch Meteorol Geophys Bioklimatol Ser A 12:151–182
46. Maykut GA, Untersteiner N (1971) Some results from a time-dependent thermo-dynamic model of sea ice. J Geophys Res 76:1550–1575
47. Semtner AJ (1976) A model for the thermodynamic growth of sea ice in numerical investigaions of climate. J Phys Oceanogr 6:379–389
48. Washington WM, Meehl GA (1989) Climate sensitivity due to increased CO_2: experiments with a coupled atmosphere and ocean general circulation model. Clim Dyn 8:211–223
49. Manabe S, Stouffer RJ, Spellman MJ, Bryan K (1991) Transient responses of a coupled ocean–atmosphere model to gradual changes of atmospheric CO_2. Part I. Annual mean response. J Clim 4:785–818
50. McFarlane NA, Boer GJ, Blanchet J-P, Lazare M (1992) The Canadian climate centre second-generation general circulation model and its equilibrium climate. J Clim 5:1013–1044
51. Flato GM, Hibler WD III (1992) Modeling pack ice as a cavitating fluid. J Phys Oceanogr 22:626–651
52. Hunke EC, Dukowicz JK (1997) An elastic-viscous-plastic model for sea ice dynamics. J Phys Oceanogr 27:1849–1867

53. Zhang J, Hibler WD III (1997) On an efficient numerical method for modeling sea ice dynamics. J Geophys Res 102:8691–8702
54. Bitz CM, Holland MM, Weaver AJ, Eby M (2001) Simulating the ice-thickness distribution in a coupled climate model. J Geophys Res 106:2441–2464
55. Holland MM, Bitz CM, Hunke EC, Lipscomb WH, Schramm JL (2006) Influence of the sea ice thickness distribution on polar climate in CCSM3. J Clim 19:2398–2414
56. Goodrich LE (1978) Some results of a numerical study of ground thermal regimes, vol 1. National Research Council Canada, Ottawa, pp 29–34
57. Goodrich LE (1978) Efficient numerical technique for one-dimensional thermal problems with phase change. Int J Heat Mass Transf 21:615–621
58. Jordan R (1991) A one-dimensional temperature model for a snow cover. Special Report 91–16, Technical report. Cold Regions Research and Engineering Laboratory, Hanover, 49pp
59. Loth B, Graf H-F, Oberhuber JM (1993) Snow cover model for global climate simulations. J Geophys Res 98:10451–10464
60. Douville H, Royer JF, Mahfouf JF (1995) A new snow parameterization for the Météo-France climate model. 1. Validation in stand-alone experiments. Clim Dyn 12:21–35
61. Robock A, Vinnikov KY, Schlosser CA, Speranskaya NA, Xue Y (1995) Use of midlatitude soil moisture and meteorological observations to validate soil moisture simulations with biosphere and bucket models. J Clim 8:15–35
62. Yang ZL, Pitman AJ, McAvaney B, Sellers AH (1995) The impact of implementing the bare essentials of surface transfer land surface scheme into the BMRC GCM. Clim Dyn 11:279–297
63. Slater AG, Pitman AJ, Desborough CE (1998) The validation of a snow parameterization designed for use in general circulation models. Int J Clim 18:595–617
64. Slater AG, Pitman AJ, Desborough CE (1998) Simulation of freeze-thaw cycles in a general circulation model land surface scheme. J Clim 103:11,303–11,312
65. Dery S, Tremblay L-B (2004) Modelling the effects of wind redistribution on the snow mass budget of polar sea ice. J Phys Oceanogr 34:258–271
66. Oleson KW et al (2010) Technical description of version 4.0 of the community land model version (CLM). NCAR/TN-478+STR. National Center for Atmospheric Research, Boulder, CO, USA. http://www.cesm.ucar.edu/models/cesm1.0/clm/CLM4_Tech_Note.pdf
67. Clarke GKC (2005) Subglacial processes. Ann Rev Earth Planet Sci 33:247–276
68. Paterson WSB (1994) The physics of glaciers, vol 3. Elsevier, Amsterdam
69. Duval P (1981) Creep and fabrics of polycrystalline ice under shear and compression. J Glaciol 27:129–140
70. Mellor M, Cole DM (1982) Deformation and failure of ice under constant stress or constant strain-rate. Cold Reg Sci Technol 5:201–219
71. Hooke RL (1981) Flow law for polycrystalline ice in glaciers: comparison of theoretical predictions, laboratory data, and field measurements. Rev Geophys Space Phys 19:664–672
72. Adalgeirsdóttir G, Gudmundsson G, Bjornsson H (2000) The response of a glacier to a surface disturbance: a case study on Vatnajkull ice cap, Iceland. Ann Glaciol 31:104–110
73. Alley RB (1992) Flow-law hypotheses for ice-sheet modelling. J Glaciol 38:245–256
74. Shoji H, Langway CC (1988) Flow-law parameters of the dye 3, Greenland, deep ice core. Ann Glaciol 10:146–150
75. Thorsteinsson T, Waddington ED, Taylor KC, Alley RB, Blankenship DD (1999) Strain-rate enhancement at dye 3, Greenland. J Glaciol 45:338–345
76. Thorsteinsson T, Waddington ED, Fletcher RC (2003) Spatial and temporal scales of anisotropic effects in ice-sheet flow. Ann Glaciol 37:40–48
77. Cuffey KM, Thorsteinsson T, Waddington ED (2000) A renewed argument for crystal size control of ice sheet strain rates. J Geophys Res 105(B12):27,889–27,894
78. Durham W, Stern L, Kirby S (2001) Rheology of ice i at low stress and elevated confining pressure. J Geophys Res 106(B6):11,031–11,042

79. Paterson WSB (1991) Why ice-age ice is sometimes 'soft'. Cold Reg Sci Technol 20(1):75–98
80. Bindschadler RA (1983) The importance of pressurised subglacial water in separation and sliding at the glacier bed. J Glaciol 29:3–19
81. Clarke GKC (1987) Fast glacier flow: ice streams, surging, and tidewater glaciers. J Geophys Res 92:8835–8841
82. Iken A, Bindschadler RA (1986) Combined measurements of subglacial water pressure and surface velocity of Findelengletscher, Switzerland: conclusions about drainage system and sliding mechanism. J Glaciol 32(110):101–119
83. Kamb B (1987) Glacier surge mechanism based on linked cavity configuration of the basal water conduit system. J Geophys Res 92(B9):9083–9100
84. Copland L, Sharp MJ, Nienow P (2003) Links between short-term velocity variations and the subglacial hydrology of a predominantly cold polythermal glacier. J Glaciol 49:337–348
85. Zwally HJ, Herring T, Abdalati W, Larson K, Saba J, Steffen K (2002) Surface melt-induced acceleration of Greenland ice-sheet flow. Science 297:218–222
86. Joughin I, Das SB, King MA, Smith BE, Howat IM, Moon T (2008) Seasonal speedup along the western flank of the Greenland ice sheet. Science 320(5877):781–783
87. Alley RB (2000) Water pressure coupling of sliding and bed deformation. Space Sci Rev 92:295–310
88. Kamb B (1991) Rheological nonlinearity and flow instability in the deforming bed mechanism of ice stream flow. J Geophys Res 96(B10):16,585–16,595
89. Tulaczyk SM, Kamb B, Engelhardt H (2001) Basal mechanics of ice stream b, west Antarctica 1. Till mechanics. J Geophys Res 105(B1):463–481
90. Clark PU, Alley RB, Pollard D (1999) Northern hemisphere ice-sheet influences on global climate change. Science 286:1104–1111
91. Marshall SJ, Björnsson H, Flowers GE, Clarke GKC (2005) Modeling Vatnajökull ice cap dynamics. J Geophys Res 110(F03009). doi:10.1029/2004JF000262
92. Flowers GE, Marshall SJ, Björnsson H, Clarke GKC (2005) Sensitivity of Vatnajökull ice cap hydrology and dynamics to climate warming over the next two centuries. J Geophys Res 110(F02011). doi:10.1029/2004JF000200
93. Ritz C, Fabre A, Letréguilly A (1997) Sensitivity of a Greenland ice sheet model to ice flow and ablation parameters: consequences for the evolution through the last glacial cycle. Clim Dyn 13:11–24
94. Kamb B, Echelmeyer KA (1986) Stress-gradient coupling in glacier flow: I. Longitudinal averaging of the influence of ice thickness and surface slope. J Glaciol 32:267–279
95. Marshall SJ, Clarke GKC (1997) A continuum mixture model of ice stream thermomechanics in the Laurentide ice sheet, 2. Application to the Hudson strait ice stream. J Geophys Res 102(B9):20,615–20,638
96. MacAyeal DR, Bindschadler RA, Scambos TA (1995) Basal friction of Ice stream E, west Antarctica. J Glaciol 41:247–262
97. Vieli A, Payne AJ (2003) Application of control methods for modelling the flow of pine island glacier, west Antarctica. Ann Glaciol 36:197–203
98. Flowers GE, Clarke GKC (2002) A multicomponent coupled model of glacier hydrology. 1. Theory and synthetic example. J Geophys Res 107(B11). doi:10.1029/2001JB001122
99. Johnson J, Fastook JL (2002) Northern hemisphere glaciation and its sensitivity to basal melt water. Quat Int 95–96:65–74
100. Arnold NS, Sharp MJ (2002) Flow variability in the Scandinavian ice sheet: modelling the coupling between ice sheet flow and hydrology. Quat Sci Rev 21:485–502
101. Conway H, Hall BL, Denton GH, Gades AM, Waddington ED (1999) Past and future grounding-line retreat of the west Antarctic ice sheet. Science 286:280–283
102. Paul F, Kotlarski S (2010) Forcing a distributed glacier mass balance model with the regional climate model REMO. Part II: downscaling strategy and results for two Swiss glaciers. J Clim 23(6):1607–1620

103. Toniazzo T, Gregory JM, Huybrechts P (2004) Climate impact of a Greenland deglaciation and its possible irreversibility. J Clim 17:21–33
104. Ridley JK, Gregory J, Huybrechts P, Lowe J (2010) Thresholds for irreversible decline of the Greenland ice sheet. Clim Dyn 35(6):1049–1057
105. Vizcaíno M, Mikolajewicz U, Gröger M, Maier-Reimer E, Schurgers G, Winguth AME (2008) Long-term ice sheet-climate interactions under anthropogenic greenhouse forcing simulated with a complex earth system model. Clim Dyn 31:665–690
106. Holland DM, Thomas RH, Young BD, Ribergaard MH, Lyberth B (2008) Acceleration of Jakobshavn Isbræ triggered by warm subsurface ocean waters. Nat Geosci 1:659–664
107. Walsh JE, Hibler WD III, Ross B (1985) Numerical simultion of northern hemisphere sea ice variability 1951–1980. J Geophys Res 90:4847–4865
108. Hibler WD (1979) A dynamic thermodynamic sea ice model. J Phys Oceanogr 9:815–846
109. Lipscomb WH, Hunke EC, Maslowski W, Jakacki J (2007) Improving ridging schemes for high-resolution sea ice models. J Geophys Res 112:C03S91. doi:10.1029/2005JC003355
110. Rothrock DA (1975) The energetics of the plastic deformation of pack ice by ridging. J Geophys Res 80:4514–4519
111. Hopkins MA, Hibler WD III (1991) On the ridging of a thin sheet of lead ice. Ann Glaciol 15:81–86
112. Flato GM, Hibler WD III (1995) Ridging and strength in modelling the thickness distribution of Arctic sea ice. J Geophys Res C9:18,611–18,626
113. Dennis JM, Tufo HM (2008) Scaling climate simulation applications on the IBM blue gene/L system. IBM J Res Dev Appl Massively Parallel Syst 52(1/2):117–126
114. Box JE, Bromwich D, Veenhuis BA, Bai L-S, Stroeve JC, Rogers JC, Steffen K, Haran T, Wang S-H (2006) Greenland ice sheet surface mass balance variability (1988–2004) from calibrated polar MM5 output. J Clim 19(12):2783–2800
115. Fettweis X, Hanna E, Gallee H, Huybrechts P, Erpicum M (2008) Estimation of the Greenland ice sheet surface mass balance for the 20th and 21st centuries. The Cryosphere 2:117–129
116. Wake LM, Huybrechts P, Box JE, Hanna E, Janssens I, Milne GA (2009) Surface mass-balance changes of the Greenland ice sheet since 1866. Ann Glaciol 50:178–184
117. Hopkins MA (1996) On the mesoscale interaction of lead ice and floes. J Geophys Res 101:18315–18326
118. Kwok R (2001) Deformation of the Arctic ocean sea ice cover between November 1996 and April 1997: a qualitative survey. In: Dempsey J, Shen H, Shapiro L (eds) IUTAM scaling laws in Ice mechanics and ice dynamics. Kluwer Academic, Dordrecht, pp 315–322
119. Hutchings J, Heil P, Hibler W (2005) Modeling linear kinematic features in sea ice. Mon Weather Rev 133:3481–3497
120. Hibler WD, Schulson M (2000) On modeling the anisotropic failure and flow of flawed sea ice. J Geophys Res 105:17,105–17,120
121. Schreyer H, Sulsky D, Munday L, Coon M, Kwok R (2006) Elastic-decohesive constitutive model for sea ice. J Geophys Res 111:C11S26. doi:10.1029/2005JC003344
122. Coon M, Kwok R, Levy G, Prius M, Schreyer H, Sulsky D (2007) Arctic ice dynamics joint experiment (AIDJEX) assumptions revisited and found inadequate. J Geophys Res 112: C11S90. doi:10.1029/2005JC003393
123. Arrigo KR, Kremer JN, Sullivan CW (1993) A simulated Antarctic fast ice ecosystem. J Geophys Res 98:6929–6946
124. Arrigo KR, Worthen DL, Lizotte MP, Dixon P, Dieckmann G (1997) Primary production in Antarctic sea ice. Science 276:394–397
125. Lavoie D, Denman K, Michel C (2005) Modeling ice algal growth and decline in a seasonally ice-covered region of the arctic (resolute passage, Canadian archipelago). J Geophys Res 110: C11009. doi:10.1029/2005JC002922

126. Jin M, Deal CJ, Wang J, Tanaka N, Ikeda M (2007) Vertical mixing effect on the phytoplankton bloom in the southeastern Bering Sea midshelf. J Geophys Res 111:C03002. doi:10.1029/2005JC002994
127. Vancoppenolle M, Goosse H, de Montety A, Fichefet T, Tremblay B, Tison J-L (2010) Modelling brine and nutrient dynamics in Antarctic sea ice: the case of dissolved silica. J Geophys Res 115:C02005. doi:10.1029.2009JC005369
128. Bitz CM, Ridley JK, Holland MM, Cattle H (in press) Global climate models and 20th and 21st century arctic climate change. In: Lemke P (ed) Arctic climate change – the ACSYS decade and beyond. Springer
129. Greve, Blatter (2009) Dynamics of ice sheets and glaciers, Springer-Verlag, Berlin, pp 287
130. MacAyeal DR, Bindschadler RA, Scambos TA (1995) Basal friction of Ice Stream E, West Antarctica. J Glacio 41:247–262

Chapter 4
Oceanic General Circulation Models

Jin-Ho Yoon and Po-Lun Ma

Glossary

Bathymetry	The depth of ocean floor. Bathymetric map represents the terrain of the seafloor.
Biogeochemical cycle	A pathway that is described by the physical, biological, and chemical processes that control the evolution of elements found in the Earth system.
Boussinesq approximation	In Boussinesq approximation, a fluid parcel is assumed to maintain the same volume or density during its transport because of near incompressibility. This approximation was named after French physicist and mathematician, Joseph Valentin Boussinesq. By adopting this approximation, sound waves that propagate through a density change can be eliminated in numerical model.
Diapycnal mixing	Mixing of a fluid across different density surfaces. To be contrasted with isopycnal mixing which occurs along the same density surfaces.
El Niño-Southern Oscillation (ENSO)	A quasi-periodic change of the ocean and atmospheric conditions along the equatorial Pacific Ocean. The change in the sea surface temperature (SST) can be as large as $\pm 2°C$ during its extreme phases: anomalously warm over the tropical Pacific (El Niño) and cold (La Niña). Surface

This chapter was originally published as part of the Encyclopedia of Sustainability Science and Technology edited by Robert A. Meyers. DOI:10.1007/978-1-4419-0851-3

J.-H. Yoon (✉) • P.-L. Ma
Atmospheric Science and Global Change Division, Pacific Northwest National Laboratory, 902 Battelle Boulevard, P. O. Box 999, MSIN K9-24, Richland, WA 99352, USA
e-mail: Jin-Ho.Yoon@pnnl.gov; Po-Lun.Ma@pnnl.gov

P.J. Rasch (ed.), *Climate Change Modeling Methodology: Selected Entries from the Encyclopedia of Sustainability Science and Technology*,
DOI 10.1007/978-1-4614-5767-1_4, © Springer Science+Business Media New York 2012

	air pressures measured at both ends of the tropical Pacific basin vary closely with the change of SST.
Geostrophic approximation	The angular momentum is balanced by the Coriolis force and the horizontal pressure gradient force. It is generally true when the spatial and temporal scales are large, roughly over 100 km and a few days in the deep ocean.
Hydrostatic approximation	The equation describing vertical motion of the ocean column is simplified to assume that the vertical pressure at any level is due to the weight of the air and water above it. Variation of density is considered only in vertical direction when gravitational acceleration term (g) exists. This is valid when the vertical scale of a feature is small compared to the horizontal scale for both the atmosphere and ocean.
Isopycnal coordinate	Vertical coordinate that follows a constant density surface.
Meridional overturning circulation (MOC)	This has often been assumed to be the same as the thermohaline circulation. However, the MOC explicitly describes the ocean circulation system with the upwelling/downwelling and associates the northward/southward transport.
Ocean gyre	A large-scale rotating circulation in the ocean primarily forced by the atmospheric wind and the Coriolis force. These include the North Atlantic Gyre, South Atlantic Gyre, Indian Ocean Gyre, North Pacific Gyre, and South Pacific which tend to be more elongated in the east-west direction. There are also other types of Gyre forced by different mechanisms such as baroclinicity.
(Oceanic) Mesoscale eddy	A vigorous rotational circulation or vortex at spatial scales roughly 100 km and smaller, existing for weeks to months.
Rossby radius (of deformation)	The horizontal scale at which rotational effect becomes as important as buoyancy or gravity wave effects. Mathematically, this can be computed in terms of potential temperature, temperature, wind speed, or the depth of the boundary layer. This radius is important in determining the phase speed and wavelength of Rossby waves.
Salinity	Dissolved content of the salt in the ocean. Traditionally, salinity is represented in the unit of either g/Kg or PSU (Practical Salinity Unit).
Shallowness approximation	This approximation can be applied when the vertical-to-horizontal aspect ratio is very small.
Structured (regular)/ unstructured (irregular) grid	A structured (unstructured) grid has regular (irregular) connectivity with neighboring points. In structured (unstructured) grid, its connectivity can (cannot) be easily represented with a two- or three-dimensional array.
Thermocline	A distinct ocean layer where the temperature changes greatly with its depth compared to the layers above and

	below. It is often thought of as a boundary separating the well-mixed upper ocean and the deep ocean.
Thermohaline circulation	The global oceanic circulation driven by the density gradients, primarily determined by salinity and temperature.
(Atmospheric) Wind stress	The horizontal force exerted by the atmospheric wind on the ocean surface. This can also be interpreted as the vertical transfer of horizontal momentum from the atmosphere to the ocean surface. Wind stress is a function of the square of the wind speed.

Definition of the Subject

The purpose of this text is to provide an introduction to aspects of oceanic general circulation models (OGCMs), an important component of Climate System or Earth System Model (ESM). The emerging need for understanding the Earth's climate system and especially projecting its future evolution has encouraged scientists to explore the dynamical, physical, and biogeochemical processes in the ocean. Understanding the role of these processes in the climate system is an interesting and challenging scientific subject. For example, a research question exploring how much extra heat or CO_2 generated by anthropogenic activities can be stored in the deep ocean is not only of scientific interest but also important in projecting future climate of the Earth. Thus, OGCMs have been developed and applied to investigate the various oceanic processes and their role in the climate system.

Coupled climate models incorporating some representation of ocean circulations have been used since the pioneering work of Manabe and Bryan [38]. Because of computational limits at that time, an extremely simplified Earth geography – composed of only one continent and ocean basin (Fig. 4.1) – was used. Despite its simplicity, this was the first numerical model to include reasonably complex formulations for the atmosphere, land, ocean, and more importantly feedbacks among these components. Later it was demonstrated that deep ocean circulation in this simplified coupled model was not at an equilibrium even after a century-long integration [64]. Like this case, processes that control the adjustment of deep ocean circulation toward an equilibrium remain an important issue in contemporary climate models. It took another 6 years for Manabe and Bryan to produce a coupled climate model with more realistic geography of the Earth [39]. Now contemporary OGCMs can resolve complex regional flows in a much more realistic way. However, there are still many issues to be solved and large room for improvement [48].

The goal in this text is to outline the basics of Ocean General Circulation modeling, reviewing the mathematical representation of the ocean circulation

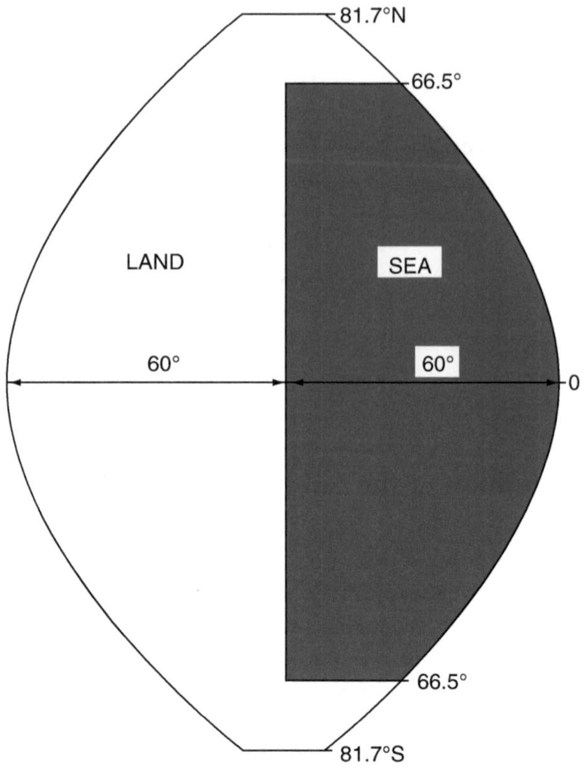

Fig. 4.1 Land-ocean configuration of the first coupled climate model of [38] (From Manabe and Bryan, J. Atmospheric Sciences 26 (1969): 786–789. © American Meteorological Society. Reprinted with permission)

(section Equations of Motion), and the discretization method with horizontal and vertical coordinate systems (section Horizontal and Vertical Grid System). A brief discussion is provided on how the sub-grid scale processes are represented in relatively coarse-resolution OGCMs (section Sub-grid Scale Parameterization). Then, using a couple of relatively simple examples, the coupling between the ocean and other climate components, especially the atmosphere by various physical processes will be explained (section Simple Conceptual Models). Also, how the ocean is coupled in terms of biogeochemical cycle will be illustrated in the section Biogeochemical Cycle Modeling. The entry concludes with a brief description of contemporary OGCMs and its future development (section Future Directions).

Introduction

An oceanic general circulation model (OGCM) is used to simulate the physical and dynamical features of the global ocean using equations representing the conservation of momentum, energy, and mass for water and salt. These equations are usually called the equations of motion, and they describe how the temperature, salinity, and currents evolve with time in the ocean. OGCMs also need appropriate boundary and initial conditions that describe how the ocean interacts with the atmosphere, land and sea ice and the state of the ocean when the model is started. An OGCM has a great similarity with its atmospheric counterpart in the entry Atmospheric General Circulation Modeling. However, there are important and non-negligible differences. First of all, the ocean is primarily forced at its surface boundaries by both thermal and mechanical forcings, while the atmosphere is forced throughout its entire volume mainly by thermal forcing. The global ocean is more strongly constrained by its complex terrain which results in oceanic gyres confined by lateral boundaries (continents and shelves), while the atmosphere is relatively less restricted. Also, very narrow boundary layers exist on almost every ocean surface and within its interior. Finally, observing the ocean is much more difficult than its atmospheric counterpart. The amount of ocean observation data is far less available than in the atmosphere. Thus, boundary and initial conditions are difficult to determine and the lack of long term and uniform observation imposes a challenge in validating the OGCMs' performance. Although a great deal of additional measurements has become available since satellite and buoys which have been deployed in the 1980s, it is estimated that data available to oceanic scientist in the 1990s is still an order of magnitude fewer than in the atmosphere [21].

The role played by the ocean in the global climate system is relatively well understood. For example, the ocean regulates the Earth surface heat budget through its large heat capacity and heat transport by the oceanic circulation from tropics to high latitudes. Large portions of this heat transport is carried out by both strong western boundary currents and oceanic mesoscale eddies (a dominant dynamical feature of the global ocean circulation at spatial scales from roughly 100 km and smaller with time scales of weeks to months). One example of a mesoscale eddy can be seen near the Atlantic coast of the USA in Fig. 4.2. The figure shows satellite-retrieved sea surface temperature from the NASA Aqua/MODIS mission over the western Atlantic where gray areas represent the land and color indicates temperature over the ocean surface. One can observe that generally warmer (colder) water exists in the south (north). Along the boundary that warm and cold waters are interwined, small-scale meanders and eddies are evident. These eddies are generated by strong shear instabilities in ocean currents, occurring particularly from currents along western boundaries and by horizontal density gradients. Oceanic kinetic energy is largely redistributed and transported by these eddies, and they are very important in the poleward transport of energy in the Earth's climate system.

In order for OGCMs to properly simulate these mesoscale eddies, they must operate at a high spatial resolution, resolving features that can be a few tens of kilometers and smaller. OGCMs divide the ocean up into small "cells" where temperature, salinity, and water velocity are calculated (discussed in more detail later). Ocean models that are designed to reproduce the behavior of eddies accurately (called Eddy-Permitting models) contain many cells and are extremely expensive in terms of computation. As of 2007, only a couple of the coupled

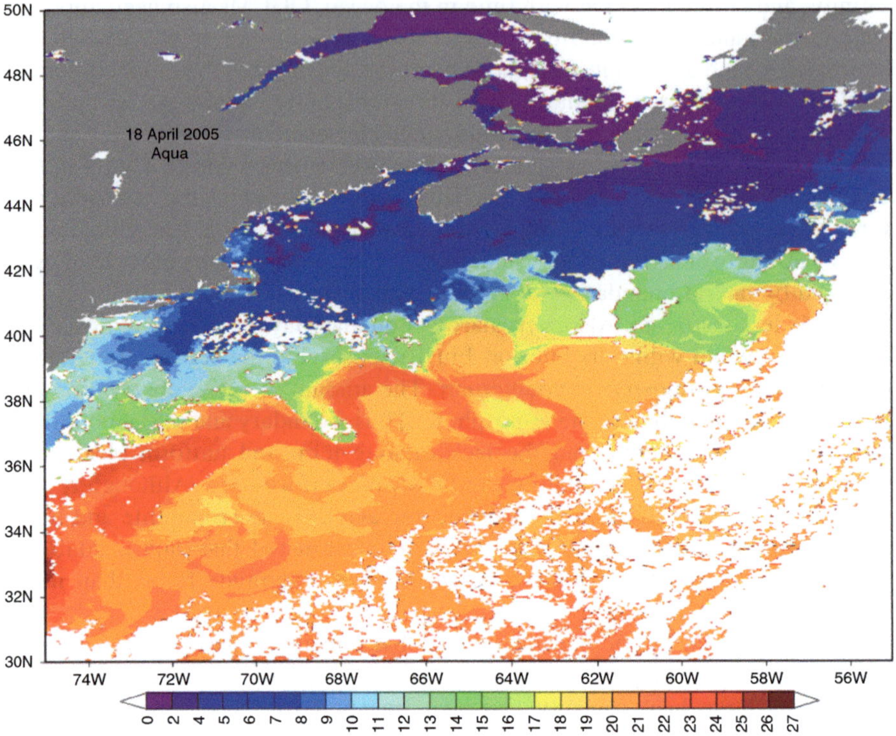

Fig. 4.2 Sea surface temperature at 18 April 2005 retrieved from Aqua MODIS [63]. Level 3 mapped data of MODIS SST was obtained from NASA JPL

climate models with eddy-permitting resolution (around 1/6° to 1/3°) successfully simulated these features explicitly [48].

OGCMs have historically been run at a much lower resolution to reduce the high computational burden. But since these lower resolution models are not able to explicitly represent these eddies, they must use other ways to represent the eddy effects. These effects are frequently treated through "parameterization" (more information on parameterization is provided below Sub-grid Scale Parameterization).

Recently ocean models now include representations of other important ocean components such as nutrient transports, ocean chemistry, and ecosystem evolution in order to calculate the evolution of biogeochemical cycles of the planet, including CO_2. The ocean regulates the amount of CO_2 stored through both biological and physical processes, which in general are similar to their terrestrial counterpart. The ocean absorbed about half of the CO_2 released in the last few centuries (e.g., [31]). And it becomes more important to understand how much of extra carbon can the ocean absorb in the near future.

This document has outlined the basis of the dynamical and physical properties of the ocean. More information on ocean processes can be found in various textbooks or other review articles (e.g., [24, 29]). In the next section, more emphasis will be given on the numerical formulation, and discretization of these processes will be provided, along with some information on the coupling between atmosphere and the ocean in a numerical form.

Equations of Motion

Ocean models solve the equations governing oceanic flow using computational fluid dynamics (CFD) methods with specific assumptions and approximations appropriate to the ocean. The equations for global ocean models are, like atmospheric models, based on Newtonian mechanics and irreversible thermodynamics applied to a particular fluid. Conservation of momentum, heat, and mass of constituents comprise the equations used for an OGCM. However, a wide range of mathematical formulation has been used to represent the numerous oceanic processes and phenomena occurring at various spatial and temporal scales, in the ocean. Only the primitive equations of motion that are the core of OGCM numerical formulations are described. A more complete derivation can be found in Griffies [24] and Vernois [62].

Hydrostatic Primitive Equations

A number of approximations are used in simplifying the fluid equations when used to represent ocean circulations, namely, the hydrostatic, shallowness, Boussinesq, and rigid lid approximations. The hydrostatic approximation assumes that the vertical momentum equation can be simplified by neglecting all terms except those resulting in the vertical pressure gradient being balanced by the gravitational acceleration (equivalently that the vertical pressure at any level is due to the weight of the air and water above it). The shallowness approximation assumes the ocean is thin compared to the Earth's radius. The Boussinesq approximation neglects variations in density except where density appears multiplied with gravity. The advantage of the Boussinesq approximation is to eliminate acoustic waves [7]. Although this approximation has been applied in many OGCMs, a couple of limitations seem to be found. One of the most noticeable limitations is that it cannot account for steric effect, i.e., thermal expansion and salinity-density compensation of sea water. Thus, sea-level change due to global warming cannot be properly predicted by this kind of model. Finally, the surface elevation of the global ocean is assumed static in the rigid lid approximation. With these approximations, the temporal and spatial variations of density are small compared to its mean value. Thus, it can be described in the following form:

$$\rho_{\text{tot}} = \bar{\rho} + \rho(x, y, z, t), \tag{4.1}$$

where $\bar{\rho}$ represents the mean density of the ocean and $\bar{\rho} \gg |\rho|$ is true and ρ can be replaced by $\bar{\rho}$ except the buoyancy term, ρg and in the thermodynamic density conservation equation.

With these approximations, the hydrostatic primitive equation in Cartesian coordinates is described in the following format:

$$\frac{du}{dt} = fv - \frac{1}{\bar{\rho}} \frac{\partial p}{\partial x} \tag{4.2}$$

$$\frac{dv}{dt} = -fu - \frac{1}{\bar{\rho}} \frac{\partial p}{\partial y} \tag{4.3}$$

$$\frac{\partial p}{\partial z} = -\rho g \tag{4.4}$$

$$\frac{dT}{dt} = 0 \tag{4.5}$$

$$\frac{dS}{dt} = 0 \tag{4.6}$$

and

$$\frac{\partial u}{\partial x} + \frac{\partial v}{\partial y} + \frac{\partial w}{\partial z} = 0, \tag{4.7}$$

where (x, y, z) represents Cartesian coordinates and (u, v, w) as current in the direction of x, y, and z, respectively. $\bar{\rho}$ is the mean density of the ocean as in Eq. 4.1, f is the Coriolis frequency, and T and S are temperature and salinity of the ocean. And the total derivative and local derivatives are related in the following way:

$$\frac{d}{dt} = \frac{\partial}{\partial t} + u\frac{\partial}{\partial x} + v\frac{\partial}{\partial y} + w\frac{\partial}{\partial z} \tag{4.8}$$

To close the set of equations (Eqs. 4.2–4.6), an equation of state, another equation of state to link density to ocean temperature, salinity, and pressure is added:

$$\rho = \rho(T, S, p) \tag{4.9}$$

It is also noted here that these sets of equation are written for spherical geometry in a global ocean model. Details can be found in Haidvogel and Bryan [29] or Bryan [5].

The above sets of equation are derived from the fluid dynamics with the help of approximations or assumptions. Some of the approximations, like Boussinesq approximation, have been still used in OGCM. Others, like the rigid lid approximation are not used any longer. There is a trend toward more explicitly representing the ocean status and processes. More realistic simulation of the ocean with better computational efficiency is now being achieved in contemporary OGCMs (see [27]).

Boundary Conditions

To solve the equations of motions, boundary conditions are needed. Upper boundary conditions are governed by interactions with the atmosphere or sea ice. The surface boundary condition on momentum below an ice-free atmosphere is governed by the wind stress:

$$\tau_x = \bar{\rho} \nu_x \frac{\partial u}{\partial z}\bigg|_{z=0}, \tag{4.10}$$

$$\tau_y = \bar{\rho} \nu_y \frac{\partial v}{\partial z}\bigg|_{z=0}, \tag{4.11}$$

where ν_x and ν_y are the turbulent eddy viscosity coefficients. The wind stress is sometimes provided by observational estimates (from a measurement climatology) and other times is provided by an atmospheric model. More discussion will be given later in Atmospheric forcing section.

The boundary condition for temperature is also treated in a variety of ways. When ocean models are coupled with a climate model, they calculate turbulent fluxes of heat and radiative fluxes based upon the state of the atmosphere and ocean. When ocean models are driven with observations, other choices are possible. Sometimes sea surface temperature (SST) is specified. In other situations surface fluxes are calculated in a similar fashion to the way they are done when coupled to an atmospheric model.

Salinity boundary condition depends upon evaporation, and the freshwater flux from precipitation, river runoff, and sea ice melt. In simple models where feedback between salinity and its forcing is not considered, one can use the prescribed surface salinity flux. However, it is found that there is important feedback between salinity and its forcing even in the case when ocean models are not coupled to atmosphere and sea ice models (e.g., [6]).

The Surface Mixed Layer

In the ocean, the heating term can be neglected generally except in the regions where solar radiation can play a role and heat exchange with the overlaying atmosphere is active, which is called the surface mixed layer. Both vigorous mixing and surface heating play important roles. Therefore, the surface heating term has to be included in the temperature equation:

$$\frac{\partial T}{\partial t} = -\frac{\partial}{\partial z}(Tw) + \frac{1}{\bar{\rho} c_p} Q(z,t), \tag{4.12}$$

where the first term in the right hand side represents the heat exchange due to vertical mixing, the second term is the solar heating absorbed at depth in the water column, and c_p is the specific heat of water. Although the equation looks simple, complication arises how to close. Two general approaches are taken: differential and bulk models [33]. The Mellor Yamada level-2 model is an example of the former [43], in which the stress terms now include both eddy coefficients for both within and below the mixed layer. On the other hand, the Niler model [41] uses a bulk model in which the mixed layer is treated as a well-defined homogeneous layer.

Horizontal and Vertical Grid System

To solve the governing equations of the oceanic general circulation, these equations have to be discretized with appropriate horizontal and vertical grid systems. Various methodologies have been developed and applied in the different OGCMs. In this section, both horizontal and vertical grid systems will be reviewed.

Horizontal Grid System

OGCMs have widely adopted the finite volume and finite elements methods. These methods will be discussed here for brevity, but a few distinguishing characteristics be mentioned. Discretized governing equations are applied in flux form so scalars such as temperature, density, salinity, etc., are updated through fluxes at the boundaries of each grid, and therefore, these quantities are conserved. Structured meshes with regular distributions of neighboring grids throughout the entire domain and unstructured meshes where the number of neighboring grids varies throughout the domain are both utilized with finite volumes and finite element discretizations. Successful model implementations include orthogonal meshes (e.g., [58]), cubed sphere meshes (e.g., [1]), icosahedral meshes (e.g., [47, 51, 52]), and unstructured triangular meshes (e.g., [10, 12]). Figure 4.3 shows an example of a structured and unstructured grid.

Structured grids are computationally efficient, accurate, and convenient. However, the major shortcoming with structured grids is that the grid-spacing is not uniform. Lack of flexibility to increase resolution in coastal regions or regions with complex topography make the computation expensive if the resolution of the entire domain has to be

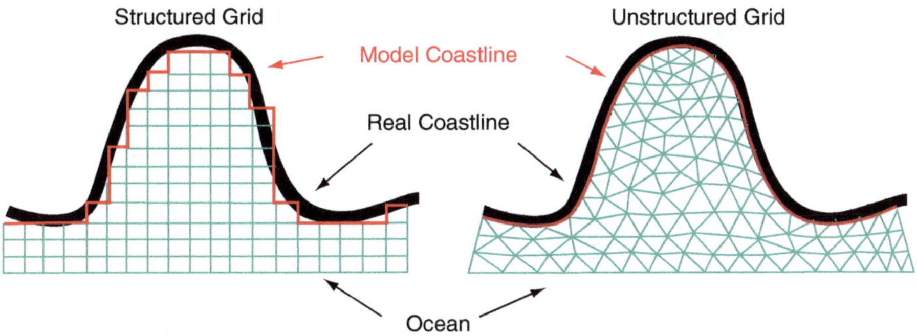

Fig. 4.3 An example of fitting a structured grid (*left*) and an unstructured grid (*right*) to a simple coastal embayment. The true coastline is shown in *black*, the model coastline in *red*. Note how the unstructured triangular grid can be adjusted so that the model coastline follows the true coastline, while the structured grid coastline is jagged – which can result in unrealistic flow disturbance close to the coast (Figure taken from Fig. 4.1 in Chen et al. [13] in *Oceanography*, reproduced with permission)

increased at the same time. An irregular gridding system that can increase resolution near the coast while keeping lower resolution over the open ocean coarse provides better computational efficiency more easily. In addition, because the unstructured grid can better represent the coastal area, it is advantageous over structured grid in simulating coastal currents and eddies.

To avoid a singular point at the North Pole, where grid lines converge, the ocean models either rotate the global gridding system, or create a tripolar grid with poles located over Canada, Russia, and Antarctica, such that the singular points of the grids are positioned over land instead of ocean.

An unresolved issue is the resolution–dependent physics in ocean models. Parameterizations of sub-grid scale processes inevitably produce some "free" (undefined) parameters (hence the name) and need to be in accordance with observations. These parameters are at varying horizontal grid sizes to capture the effects of processes from eddies with different sizes. The diffusion and viscosity terms will need to be adjusted accordingly.

Vertical Coordinate

Ocean models have developed several different vertical coordinate systems: (1) geopotential coordinate (or *z*-coordinate), (2) terrain-following coordinate, and (3) isopycnal coordinate, as shown in Fig. 4.4. Also, there is hybrid coordinate which combines various coordinate systems. Each one has its own advantages and disadvantages.

The most popular coordinate system is the geopotential coordinate [42]. For example, the first coupled climate model also utilized this vertical coordinate

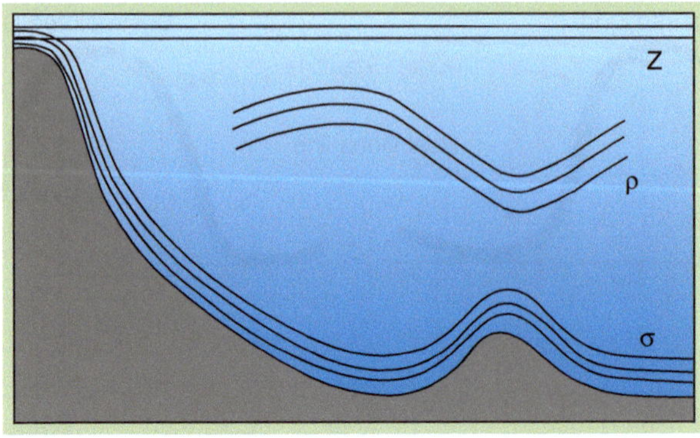

Fig. 4.4 A schematic diagram of an ocean basin illustrating the three regimes of the ocean germane to the considerations of an appropriate vertical coordinate. The surface mixed layer is naturally represented using fixed-depth z (or pressure p) coordinates, the interior is naturally represented using isopycnic ρ (density tracking) coordinates, and the bottom boundary is naturally represented using terrain-following σ coordinates (after [25]) (Figure taken from Box 1 in Chassignet et al. [11] in *Oceanography*, reproduced with permission)

system with the hydrostatic and Boussinesq approximations [26, 38]. The geopotential coordinate adopts a straightforward concept that the ocean water body is divided into regular boxes which are invariant with time. The vertical coordinate is based on the geopotential or depth of the ocean. Governing equations using lower-order finite difference methods are frequently used. In ocean models that adopt the geopotential coordinate (e.g., [35, 56]), the vertical resolution decreases as a function of depth (i.e., higher vertical resolution in the upper ocean and lower vertical resolution in the deep ocean) in order to better resolve processes around the thermocline. However, some studies have shown that the geopotential coordinate models have problems simulating the topographical effects on the ocean circulation because of their "stepwise" representation of the topography. They also have mesoscale eddy-induced vertical diffusion much greater than the observations.

The discontinued representation of the bathymetry and sidewall geometry in the geopotential coordinate system has been related to spurious topographic effects on the ocean circulation. Terrain-following (or sigma) coordinate models have been developed to handle complex lateral and bottom boundaries and avoid some of these problems. The sigma coordinate system is based on a fractional depth scaled from 0 to 1, with 0 being the surface and 1 being the bottom of the ocean. However, because the pressure gradient term is projected to be along the sigma surfaces, the horizontal pressure gradient produces considerable errors as a function of topographical slope and the stratification near the sea floor [2, 16, 32, 57]. Due to this problem, studies have shown that models that adopt the terrain-following coordinate are not suitable for global-scale simulations, unless the horizontal resolution is very high (the order of 10 km).

The isopycnal coordinate system uses isopycnal surfaces to be the vertical levels. Its advantage is due to the fact that the ocean currents and the transport of tracers generally proceed along isopycnal surfaces, so that the model that adopts this coordinate can ease the treatment of advection. The diapycnal mixing due to eddies can be added to the model as a parameterization. Furthermore, since isopycnal surfaces evolve over time, the isopycnal coordinate is, by nature, an adaptive system. It can better resolve the frontal region and thermocline where sharp vertical gradients of density exist [4]. However, for areas with a low-density gradient such as the well-mixed shelf area or deepwater formation areas, the model has a lower resolution than it would with other vertical coordinate choices. Because it avoids some of the shortcomings of the two other coordinate systems, isopycnal models, or "hybrid" models using a combination of these coordinate systems have a great potential in becoming more and more viable for global climate simulations in the future. Examples of models that use hybrid coordinates include the Global Navy Coastal Ocean Model that uses sigma-z coordinate, and the experimental Hybrid Coordinate Ocean Model that uses the z- (or sigma) coordinate in the upper ocean and the isopycnal coordinate in the stratified ocean.

Sub-grid Scale Parameterization

With spatial discretization, one can solve the governing equation discussed in the section Equations of Motion. But, representing continuous fluid with limited number of grid points imposes some new challenges. For example, the global OGCM with horizontal grid scale of 200 km cannot explicitly simulate oceanic mesoscale eddies (Fig. 4.2). To overcome this, spatial resolution of the OGCM has been increasing greatly. Indeed, increasing the ocean resolution in the coupled climate model has resulted in many improvements in the simulation of ocean circulation features. However, the impact of higher ocean resolution on the atmospheric simulation in coupled models has been found to be relatively small and localized unless the resolution of the atmosphere is also changed [30]. Further it has been shown that increasing spatial resolution plays a secondary role in simulating ENSO properly [28]. To achieve realistic representation of the ocean circulation and physical status requires not only increasing resolution but also physical parameterization. In this section, these parameterizations are discussed.

There are many processes occurring in the ocean that are not expressed explicitly in the equations of motion, or that take place at time and space scales that are too small to be treated explicitly in a global ocean model. No OGCM can simulate process in the ocean explicitly due to computational constraints or lack of knowledge about these processes. However, these can be very important in influencing oceanic circulations. Sub-grid processes such as vertical mixing and eddies with a scale smaller than the grid that cannot be resolved are "parameterized" based on theoretical considerations, observational data, and/or results from finer-resolution models. In "non-eddy" resolving (or permitting) OGCMs, the effects of mesoscale eddies are represented by parameterizations called neutral diffusion [49, 59] and eddy-induced advection [19, 20]. Maltrud and McClean [37] have suggested that a horizontal grid-spacing of 50 km in mid-latitudes and 10 km in high latitudes is sufficient to resolve eddies. Higher resolution models are believed to be able to resolve these processes.

Turbulent mixing at the upper ocean surface must also be represented (e.g., [18, 34, 45]). The stirring and mixing at the ocean surface can penetrate to the interior of the ocean and trigger further vertical mixing. These waves can be reflected, scattered, or transformed by the seafloor at the bottom and lateral boundary of the ocean. The diapycnal mixing takes place as a result of baroclinic instability, and is brought to balance by geostrophic adjustment via radiation of gravity waves. These processes interact with the topography and the mean flow. Depending on the stability of the ocean, the diapycnal mixing can affect the general circulation or dissipate without having any effect.

Many sub-grid scale processes are recognized to be important but remain poorly understood. The lack of the four-dimensional observational data of ocean flows as well as the state variables of the ocean in the oceanic interior make validation of the parameterizations against observations difficult, so validations are usually performed by evaluating their bulk effect against a high-resolution model. Recently satellite measurements of sea surface height, sea surface temperature, winds, etc. that cover a large scale of domain with a reasonable resolution have proved very helpful for understanding the features of mesoscale eddies but more in situ measurements are necessary to validate the model and to calibrate satellite observations. With the coordinated efforts from both modeling and measurement communities, parameterizations in ocean models can be scientifically evaluated and improved accordingly.

Simple Conceptual Models

Continuing atmospheric-ocean coupling process, a brief review is provided in this section on two simple ocean models which describe two important oceanic phenomena that have a large impact on the entire climate system: (1) the El Niño Southern Oscillation and (2) the thermohaline circulation. The former feature produces the largest signal in interannual climate variability in the Earth system, and the latter is a feature that is very important in longer-term climate variability.

These features are most easily described using simple models based on the full sets of mathematical equations but with their complexity reduced to make it easier to describe the basic features of particular phenomena. For example, idealized geometries with simplified boundary conditions may be applied or several approximations can be made for some regions, which are not uniformly applicable to the global ocean. However, these simple models provide very useful insight on how a specific physical process is maintained and respond to individual forcing. For example, the ocean thermocline and wind-stress effects on El Niño Southern

Oscillation can be easily understood and modeled in a Cane–Zebiak model. In this subsection, a couple of simple models will be reviewed. One is this Cane–Zebiak model and the other describes the oceanic thermohaline circulation [15].

Cane–Zebiak Model

El Niño Southern Oscillation (ENSO) is a climate pattern found in nature. The pattern is centered in the tropical pacific, but its signal can be detected globally. It is frequently described through an index measuring surface pressure differences between Tahiti and Darwin, or ocean surface temperature over the eastern Pacific but its signatures can be found in precipitation changes over tropics as well as mid-latitudes and in many other components of the Earth climate system. The feature has a very large interannual variability. On average there is an event every 3–6 years but it occurs irregularly. Various degrees of simple conceptual models have been proposed to explain and model ENSO. All these efforts can be summarized by the Bjerknes–Wyrtki model [3, 65], in which it was suggested that ENSO is an internal mode of oscillation of the coupled atmosphere-ocean system, driven by a continuous imbalance between the tightly coupled surface winds and temperatures on the one hand, and the more sluggish subsurface heat reservoir on the other. In the early 1980s, several highly idealized coupled models were developed based on the Bjerknes–Wyrtki model. The first one was that of Cane and Zebiak [9] and Zebiak and Cane [67]. Both the atmosphere and the ocean are highly simplified. These components are outlined in the next subsection.

The Atmosphere Model

Tropical atmosphere exhibits a couple of unique characteristics. One is that the wind is driven by latent heat release due to convection, and the other is a reversed polarity in the vertical structure, for example, low-level convergence with upper-level divergence in the divergent circulation. Therefore, linear dynamics with a single degree of freedom in the vertical coordinate has proven surprisingly accurate in reproducing the atmospheric circulation, classified as a Gill-type model [22, 44]. In ZC model, steady-state, linear shallow-water equation on an equatorial beta plane is used, just like a Gill-type model as follows:

$$\frac{\partial u}{\partial t} - \beta y v = -\frac{\partial \Phi}{\partial x} \tag{4.13}$$

$$\frac{\partial v}{\partial t} - \beta y u = -\frac{\partial \Phi}{\partial y} \tag{4.14}$$

$$\frac{\partial \phi}{\partial t} + c_a^2 \nabla \cdot \vec{V} = \frac{J}{C_p} \tag{4.15}$$

where $\vec{V} = (u, v)$ are the surface wind vector, ϕ is the geopotential, $\beta = df/dy$, with f being the Coriolis parameter, and c_a is a constant related to momentum and mass in the form of Newtonian cooling. Earlier, Matsuno [8] solved this equation for free tropics diabatic heating, and later Gill [22] found the steady solution for both the forced and damped versions. Lindzen and Nigam [36] introduced an alternative view on how surface wind was forced. They proposed the surface temperature gradient causes overlaying atmospheric pressure gradient and in turn surface wind anomalies.

The atmospheric heating anomaly Q is contributed by local evaporation (Q_S) and low-level moisture convergence (Q_L). It has been shown that the atmospheric moisture convergence is important for tropical heat budget and to maintain divergent circulation (e.g., [14]).

The Ocean Model

The ZC model represents the ocean as a single basin, the tropical Pacific, consisting of two vertical layers with a simplification of the ocean dynamics called a linear reduced gravity model. Vertically, the change of the thermocline needs to be considered, which is assumed motionless and linear. The set of equations is as follows:

$$\frac{\partial u}{\partial t} - \beta y v = -\frac{\Delta\rho}{\rho}\frac{\partial h}{\partial x} + \frac{\tau_{S_x}}{\rho \bar{h}} - ru\,, \tag{4.16}$$

$$\beta y u = -\frac{\Delta\rho}{\rho}\frac{\partial h}{\partial y} + \frac{\tau_{S_y}}{\rho \bar{h}} - rv\,, \tag{4.17}$$

$$\frac{\partial h}{\partial t} + h \nabla \cdot \vec{V} = -rh\,, \tag{4.18}$$

$$\frac{\partial T}{\partial t} = -\vec{V} \cdot \nabla T - M w_s \frac{\partial T}{\partial z} - \alpha_S T\,, \tag{4.19}$$

where (u,v) is the horizontal velocity in the layer, h is the derivation of the layer thickness from its mean value H, r is a Rayleigh friction parameter, τ_S is the surface wind stress, and $\Delta\rho$ a characteristic density difference between the layers. The upwelling is represented as follows:

$$w_S = H_1 \nabla \cdot \vec{V}_1. \tag{4.20}$$

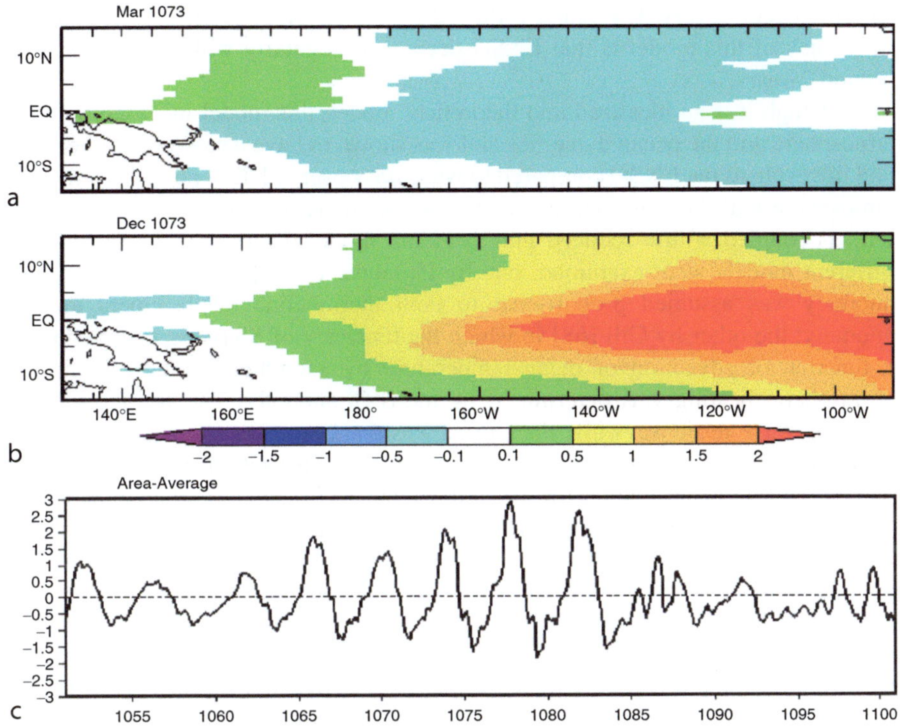

Fig. 4.5 An example of El Nino simulated by ZC model. (**a**) SST anomalies at March 1073, (**b**) same as (**a**) except December 1073, and (**c**) area averaged over the eastern equatorial Pacific (90° W to 150° W, 5° S to 5° N). The entire simulation was for the period of 1001–1200

The function M accounts for the fact that surface temperature is affected by upwelling. The anomalous temperature gradient, $\partial T/\partial z$, is defined by

$$\frac{\partial T}{\partial x} = \frac{(T - T_\mathrm{e})}{H_1}, \tag{4.21}$$

$$T_\mathrm{e} = (1-\gamma)T + \gamma T_\mathrm{d}, \tag{4.22}$$

where H_1 is the surface layer depth, T_e measures temperature entrained into the surface layer, with γ as a mixing parameter (= 0.5), and T_d relates the subsurface temperature anomaly to the mean and anomalous thermocline depths.

In summary, the ocean model is essentially a reduced gravity model with one and half layer of ocean which includes the surface mixed layer and the layer below as one layer. Thus, the surface temperature is affected by both the atmospheric wind stress at the surface and the temperature entrained into the surface layer from the deep layer.

With reasonable choice of parameters, this coupled model has produced anomalous SST pattern and wind that resembles the observed ENSO anomalies (Fig. 4.5).

El Niño occurs every 3–4 years, followed by the La Niña events. A well-known deficiency of this model is that ENSO occurs too regularly with the fixed period around 2 years.

Although it is an idealized and theoretical model, this model included both the atmosphere and the ocean dynamics and was shown to have a reasonable evolution and life cycle of the ENSO compared to what was observed. Essentially, a Gill-type atmosphere with heating mainly tied to SST anomaly and convergence feedback term is coupled with a reduced gravity ocean model. In the mid-1980s, slightly different models were examined: one by Philander et al. [46], in which the SST anomaly was assumed proportional to both thermocline displacement and its anomaly; the other by Gill [68] in which the tendency of SST was assumed to be a function of advection of mean temperature by anomalous zonal current and atmospheric heating was determined by the SST anomaly. The former resulted a Kelvin wave response, while the latter a westward growing coupled mode.

These model variants, although far simpler than an OGCM, were constructed with careful thought on the processes influencing SST changes that are essential in producing the coupled ENSO phenomena. Simulating realistic ENSO remains a challenge for developers of climate models. Because ENSO is a dominant feature in interannual variability, and El Niño (La Niña) induces anomalous wet (dry) conditions all over the worlds, it is critical to simulate right. Understanding ENSO, and how it might change in an era of the global warming is an important research topic. For example, a recent study using the IPCC's AR4 model projections showed the El Niño with its center over the central Pacific may dominate in the future [66].

A Box Model for Thermohaline Circulation

The ZC model is primarily forced by surface wind stress and buoyancy plays a minor role. However, in climate application, ocean circulation can be driven by salinity and density gradients, which contributes much longer-term variability than ENSO which is predominately in interannual timescales. Although state-of-the-art OGCMs can now simulate the thermohaline circulation relatively well, it has been one of the long-standing challenges for modelers. Therefore, different types of simple conceptual models have been developed and used. Difference between the ZC model and the thermohaline models are not limited to the physical forcing mechanisms, but also through the introduction of nonlinearity in the thermohaline circulation and different mathematical treatment associated with the vertical stratification of buoyancy.

Therefore, an alternative to the formal mathematical derivation of simplified models from the full equations of motion is used to pose a conceptual model or simple physical analog for the ocean system. In this entry, an idealized model by Marotzke [40] will be briefly described. Two boxes which represent tropical and polar regions of the ocean are considered. Several assumptions are made: (1) ocean temperature is very close to atmospheric temperature in each box, and (2) salinity is forced by

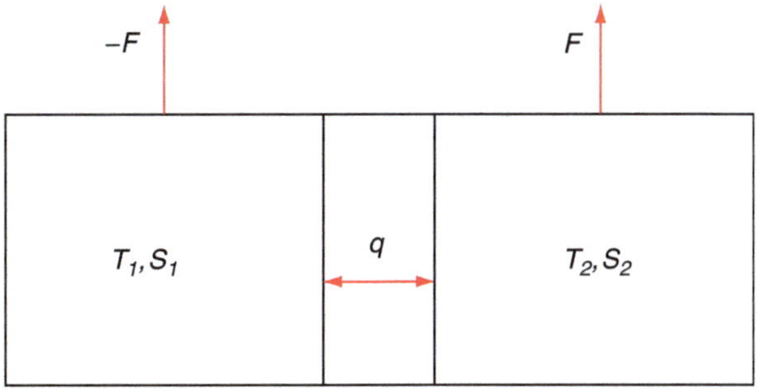

Fig. 4.6 A schematic diagram of the two-box model of the thermohaline circulation (modified from [29])

a flux of moisture through the atmosphere from the low-latitude to the high-latitude box. Salinity forcing is used to make the low-latitude box saltier and denser, and the high-latitude box fresher and less dense (Fig. 4.6). The governing equation of salt can be described as follows following Haidvogel and Bryan [29]:

$$V\frac{dS_1}{dt} = -FS_0 + |q|(S_2 - S_1), \tag{4.23}$$

$$V\frac{dS_2}{dt} = -FS_0 + |q|(S_1 - S_2), \tag{4.24}$$

where V is the volume of the boxes which is assumed to be equal, S_0 is a constant reference salinity, and q is the rate of volume exchanged in the pipes which connect the low-latitude and polar boxes. Pressure differences between the two boxes drive the system, which can be modeled using resistivity k^{-1}:

$$q = -\frac{k}{\rho_0}(\rho_2 - \rho_1)\,. \tag{4.25}$$

These are completed using a linear equation of state:

$$\rho = \rho_0\,(1 - \alpha T + \beta S)\,, \tag{4.26}$$

where α is the thermal expansion coefficient and β is the haline contraction coefficient.

As mentioned above, the state-of-the-art climate models now can simulate decent thermohaline circulation. However, it is still difficult to observe this circulation and bias exists in its location and intensity simulated by OGCM.

Biogeochemical Cycle Modeling

The oceans play a role in not only through physical quantities, such as wind stress and surface energy exchange, but also through chemical species, especially

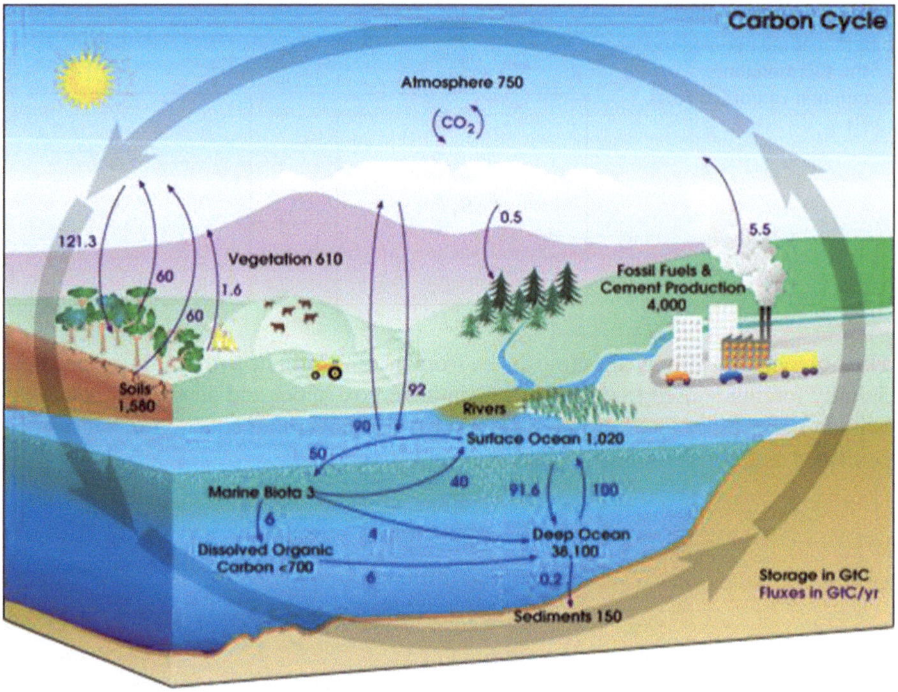

Fig. 4.7 A schematic diagram of the global carbon cycle, obtained from http://earthobservatory.nasa.gov/Features/CarbonCycle/carbon_cycle4.php

greenhouse gases, such as CO_2 and N_2O. Also, sea salt from the ocean surface is an important source of natural aerosols which have impacts on climate through aerosol–climate interaction. In this section, a few aspects on modeling of the biogeochemical cycle in the OGCM are briefly reviewed.

Currently, approximately 6 Gt/year of carbon is released in the atmosphere due to anthropogenic activity. Only about half of carbon remains in the atmosphere and the rest is stored either in the vegetation and soil, or in biota and dissolved trace species in the ocean. This stored carbon is typically identified as terrestrial and ocean carbon pools. These fluxes and reservoirs for carbon are summarized in Fig. 4.7. The size of the terrestrial and atmospheric carbon pools are about 2,200 and 750 Gt, while the Ocean carbon pool is estimated to hold about 40,000 Gt, the largest carbon pool except sediments and rocks.

The first generation of the Ocean Carbon model to treat the exchange process between atmosphere and ocean was developed more than 50 years ago by Revelle and Suess [50], which used two boxes of oceanic carbon pools with exchange processes between atmosphere and ocean. The rationale of this simple box model has been employed consistently through later generations of the ocean carbon model (e.g., [31]). Research questions about how much of the extra CO_2 released by human activity in the atmosphere can be stored in the ocean carbon pools resides at the core of climate change science, and how these large carbon pools will respond to global warming is also critical to understand.

The conservation equation for any trace constituent (like CO_2) in the ocean or atmosphere can be written as follows:

$$\frac{\partial C}{\partial t} = -\vec{V} \cdot \nabla C - C\nabla \cdot \vec{V} + C_{\text{Source}} - C_{\text{Sink}}, \tag{4.27}$$

where C is a mixing ratio of any trace constituent (units mass of trace constituent to mass of sea water).

There exist two important source and sink processes for carbon in the ocean, that are typically labeled as the biological and solubility pumps. The biological pump refers to the transport of carbon from the oceanic surface to its interior in the form of carbon bound to other elements and labeled organic or inorganic carbon, according to whether the carbon resides in chemical compound that arises primarily through biological activity (organic) or other chemical reactions (inorganic). Various routes for the transport of carbon from the atmosphere into different parts of the ocean have been discovered and are treated in modern climate models treating biogeochemistry. The strength of this biological pump is measured by the so-called f-ratio, a fraction of total primary production fueled by nitrate [17].

Second, the solubility pump is a transport process that takes place through physical-chemical interactions. Both temperature in the ocean layers and the aforementioned thermohaline circulation are two important players in determining the strength of the solubility pump. The most of the climate models in the IPCC 4th Assessment Report (AR4) does not have representation of carbon cycle. However, the next generation of OGCMs to be used in the IPCC 5th Assessment Report (AR5) has some degree of numerical representation of these two pumps.

Future Directions

The simulations of the oceans by contemporary OGCM in coupled climate model still exhibit various degrees of biases compared to observations despite improvements. Further reduction of these biases is required. Some dynamical, physical, and even biological processes are poorly represented, and others are entirely missing. Thus, much emphasis has been weighted on developing more comprehensive and computationally efficient ocean climate model at the same time. As a first step, increasing spatial resolution has been pursued as a way forward by many modeling groups in the world and reduces the need for some sub-grid scale processes which must currently be parameterized. Also, better numerical algorithm suitable for extreme computing environment has been sought at the same time.

In summary, several key aspects in the development of the next generation of the OGCM are as follows:

1. Multi-decadal natural variability-Atlantic Meridional Overturning Circulation (AMOC): This phenomenon is important for climate not only in the pan-Atlantic regions, but also for other regions in longer-term climate variability [60]. With limited availability of observational data, the OGCM has been playing an important role in understanding many processes involved with the AMOC and in simulating its historical evolution as well as current status. For example, most of climate simulations produce a weakening of the AMOC under the global warming due to increasing temperature [23, 55]. However, a role played by the fresh water budget on the AMOC has large uncertainty [61]. Many regional and international efforts have been made to collect more data and to improve model simulation of the AMOC [53].
2. Sea-level change in the warming world: Sea-level change in the global ocean is not only a function of temperature increase, but also many regional processes. As reviewed earlier, the OGCM with the Boussinesq approximation has its limitation in suitably simulating this sea-level change. Also, the ice sheet and sea ice variability and change have to be simulated and incorporated into the OGCM.
3. Simulation of marine ecosystem in the OGCM: The OGCM was typically for description of ocean circulation and physical properties. However, as our planet changes due to the anthropogenic activities, there are increasing demands on what would be the impact of climate change on marine ecosystem in a very regional scale (such as the coastal region of the Gulf of Mexico). On the other hand, proper simulation of marine ecosystem benefits better representation of the biological pump in the ocean biogeochemical cycle.

This entry introduces some of the basic concepts and examples of the OGCM. More details on each subject can be found in the following articles or textbooks. For example, general introductory and review on the contemporary OGCM can be found in Randall et al. [48], Griffies et al. [27], and Haidvogel and Bryan [29]; numerical formulation can be found in Griffies [24], Griffies et al. [26], and Bryan [5]; and biogeochemistry in the OGCM can be described in much more detail in Sarmiento [54].

Bibliography

1. Adcroft A, Campin J-M, Hill C, Marshall J (2004) Implementation of an atmosphere–ocean general circulation model on the expanded spherical cube. Mon Weather Rev 132:2845–2863
2. Beckmann A, Haidvogel D (1993) Numerical simulation of flow around a tall isolated seamount. Part I: Problem formulation and model accuracy. J Phys Oceanogr 23:1736–1753
3. Bjerknes J (1969) Atmospheric teleconnections from the equatorial Pacific. Mon Weather Rev 97:163–172
4. Bleck R (1998) Ocean modeling in isopycnic coordinates. In: Chassignet EP, Verron J (eds) Ocean modeling and parameterization. Kluwer, Dordrecht, pp 423–448

5. Bryan K (1989) The design of numerical model of the ocean circulation. In: Anderson DLT, Willebrand J (eds) Oceanic circulation models: combining data and dynamics. Kluwer, Dordrecht, pp 465–500
6. Breugem W.-P, Chang P, Jang CJ, Mignot J, Hazeleger W (2008) Barrier layers and tropical Atlantic SST biases in coupled GCMs. Tellus A 60:885–897. doi: 10.1111/j.1600-0870.2008.00343.x
7. Gill AE (1982) Atmosphere-ocean dynamics. Academic Press, 662 pp
8. Matsuno T (1966) Quasi-geostrophic motions in the equatorial area. J Meteorol Soc Jpn 44:25–43
9. Cane MA, Zebiak S (1985) A theory for El Niño and the southern oscillation. Science 228:1085–1087
10. Casulli V, Walters RA (2000) An unstructured grid, three-dimensional model based on the shallow water equations. Int J Numer Methods Fluids 32:331–346
11. Chassignet EP, Hurlburt HE, Smedstad OM, Halliwell GR, Wallcraft AJ, Metzger EJ, Blanton BO, Lozano C, Rao DB, Hogan PJ, Srinivasan A (2006) Generalized vertical coordinates for eddy resolving global and coastal ocean forecasts. Oceanography 19(1):118–129
12. Chen C, Liu H, Beardsley RC (2003) An unstructured grid, finite-volume, three-dimensional, primitive equations ocean model: applications to coastal ocean and estuaries. J Atmos Ocean Technol 20:159–186
13. Chen C, Beardsley RC, Cowles G (2006) An unstructured-grid finite-volume coastal ocean model (FVCOM) system. Oceanography 19(1):78–89
14. Cornejo-Garrido AG, Stone PH (1977) On the heat balance of the Walker circulation. J Atmos Sci 34:1155–1162
15. Cox MD (1989) An idealized model of he world ocean. Part I: The global-scale water masses. J Phys Oceanogr 19:1730–1752
16. Deleersnijder E, Beckers J-M (1992) On the use of the σ-coordinate system in regions of large bathymetric variations. J Mar Syst 3:381–390
17. Eppley RW, Peterson BJ (1979) Particulate organic matter flux and planktonic new production in the deep ocean. Nature 282:677–680
18. Gaspar P, Gregoris Y, Lefevre J (1990) A simple eddy kinetic energy model for simulations of the oceanic vertical mixing: Tests at station Papa and long-term upper ocean study site. J Geophys Res 95:16179–16193
19. Gent PR, McWilliams JC (1990) Isopycnal mixing in ocean circulation models. J Phys Oceanogr 20:150–155
20. Gent PR, Willebrand J, McDougall TJ, McWilliams JC (1995) Parameterizing eddy induced tracer transports in ocean circulation models. J Phys Oceanogr 25:463–474
21. Ghil M, Malanotte-Rizzoli P (1991) Data assimilation in meteorology and oceanography. Adv Geophys 33:141–266
22. Gill AE (1980) Some simple solutions for heat induced tropical circulation. Q J R Meteorol Soc 106:447–462
23. Gregory JM et al (2005) A model intercomparison of changes in the Atlantic thermohaline circulation in response to increasing atmospheric CO_2 concentration. Geophys Res Lett 32: L12703. doi:10.1029/2005GL023209
24. Griffies SM (2004) Fundamentals of ocean climate models. Princeton University Press, Princeton, 518 pp
25. Griffies SM et al (2000) Developments in ocean climate modeling. Ocean Model 2:123–192
26. Griffies SM, Ganadesikan A, Dixon KW, Dunne JP, Gerdes R, Harrison MJ, Rosati A, Russell JL, Samuels BL, Spelman MJ, Winton M, Zhang R (2005) Formulation of an ocean model for global climate simulations. Ocean Sci 1:45–79
27. Griffies SM, Adcroft AJ, Banks H, Böning CW, Chassignet EP, Danabasoglu G, Danilov S, Deleersnijder E, Drange H, England M, Fox-Kemper B, Gerdes R, Gnanadesikan A, Greatbatch RJ, Hallberg RW, Hanert E, Harrison MJ, Legg S, Little CM, Madec G, Marsland SJ, Nikurashin M, Pirani A, Simmons HL, Schröter J, Samuels BL, Treguier A-M, Toggweiler

JR, Tsujino H, Valllis GK, White L (2009) Problems and prospects in large-scale ocean circulation models. Community White Paper for OceanObs09. https://abstracts.congrex.com/scripts/jmevent/abstracts/FCXNL-09A02a-1672431-1-cwp2a07.pdf
28. Guilyardi E et al (2004) Representing El Niño in coupled ocean-atmosphere GCMs: the dominant role of the atmospheric component. J Climate 17:4623–4629
29. Haidvogel DB, Bryan FO (1992) Ocean general circulation modeling. In: Trenberth KE (ed) Climate system modeling. Cambridge University Press, New York, pp 371–412
30. Roberts MBH, Gedney N, Gregory J, Hill R, Mullerworth S, Pardaens A, Rickard G, Thorpe R, Wood R (2004) Impact of an eddy-permitting ocean resolution on control and climate change simulations with a global coupled GCM. J Clim 17:3–20
31. Sarmiento JL, Gruber N (2002) Sinks for anthropogenic carbon. Physics Today 55(8):30–36
32. Haney RL (1991) On the pressure gradient force over steep topography in sigma-coordinate ocean models. J Phys Oceanogr 21:610–619
33. Henderson-Sellers B, Davies AM (1989) Thermal stratification modeling for oceans and lakes. Annu Rev Numer Fluid Mech Heat Transfer 2:86–156
34. Jackson L, Hallberg R, Legg S (2008) A parameterization of shear-driven turbulence for ocean climate models. J Phys Oceanogr 38:1033–1053
35. Lehmann A (1995) A three-dimensional baroclinic eddy-resolving model of the Baltic Sea. Tellus 47A:1013–1031
36. Lindzen RS, Nigam S (1987) On the role of sea surface temperature gradients in forcing low level winds and convergence in the tropics. J Atmos Sci 44:2418
37. Maltrud M, McClean J (2005) An eddy resolving global 1/10° ocean simulation. Ocean Model 8:31–54
38. Manabe S, Bryan K (1969) Climate calculations with a combined ocean-atmosphere model. J Atmos Sci 26:786–789
39. Manabe S, Bryan K, Spelman MJ (1975) A global ocean-atmosphere climate model. Part I. The atmospheric circulation. J Phys Oceanogr 5:3–29
40. Marotzke J (1989) Instabilities and multiple steady states of the thermohaline circulation. In: Anderson DLT, Willebrand J (eds) Oceanic circulation models: combining data and dynamics. Kluwer, Dordrecht, pp 501–511
41. Martin PJ (1985) Simulation of the mixed layer at OWS November and Papa with several models. J Geophys Res 90:903–916
42. Meehl G et al (2007) The WRCP CMIP3 Multimodel Dataset: A new era in climate change research. Bull Am Meteorol Soc 88:1383–1394
43. Mellor GL, Yamada T (1982) Development of a turbulence closure model for geophysical fluid problems. Rev Geophys Space Phys 20:851–875
44. Neelin JD (1989) On the interpretation of the Gill model. J Atmos Sci 46:2466–2468
45. Noh Y, Kim H-J (1999) Simulations of temperature and turbulence structure of the oceanic boundary layer with the improved near-surface process. J Geophys Res 104:15621–15634
46. Philander SGH, Yamagata T, Pacanowski RC (1984) Unstable air-sea interaction in the tropics. J Atmos Sci 41:604–613
47. Randall D, Ringler T, Heikes R, Jones P, Baumgardner J (2002) Climate modeling with spherical geodesic grids. Comput Sci Eng 4:32–41
48. Randall DA, Wood RA, Bony S, Colman R, Fichefet T, Fyfe J, Kattsov V, Pitman A, Shukla J, Srinivasan J, Stouffer RJ, Sumi A, Taylor KE (2007) Climate models and their evaluation. In: Solomon S, Qin D, Manning M, Chen Z, Marquis M, Averyt KB, Tignor M, Miller HL (eds) Climate Change 2007: The Physical Science Basis. Contribution of Working Group I to the Fourth Assessment Report of the Intergovernmental Panel on Climate Change. Cambridge University Press, Cambridge, UK
49. Redi MH (1982) Oceanic isopycnal mixing by coordinate rotation. J Phys Oceanogr 12:1154–1158
50. Revelle R, Suess HE (1957) Carbon dioxide exchange between atmosphere and ocean and the question of increasing of atmospheric CO_2 during the past decades. Tellus 9:18–27

51. Ringler T, Ju L, Gunzburger M (2008) A multiresolution method for climate system modeling: application of spherical centroidal voronoi tessellations. Ocean Dyn 58:475–498
52. Sadourny R, Arakawa A, Mintz Y (1968) Integration of the non-divergent barotropic vorticity equation with an icosahedral-hexagonal grid for the sphere. Mon Weather Rev 96:351–356
53. Sanford TB, Kelly KA, Former DM (2011) Sensing the ocean. Phys Today 64(2):24–28
54. Sarmiento JL (1992) Biogeochemical ocean models. In: Trenberth KE (ed) Climate system modeling. Cambridge University Press, New York, pp 519–551
55. Schmittner A, Latif M, Schneider B (2005) Model projections of the North Atlantic thermohaline circulation for the 21st century assessed by observations. Geophys Res Lett 32:L32710. doi:10.1029/2005GL024368
56. Semtner AJ, Chervin RM (1992) Ocean general circulation from a global eddy-resolving model. J Geophys Res 97:5493–5550
57. Shchepetkin A, McWilliams J (2002) A method for computing horizontal pressure-gradient force in an ocean model with a non-aligned vertical coordinate. J Geophys Res 108:35.1–35.34
58. Smith R, Gent P (2004) Reference Manual for the Parallel Ocean Program (POP)—ocean component of the community climate system model. Los Alamos National Laboratory, LAUR-02-2484
59. Solomon H (1971) On the representation of isentropic mixing in ocean models. J Phys Oceanogr 1:233–234
60. Sutton RT, Hodson LR (2005) Atlantic forcing of North American and European summer climate. Science 309:115–118
61. Swingedouw D, Braconnot P, Marti O (2006) Sensitivity of the Atlantic Meridional overturning circulation to the melting from northern glaciers in climate change experiments. Geophys Res Lett 33:L07711. doi:10.1029/2006GL025765
62. Vernois G (1973) Large scale ocean circulation. In: Yih C-S (ed) Advances in applied mathematics, vol 13. Academic, New York, pp 1–92
63. Walton CC, Pichel WG, Sapper JF (1998) The development and operational application of nonlinear algorithms for the measurement of sea surface temperatures with the NOAA polar-orbiting environmental satellites. J Geophys Res 103(C12): 27999–28012
64. Weart SR (2008) The discovery of global warming: revised and expanded edition. Harvard University Press, Cambridge, MA. 240 p. ISBN: 978–0674031890
65. Wyrtki K (1975) El Niño – The dynamic response of the equatorial Pacific Ocean to atmospheric forcing. J Phys Oceanogr 4:91–103
66. Yeh S-W, Kug J-S, Dewtte B, Kwon M-H, Kirtman B, Jin F-F (2009) El Niño in a changing climate. Nature 461:511–514
67. Zebiak S, Cane MA (1987) A model El Niño-southern oscillation. Mon Weather Rev 115:2262–2278
68. Gill AE (1985) Elements of coupled ocean-atmosphere models for the tropics. In: Nihoul JCJ (ed) Coupled ocean-atmosphere models, vol 40. Elsevier Oceanogr. Ser., Amsterdam, pp 303–327

Chapter 5
Weather Prediction Models

Julio T. Bacmeister

Glossary

Assimilation	The process of combining observations of the atmosphere with a "first guess" (usually a model forecast) to define the atmospheric state on a forecast model grid.
Geostrophic balance	A possible state of rotating fluids in which flow is directed along pressure gradients rather than across them.
Gravity waves	Rapidly moving atmospheric disturbances driven by gravity acting on vertical density gradients. Often arise as a consequence of spurious geostrophic imbalance in initial conditions.
Hydrostatic balance	State in which the vertical pressure gradient force cancels the downward accleration of gravity. Approximately obeyed in atmospheric flows with horizontal scales larger than several km.
Instabilities (or unstable modes)	Spatial patterns in a flow that are able to extract energy from the background flow and grow in amplitude.
Primitive equations	Complete set of equations describing flow of a thin envelope of fluid or gas surrounding a sphere.
Resolution	Separation in space of notional points at which quantities are defined in a numerical model.

This chapter was originally published as part of the Encyclopedia of Sustainability Science and Technology edited by Robert A. Meyers. DOI:10.1007/978-1-4419-0851-3

J.T. Bacmeister (✉)
NCAR Earth System Laboratory, National Center for Atmospheric Research,
1850 Table Mesa Drive, 80305 Boulder, CO, USA
e-mail: juliob@ucar.edu

P.J. Rasch (ed.), *Climate Change Modeling Methodology: Selected Entries from the Encyclopedia of Sustainability Science and Technology*,
DOI 10.1007/978-1-4614-5767-1_5, © Springer Science+Business Media New York 2012

Definition of the Problem

Awareness of weather and concern about weather in the proximate future certainly must have accompanied the emergence of human self-consciousness. Although weather is a basic idea in human existence, it is difficult to define precisely. Weather intuitively refers to a set of atmospheric conditions prevailing over a relatively small area, and even more emphatically, over a relatively short time. The immediacy contained in our notion of weather may be reflected in the fact that in many languages the same root appears in the words for time and weather, for example, Spanish (*tiempo*) or Hungarian (*idő*). Thus, weather is to be distinguished from the notion of climate, or more subtly from the notion of "spells" which imply a time window anywhere from a week to several years. Our experience of weather does involve quantities which can be defined with reasonable precision. These include air temperature, wind speed, precipitation rates and types, cloud cover, and also humidity, air-quality, and barometric pressure. Numerical weather prediction (NWP) is the attempt to predict the evolution of such quantities by solving a set of partial differential equations which describe the dynamics of a fluid like the atmosphere [1, 2]. These equations must be solved using approximate or "numerical" techniques using computers. They are integrated forward in time from a set of initial conditions, which are derived from an optimized combination of observations and previous model forecasts. This bootstrapped procedure is known as the analysis cycle and is an integral part of the activities at all modern forecasting centers.

NWP and global climate simulation are closely related problems. The models used in both endeavors are essentially the same. A key difference between NWP and the climate problem is the role of atmospheric initial conditions. Initialization of the atmosphere is of secondary importance in multiyear climate simulations. However, good initialization of the atmospheric state is at the heart of the forecasting problem. Initialization must give an accurate and comprehensive representation of the state of the atmosphere that is compatible with numerical forecast model being used. The initial state must also satisfy a number of "balance constraints" to avoid spurious initial variability.

Lewis Fry Richardson (1881–1953) reported the first numerical weather forecast, performed using hand calculations, in 1922 [3]. His attempt did not succeed for reasons that are summarized below, and discussed in detail in the book by Lynch [2], but Richardson's effort marks the beginning of NWP as a field of inquiry.

Introduction: Direct Simulation of Atmospheric Flows

In order to appreciate the challenges faced by numerical models of the global atmosphere it is useful to have a sense of the complex nature of the motions which must be represented. A brief account of some of the dynamical processes

Table 5.1 Motions in atmosphere

Phenomenon	Spatial scale	Temporal scale	Role in weather
Acoustic waves	Meters	Seconds	None
Gravity waves	1–1,000 km	15 min to days	Initiating and organizing convection. Mountain flows, e.g., chinooks, Foehns . . .
Eastery waves	500–2,000 km	Days	Organizing tropical convection, tropical cyclogenesis
Baroclinic instability	Thousands of km	Days	Midlatitude cyclogenesis
Madden–Julian oscillation	10,000 km	20–60 days	Possible modulation of tropical and midlatitude weather

that play a part in weather is given below. For more details, the reader is referred to the excellent, comprehensive introduction to the dynamics of the atmosphere and ocean by A.E. Gill (1937–1986) [4].

A Zoology of Atmospheric Motion

The atmospheric flows which are responsible for creating weather span a wide range of space and time scales and are driven by rich variety of dynamical and thermodynamic processes. Atmospheric flows may be forced by features at the Earth's surface. Mountainous terrain forces atmospheric circulations with spatial scales ranging from several kilometers to several thousand kilometers. Surface temperature contrasts, most pronounced between land and ocean but also created by variations in ground cover, force sea-breeze circulations with scales of tens of kilometers, [5] as well as continental-scale monsoonal circulations (Table 5.1) [6–8].

External forcing can produce wave-like motions in the atmosphere. Gravity waves, or buoyancy waves, exist because of density stratification in the atmosphere, and are analogs to waves on the surface of water. These waves are typically one to hundreds of kilometers in scale, and have periods of minutes to hours, although in the tropics both spatial and temporal scales may be longer [4]. Gravity waves may play a role in triggering convection [9–11], as well as in organizing convection in the mesoscale [10, 12, 13]. Mountain waves are large, nonlinear, gravity waves generated by flow over mountains with horizontal scales from less than ten kilometers to several hundred kilometers. Such waves are responsible for a number of local but intense winds [14]. Gravity waves are also notorious for contaminating forecasts when errors in initial conditions are present.

Rossby waves [4, 15–17] named after the Swedish meteorologist Carl Gustav Rossby exist because of the change in the effective rotation rate experienced by a fluid parcel as it moves from equator to pole. The effective rotation rate at the

Equator is zero, since all of the effects of Earth's rotation are directed in the vertical and are felt as a slight reduction in the pull of gravity. At the poles, the effective rotation rate is as intuitively expected – one cycle or 2π radians per day. In between equator and pole, this rate varies as the sine of latitude. Rossby waves which exploit this gradient have scales of thousands to ten thousands of kilometers and periods of days to weeks. Immense planetary-scale Rossby waves or "planetary waves" generated by topographic features such as the Himalayas and Rockies, as well as, by continental-scale land-sea contrasts, dominate the tropospheric flow and are responsible for the mean position of the jet-streams, and for the mean paths of storms.

Probably more important in the overall problem of weather, and certainly more difficult to predict, is another class of flows driven by internal exchanges of energy in the atmosphere. These instabilities or unstable modes are essentially flow patterns that are able to extract energy from their surroundings and grow in amplitude. Initial growth of unstable modes is typically exponential. Familiar examples of such behavior in fluids include the growth of wind driven waves on the surface of water. The convective instability of a fluid heated from below is another example that is both familiar in everyday experience and also important in the atmosphere.

In the atmosphere, a particularly important mode of unstable growth is through baroclinic instability [18, 19]. This instability arises from a combination of thermal and inertial effects in a rotating fluid with a horizontal temperature along the lower boundary. Baroclinic instability is characterized by length scales of 1,000 km and growth times of days – making it a key factor in weather. It is fair to say that the problem of weather forecasting in midlatitudes is essentially that of predicting the evolution of baroclinically unstable modes in the atmosphere.

The tropics possess another as yet poorly understood class of motions, in which moist heating plays a key role in energizing and modifying wave motions in the atmosphere [20]. Tropical easterly waves [21] have periods of several days and scales of hundreds to thousands of kilometers and play an important role in the genesis of tropical cyclones [22–24]. The Madden–Julian oscillation or "MJO" [25, 26] is an eastward traveling disturbance in the tropics with a length scale close to 10,000 km and a period of weeks. It is thought to play a role in modulating tropical cyclone frequency in various basins [27–31] and possibly midlatitude disturbances as well [32, 33]. The dynamics behind the MJO are not yet understood. Successful forecasting of the MJO could improve prospects for accurate forecasts out to lead times of weeks [34].

Early History

Weather prediction, not climate simulation, was the original motivation for developing numerical models that describe the time evolution of the atmosphere. Richardson's 1922 attempt at NWP predates initial attempts to study climate numerical models by at least four decades. The first attempts at simulating the longer-term equilibrium state of the atmosphere did not occur until the 1960s [35].

The notion of predicting weather systematically using equations to describe the evolution of systems can probably be traced to the Norwegian meteorologist Vilhelm Bjerknes (1862–1951) [1, 2] who founded the famous "Bergen School" of meteorology [36, 37]. Bjerknes considered graphical methods to predict the motion of fronts and other features in the atmosphere, as well as numerical techniques to solve the equations themselves. However, it was L. F. Richardson who finally conceived and implemented a concrete plan to use a numerical solution of the partial differential equations describing the atmosphere (see section on "primitive equations") to make forecasts. His approach was remarkably prescient both in concept and in detail. He employed a finite-difference technique on a regularly spaced grid of points over central Europe and attempted to predict the tendency of surface pressure over 6 h. Richardson's forecast was a famous failure or "bust" (see the book by Lynch [2] for a detailed and readable account of Richardson's attempt, as well as for a comprehensive account of the development of NWP). However, the reasons for Richardson's failure lay in the initial conditions used in the forecast not in his method as Richardson himself suspected [2].

The potential for numerical prediction was clear. The major obstacle beside the question of initialization, was the sheer amount of calculation required to produce even a short forecast over a limited area. Richardson imagined computational "factories" employing thousands of people to produce weather forecasts [2]. The appearance of electronic computers soon after World War II made numerical prediction plausible. The potential application to the problem of weather prediction was recognized by one of the main intellects behind the development of electronic computers John von Neumann (1903–1957) [1]. In the first successful attempts at NWP using electronic computers so-called filtered equation sets were used [2]. Filtered equations describe a limited set of atmospheric motions, but allow large time steps to be taken in numerical integration and side step the need for well-balanced initial conditions (see sections on "Numerics and Initialization").

Development of Modern NWP Models

It was recognized early on by Jule Charney and others that models using filtered equations were not a promising long-term path for NWP [2]. As computer power increased, the limitations on the time-step length allowed by more complete equations became less important. The problem of initialization was not solved, but its tractability became apparent [38]. Development work on NWP models using the primitive equations began in the late 1950s and eventually led to the adoption of primitive equation models at all major forecasting centers by the mid 1960s [2].

The Primitive Equations

The primitive equations are essentially the complete Navier–Stokes or Euler equations for fluid motion with the hydrostatic approximation invoked. As long as the horizontal scales of interest are much larger than 10 km, the hydrostatic approximation is well satisfied. However, at the time of writing, major NWP models are reaching horizontal resolutions that test the limits of this approximation. The Euler equations using a generalized vertical coordinate η, including connections between the fully nonhydrostatic system and the hydrostatic system currently used in NWP models, are nicely described by Laprise [39].

The hydrostatic primitive equations include an equation describing the evolution of horizontal momentum or velocity V:

$$\frac{d}{dt}V + f\mathrm{k} \times V = -\alpha\nabla_{\eta}p - \nabla_{\eta}\phi + F_{phys}$$

where f is the Coriolis parameter or local apparent rotation rate and k is the unit vector in vertical direction. The symbol ∇_{η} denotes a gradient along surfaces of fixed η. As is common in the meteorological literature φ denotes the geopotential height or potential energy density of a fluid parcel along constant η. The remaining symbols α and p denote specific volume and pressure. This equation is simply the fluid dynamical form of Newton's law $F = ma$, where the right-hand side contains forces accelerating fluid parcels in the horizontal. In current meteorological literature, the individual velocity components are usually designated as u for the eastward or "zonal" component, and υ for the northward or "merdional" component.

Another equation restates the first law of thermodynamics $dE = dQ + dW$ in fluid form:

$$C_p\frac{d}{dt}T - \alpha\frac{d}{dt}p = H_{phys}$$

where T is the absolute temperature and C_p is the heat capacity of air at constant pressure. In both the momentum and energy equations the symbol $\frac{d}{dt}$ is used to denote:

$$\left(\frac{\partial}{\partial t}\right)_{\eta} + V \cdot \nabla_{\eta} + \dot{\eta}\frac{\partial}{\partial \eta}$$

the Lagrangian derivative that tracks changes in a quantity following a fluid parcel.

A prognostic equation for mass continuity is also required:

$$\left[\frac{\partial}{\partial t}\left(\frac{\partial p}{\partial \eta}\right)\right]_{\eta} + \nabla_{\eta} \cdot \left(V\frac{\partial p}{\partial \eta}\right) + \frac{\partial}{\partial \eta}\left(\dot{\eta}\frac{\partial p}{\partial \eta}\right) = 0$$

where $g^{-1}\frac{\partial p}{\partial \eta}$ is the mass per unit area in a column between surfaces of constant η.

Three diagnostic relationships are needed to complete the system. The equation of state for a gas:

$$p\alpha = RT$$

and a relationship that determines the geopotential height of η-surfaces:

$$\phi = \phi_s + \int_{\eta}^{\eta_s} \alpha \frac{\partial p}{\partial \eta'} \, d\eta$$

This relationship uses the surface geopotential height φ_s as well as the integral of the hydrostatic relationship:

$$\frac{\partial p}{\partial \eta} = \frac{1}{\alpha} \frac{\partial \phi}{\partial \eta}$$

between pressure and specific volume.

The right-hand sides of the momentum and energy equations also contain the terms F_{phys} and H_{phys}. These represent the effects of physical parameterizations on the grid scale variables, and include effects from radiative heating, friction, and other processes which will be described in more detail in the section on parameterization. The use of the generalized vertical coordinate η gives the equations a somewhat unfamiliar look. However, replacing η with the geometric height z, and noting that $\nabla_z \varphi = 0$ reduces them to a more familiar form.

Due to the complex shape of Earth's topography most NWP models do not use geometric height as their vertical coordinate. Most use a version of the so-called σ-coordinate defined by:

$$\sigma = \frac{p - p_t}{p_s(x, y, t) - p_t}$$

where $p_s(x, y, t)$ is the surface pressure and p_t is the pressure at the model top, typically a constant value. The coordinate surface $\sigma = 1$ follows the bottom of the model domain while $\sigma = 0$ follows the top. Boundary conditions on "vertical" velocity become simply $\dot{\sigma}(1) = \dot{\sigma}(0) = 0$. Thus, the difficulties of representing flow boundaries in and around topographic obstacles are replaced by the need for a prognostic equation describing π_s. This is obtained by integrating the mass continuity equation in the vertical.

The primitive equations can describe all of the motions discussed in the introduction except for fully three dimensional acoustic waves. They do allow a horizontal acoustic mode known as the Lamb wave [4] which can produce difficulties for numerical integrations. Gravity waves and convective instabilities with horizontal scales much smaller than 100 km are not well represented, and this may become a significant handicap for models in the next 10 years (see section on "Future Directions").

In addition to the equations describing the dynamics and thermodynamics of the atmosphere, modern NWP models include equations that describe the evolution of trace gases and trace species in the atmosphere. Each additional species results in an additional prognostic equation of the form:

$$\frac{d}{dt}q_i = C_{i,phys}$$

where q_i is the mixing ratio of the ith species and $C_{i,phys}$ are the sources and losses of the species. The most important of these trace quantities is water vapor. Water vapor was included in primitive equation NWP models early on [40]. As NWP model domains were extended into the tropics during the 1970s [41] strong condensational heating associated with high tropical humidities presented problems for NWP that spurred the development of deep convection parameterizations [42] (see section on "Parameterization"). More recently, NWP models have incorporated prognostic treatment of condensed water species known as prognostic cloud schemes [43, 44].

Numerics

Closed form solutions of the primitive equations do not exist. Approximate numerical techniques must be used. An illustration of how this proceeds is given here using a simple equation that describes one dimensional advection of a constituent C by a constant flow u:

$$\partial_t C + u\partial_x C = S$$

Figure 5.1 shows three time steps from a numerical integration of this equation. Finite-difference approximations of the partial derivatives in both space and time are calculated as shown in the figure. The approach illustrated is known as a "centered difference" since the approximation uses a symmetric stencil of equally weighted points. With these approximations to the derivatives in time and space, a solution for the tracer distribution at $\tau + 1$ can be obtained:

$$C(i, \tau + 1) \approx C(i, \tau - 1) - u2\Delta t\left(\frac{C(i+1, \tau) - C(i-1, \tau)}{2\Delta x}\right)$$

As a technical detail, notice that to start (or initialize) this calculation two time levels of data must be given. In practice, these can be set equal to each other.

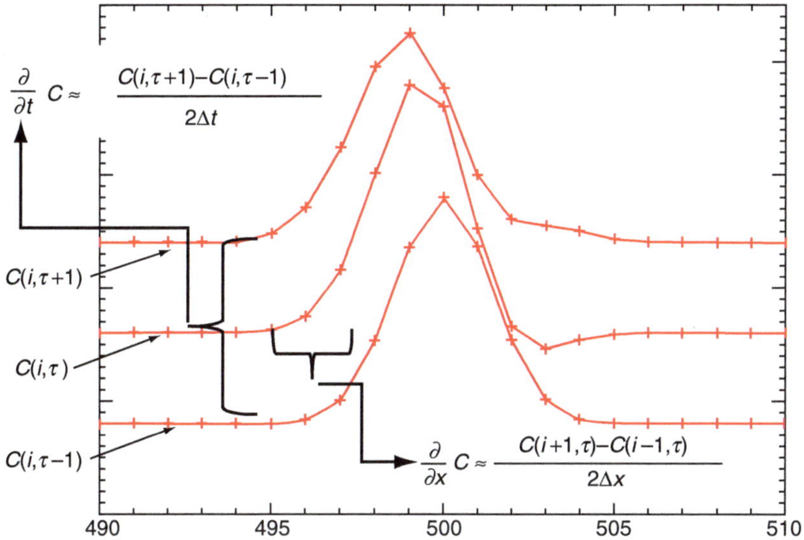

Fig. 5.1 Three time steps from numerical advection of a Gaussian tracer pulse using a second-order, space-centered, time-centered finite-difference scheme. The +'s indicate grid-point locations. A constant velocity of $u = -0.5$ is used with grid spacing $\Delta x = 1$ and time step $\Delta t = 1$

Once initial values are given, the arithmetic equation above can be repeated, or iterated, many times to give the time evolution of C.

The algorithm described above is a simple but stable scheme that was employed in early numerical models. However, it has many undesirable properties, which are described in [45]. Nevertheless, the basic concepts illustrated by this method hold for all explicit finite-difference approaches. In implicit methods, the terms in the approximation of $\partial_x C$ above are replaced by their values at time $\tau + 1$. In the case above, this leads to a tridiagonal matrix problem that can be easily solved using standard techniques [45].

The most natural coordinate system for a global atmospheric model is spherical, with latitude and longitude as the horizontal coordinates. Unfortunately, as either pole is approached, equally spaced longitude lines come arbitrarily close together. This presents problems for most finite-difference numerical schemes, which become inaccurate or unstable when information travels across multiple grid lengths in a single time step. For example, in the simple case described in Fig. 5.1 the time step must be chosen such that $\frac{u\Delta t}{\Delta x} < 1$ to avoid numerical instability. Versions of this limit, known as the Courant-Friedrichs–Levy (CFL) limit, exist for most explicit finite-difference schemes. In systems of equations that support propagating waves, as the primitive equations do, the relevant velocity in the CFL limit is typically the sum of the wave propagation speeds and the advective speed. Thus, the stability of explicit calculations is limited by the fastest wave modes in the system, which are often of little interest, for example, Lamb waves or deep gravity waves. Global models whose numerics are based on finite differences ("grid-point models") address this issue by introducing polar filters which are designed to suppress small-scale motions in the polar regions of the model.

A revolutionary innovation occurred in the late 1960s with the introduction of spectral techniques for solving the primitive equations [46]. Spectral models decompose the atmosphere into a finite sequence of spherical harmonic functions, rather than onto a grid of points. For example, the wind u is approximated by:

$$u(\lambda, \theta, z, t) = \sum_{n=0}^{N} \sum_{m=-n}^{n} U_n^m(z, t) Y_n^m(\lambda, \theta)$$

where λ is longitude and θ is latitude. The time evolution of the atmospheric flow is then represented as the time evolution of the amplitudes $U_n^m(z, t)$. Nonlinear advection terms arising in the primitive equations may be dealt with directly in spectral space, as interactions of spherical harmonics or in physical space with transformation of the tendencies back to spectral space [45].

The resolution of spectral models is determined by the truncation parameter N. The form of the finite sum used above is referred to as a triangular truncation. A model with $N = 799$ in this kind of decomposition is referred as a "T799" model. Lynch [2] gives a useful rule-of-thumb for estimating the spatial resolution Δ corresponding a particular value of N. He uses $\Delta \approx (2\pi a_e)/2\ N$ or $\Delta \approx$ (20,000 *km*)/*N* where the circumference of the Earth $2\pi a_e$ has been approximated by 40,000 km.

Spectral techniques are not only highly accurate, but also nicely sidestep the pole problem faced by grid-point models formulated in terms of latitude and longitude on the sphere. Discretization into spherical harmonics produces no special difficulties at the poles.

A disadvantage of spectral schemes is that fields with strong variation across small-spatial scales, such as most trace gas concentrations (including water vapor), precipitation, or topography cannot be represented without introducing significant spurious nonlocal oscillations in these fields. This behavior, known as the Gibbs phenomenon is a simple consequence of attempting to represent highly localized features with global basis functions. The nature of these truncation errors is such that the amplitude of spurious oscillations decreases slowly with resolution. The presence of Gibbs oscillations can lead to serious problems in global simulations, such as the formation of negative trace gas concentrations.

Due to the difficulties in spectrally representing fields with intense spatial variability, grid-point models have not been abandoned. In addition, grid-point models can be made more efficient than spectral models at very high resolution. So, while most operational forecasting centers currently use some form of spectral dynamical core in their NWP models, this may change in the next decade. Current research is focused on developing grid-point or finite element approaches on nonstandard grids such as the icosahedral, that is, "bucky-ball" or "soccer-ball," grid to bypass the pole problem encountered in latitude-longitude discretizations (see section on "Future Directions") (Table 5.2).

Table 5.2 Parameterizations in Weather Forecasting models. The first column gives the usual designation used in the meteorological community. The second column summarizes the effects of the parameterization. The third column indicates the primitive equation forcing term in which tendencies from the parameterization appear

Parameterization	Effects	Included in
Deep convection	Transports heat, moisture and momentum vertically. Damps convergence	H_{phys}, $\mathbf{F}_{phys}$, $C_{i,phys}$
Orographic gravity wave drag	Decelerates flow over mountains	$\mathbf{F}_{phys}$
Planetary boundary layer (PBL) turbulence	Transports heat, moisture and momentum vertically	H_{phys}, $\mathbf{F}_{phys}$, $C_{i,phys}$
Radiation (Solar and IR)	Calculates heating due radiative flux convergence	H_{phys}
Diagnostic cloud	Estimates cloud cover and thickness	
Prognostic cloud	Calculates cloud condensate concentrations and estimates cloud optical properties	H_{phys}, $C_{i,phys}$

Parameterizations

Representation of physical processes such as radiation, turbulence, gravity wave drag, convection, and precipitation also became more sophisticated in NWP models as they evolved. The earliest successes in NWP using a filtered barotropic model did not even formally incorporate temperature as a prognostic variable. Today, NWP models may track several condensed water species, as well radiatively active trace gases such as ozone. It was the incorporation of radiative transfer schemes and simple moist convective adjustment schemes into NWP models in the 1960s [35] that led to the first climate models capable of self-consistently representing the basic feature of Earth's atmospheric general circulation.

Parameterization development for NWP models has followed that for climate models. The suite of parameterizations currently used in global NWP models is identical to that used for climate simulations. It is arguable whether most physics parameterizations exert an appreciable effect on short-term (1–3 day) forecasts. Deep convection and orographic wave drag parameterizations have been shown to exert a significant short-term effect [47–49]. At medium range (3 days to 2 weeks), physical parameterizations are thought to have an important effect on forecast skill (Bengtsson 2000). In addition climate biases resulting from deficient parameterizations can have a negative impact on data assimilation schemes (see section on "Data Assimilation Systems"). This can indirectly affect short-term forecasts by introducing errors in the initial conditions.

How Are NWP Models (Versus Climate Models) Evaluated?

Perhaps the most significant differences between global climate models and NWP models arise from the different jobs they are expected to perform. Ideally, a solver for the primitive equations, coupled to set of physically realistic parameterizations of processes like convection, should perform both long-term and short-term simulations of the atmospheric flow with equal accuracy. However, this is not the case. Even when run at comparable resolutions, models developed for NWP and climate do not perform each other's tasks with comparable skill. The likely cause of this discordance is the process of model tuning. The need for "tuning" is widely recognized in both the NWP community and the climate modeling community as an inevitable consequence of using imperfect models [41]. The bulk of model tuning occurs in choosing empirical factors that regulate physical parameterizations. Some of this can be explained (or excused) as an attempt to represent unknown sub-grid distributions of quantities such as water vapor, temperature, or topographic roughness. Unfortunately, the process of selecting these empirical factors can be ad hoc, and parochial in character. Even if optimal tunings exist that combine high accuracy in short-term forecasting as well as unbiased climate simulations, they are unlikely to be found. Groups involved in developing climate models rarely have the computational resources to perform extensive testing of their models at high resolution in forecast mode, while NWP groups are typically under intense operational pressures, and have little time or incentive to examine their models in free-running climate simulations.

There is as yet no set of universally accepted metrics for climate models, although developments in this direction are taking place [50]. Metrics typically targeted by climate modelers include seasonal mean distributions of precipitation and seasonal mean planetary wave patterns. Other, functional, constraints exist for atmospheric models used in climate research. For example, when used in coupled climate simulations, that is, connected through boundary fluxes to ocean and land surface models, obtaining correct global budgets of energy and momentum in an atmospheric model is critical. Thus, intensive tuning of cloud parameterizations is usually conducted to ensure that seasonal and annual-mean radiation budgets at the top of the atmosphere are realistic, to minimize spurious long-term drift in extended simulations. Generally speaking, exact conservation of energy and momentum is a key concern in the design of atmospheric models for climate, while being of secondary importance in the design of NWP models.

On the other hand, operational NWP models are regularly subjected to a number of rigorous, and more-or-less universally accepted tests at each step in their development. Not surprisingly, these tests emphasize short-term simulation accuracy rather seasonal or annual-mean performance.

500 hPa Height Anomaly Correlation

At many centers including the European Center for Medium Range Weather Forecasting (ECMWF), the most important measure of global forecast model performance or skill is the 500 hPa height anomaly correlation. This measure is

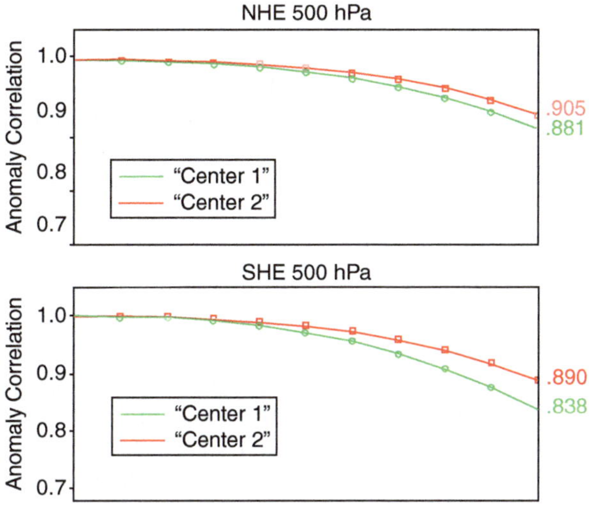

Fig. 5.2 Mean evolution of 500 hPa anomaly height correlation in two major global NWP models for the January through March

essentially the pattern correlation of two maps of geopotential height anomalies φ' interpolated to a pressure level of 500 hPa (corresponding to altitude close to 5 km). A height anomaly is defined the deviation in height from its average value along a latitude circle. One map is an analysis of height anomaly at 500 hPa $\phi'_{ana}(x,\ y,\ p_{500},\ t_a)$ (see section on "Initialization and Data Assimilation") and the second is a map of forecast height anomaly valid at the same time $\phi'_f(x,\ y,\ p_{500},\ t_a)$ where $t_a = t_i + \Delta t_f$. Here Δt_f is the forecast lead time and t_i is the initiation time. These two height fields are then used to form a correlation.

$$r_{500}(\Delta t_f) = \frac{\left\langle \phi_f'(t_i + \Delta t_f), \phi'_{ana} \right\rangle}{\sqrt{\left\langle \phi_f', \phi_f' \right\rangle \langle \phi'_{ana}, \phi'_{ana} \rangle}}$$

where <> represents the spatial covariance over some region, typically the southern hemisphere or the northern hemisphere.

Figure 5.2 shows the average evolution of r_{500} in two major global forecast models for the period January 1 through March 31, 2009. The plots illustrate the state-of-the-art in NWP as of this writing. As expected the pattern correlations decrease with time, but remain quite high, above 0.8, out to a forecast lead of 5 days. Northern hemisphere correlations are higher in both systems, probably reflecting the higher density of in situ measurements available there. Many operational forecasting centers do not allow changes to their systems that degrade this measure of performance.

Although using r_{500} as *the* single measure of forecast accuracy may seem somewhat restrictive, it should noted that that the geopotential height φ at 500 hPa is an integrated measure of the temperature in a deep layer, form the surface to around 5,000 m. So, r_{500} is a concise summary of model performance in a horizontally extensive and deep atmospheric slab.

Skill Scores

The S1 skill score [51, 52] has been used since the 1950s by forecasters at the National Meteorological Center (NMC), and after 1995 the National Center for Environmental Prediction (NCEP), to evaluate forecast performance. An S1 score can be defined for any quantity. It is calculated:

$$S1(\chi) = 100 \times \frac{\int_A |\nabla(\chi_f - \chi_o)| dA}{\int_A \max(|\nabla \chi_f|, \, |\nabla \chi_o|) dA}$$

where χ_f is forecast χ and χ_o is the observed value for verification. $S1$ then is the ratio of the integrated absolute gradient in forecast error, normalized by the integrated absolute gradient of the quantity itself, where at each location the larger of the forecast or observed values is used. A value of $S1 = 0$ represents a perfect forecast. The quantity χ used to calculate $S1$ is typically sea-level pressure or geopotential height.

The $S1$ skill score was selected by NMC from many measures of forecast quality with guidance from practicing forecasters. Forecasters in 1950s noted that values of $S1$ around 20 corresponded to very good forecasts, while values of 70 or more represented nearly worthless forecasts. As a result, it became common practice to express "skill" as $2(70 - S1)$, so that now a very good forecast $S1 = 20$ has a skill score of 100, while useless forecasts have a skill score of 0 [52].

Equitable Threat Scores

Evaluation of precipitation forecasts is difficult for a number of reasons. Precipitation is a field with high variance and sharp boundaries. In many circumstances, the important forecast parameter is whether rain or precipitation (above a certain threshold) has occurred. Such categorical forecasts are evaluated using various methods based on matrices of possible outcomes, for example, **YY** – rain is forecast and occurs, **YN** – rain is forecast but does not occur, **NY** – rain is not forecast but occurs, and **NN** – rain is not forecast and does not occur. The most commonly used method is that of Equitable Threat Scores (ETS) [53] which attempts to account for the long-term statistical probability of each category.

Initialization and Data Assimilation

A moment's reflection shows that establishing initial conditions for a global model of the atmosphere is a nontrivial task. First of all, there may be instrumental errors

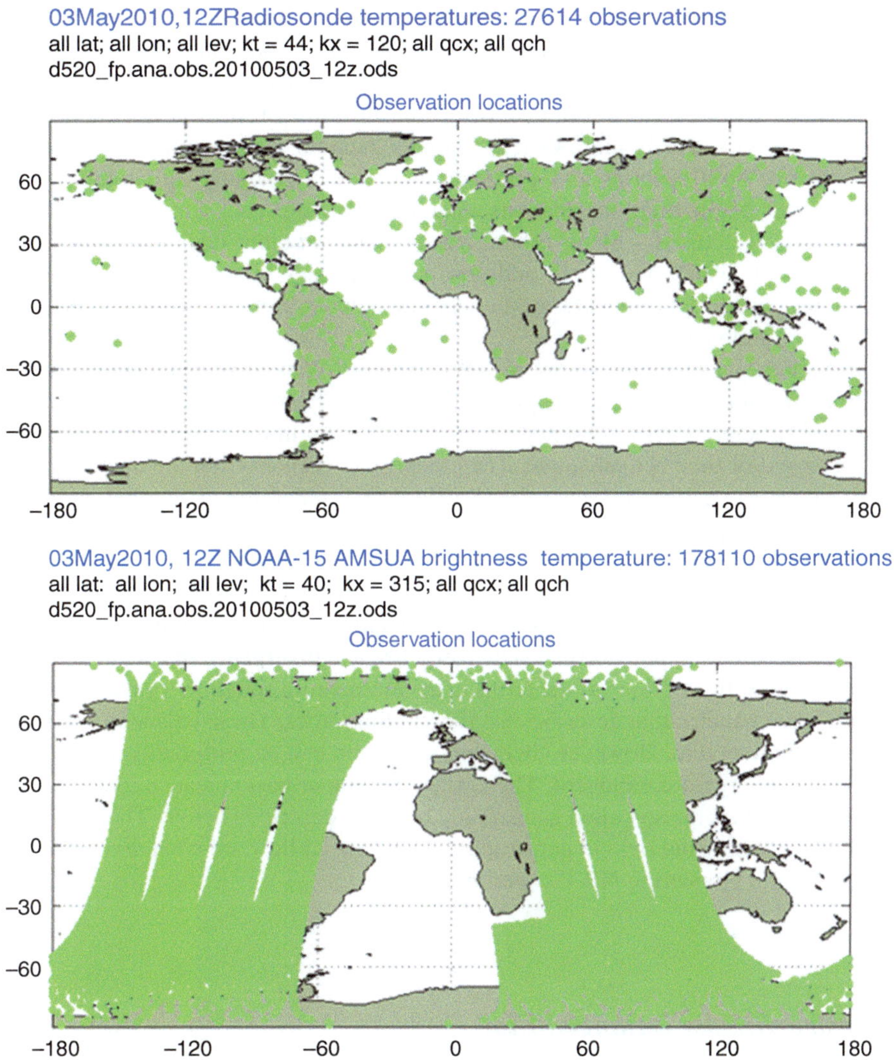

Fig. 5.3 Locations of radiosonde observations (*top*) and satellite temperature observations (*bottom*) for May 12, 2008 from NASA's GEOS-5 DAS

in the measurements of wind, temperature, humidity, or other quantities needed to specify the initial state of the atmosphere. While perhaps the most obvious problem, instrumental error may also be the least important problem faced in initialization. A more serious problem is hinted at in Fig. 5.3 which shows the current global distribution of radiosonde balloon launch sights. Radiosondes provide very accurate and reliable measurements of winds, temperatures, and humidities from the surface to around 10 km altitude. Launches are made by international agreement at either 0Z or 12Z ("Z" refers to Greenwich mean time) or at both times, depending on the station. However, as the figure shows, radiosonde launch sites are distributed in a completely unstructured way across the globe. High concentrations occur in the developed world, with sparse or no coverage over oceans and over less-developed land areas. Even in the developed world the location of sites is determined by human factors – and is more-or-less spatially random. Numerical models of the atmosphere, both spectral and grid point, require a spatially structured set of numbers to begin their integrations. Satellite data presents similar challenges. While it is structured – along orbital tracks – the structure does not conform to the needs of NWP models. In addition, satellite data is asynchronous, that is, sampling occurs continuously as the satellite travels, not at a specified time as with radiosonde data.

Thus a major challenge faced in NWP is to derive a complete model state on a structured set of points at a single instant in time, from data that may be neither spatially structured nor representative of a single point in time. Naive interpolation in space and/or time is inadequate. The model state derived from the data must not only cover the globe, it must do so while also satisfying a number of other dynamical and physical constraints.

Fluid flows, and vector fields in general, can be decomposed into a sum of divergent ($\nabla \cdot \mathbf{u}_1 \neq 0$) and nondivergent ($\nabla \cdot \mathbf{u}_2 = 0$) components. Generally speaking, the atmospheric motions of most significance in 1–5 day forecasts are characterized by "small" horizontal convergence and divergence in a relative sense. In these flows, two or three of the terms in the momentum equation form a dominant steady-state balance that describes the flow to first-order. These balanced flows are *almost* nondivergent. However, their time evolution can be profoundly affected by the small divergent component. The earliest and most basic balance identified by meteorologists is the so-called geostrophic balance, described below. The subtleties of the divergent wind field in geostrophically balanced flow are what doomed L. F. Richardson's pioneering NWP experiment.

Geostrophically Balanced Flow

The origin of geostrophic balance is most easily seen by performing a scale analysis of the momentum equation in the primitive equation system. Scale analysis is a common procedure in fluid mechanics to systematically identify the most important terms in the complicated equations describing fluid flow [4]. It begins by identifying scales of motion for the phenomenon of interest. For midlatitude weather systems, especially after considering the spacing and resolution of radiosonde and satellite data, a reasonable choice of spatial scale L is around 1,000 km. Other reasonable choices of scales are for horizontal wind $U \sim 10\ \mathrm{ms}^{-1}$, and for pressure disturbance $P \sim 10$ hPa or 1,000 Pa. These scales along with local apparent rotation rate f with values $\sim 10^{-4}\ \mathrm{s}^{-1}$ in midlatitudes are used to estimate the sizes of the terms in the equation. For example, $\frac{d}{dt}V$ will be $\sim U^2/L$ where the horizontal advective time scale has been used as the relevant time scale. It is easy to see that the ratio of this term to the Coriolis term $f\mathbf{k} \times V$ will scale as U/fL, which is a key nondimensional parameter in dynamic meteorology known as the Rossby number or Ro. For the scales of motion typical for midlatitude systems, Ro is close to 0.1. So the advective term is likely to small compared to the Coriolis term and also, it turns out, compared to the pressure gradient.

Thus, the "leading order" balance in midlatitude systems is left as:

$$f\mathrm{k} \times V = -\alpha \nabla_z p$$

where we have used height as the vertical coordinate. Equivalent expressions exist for other vertical coordinates. In component form this balance is written as:

$$-fv = -\alpha \partial_x p; \; fu = -\alpha \partial_y p$$

This balance is the dominant feature of midlatitude flow. It is also responsible for one of the most counterintuitive aspects of weather maps in midlatitudes – that the wind blows along pressure contours rather than from high pressure to low pressure. It is also the reason air flows in a counterclockwise sense around low-pressure centers in the northern hemisphere. A perplexing aspect of this balance is that it is a steady-state relation. In other words, the largest forces in the system give no information about its time evolution.

It is easy to see that this leading-order geostrophic flow is horizontally nondivergent. However, spurious divergent flow features can easily appear when constructing initial conditions from observations of horizontal. An idea of the difficulty of this challenge can be obtained by considering the following argument. The relative vorticity of the horizontal wind is determined from:

$$\zeta = \partial_x v - \partial_y u$$

and, further scale analysis of the equations of motion in midlatitudes (see [45]) shows that the ratio $\nabla \cdot V$ to ζ will typically be close to Ro or ~ 0.1. The individual horizontal derivative terms in the expressions for vorticity and divergence are of the same order. The small relative magnitude of $\nabla \cdot V$ is only possible through near cancelation of its much larger component terms. It is not trivial to maintain this cancelation during the data assimilation process. Errors in the divergence can have large effects on surface pressure tendencies since

$$\partial_t p_s \sim p_s \nabla \cdot V.$$

More complete and correct balance relationships than geostrophic balance can be derived [54]. Balanced initial conditions have been sought in a number of ways during the history of NWP. The most successful of these was perhaps nonlinear normal mode initialization introduced in the 1970s [55–57]. In this technique, fast and slow normal modes of the nonlinear equations of motion are found through an iterative procedure.

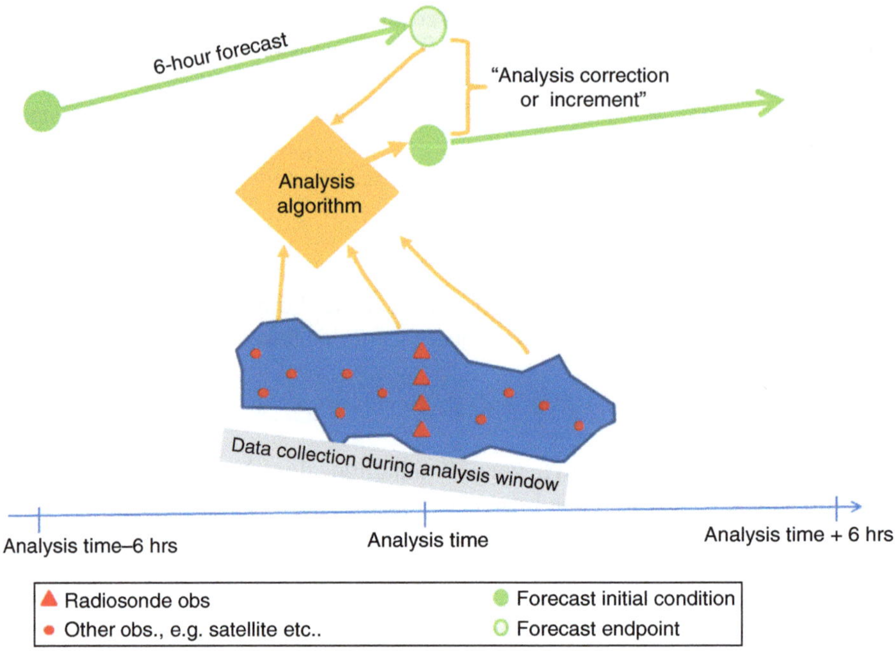

Fig. 5.4 Schematic diagram of 3D-var analysis

Data Assimilation Systems

Modern operational centers handle the problem of initializing their forecasts by using a data assimilation system (DAS). A central feature of a DAS is the forecast model itself (or a linearized version thereof). Analyses at operational forecasting centers are typically performed four times daily at 00Z, 06Z, 12Z, and 18Z. The analysis procedure combines measurements with a short model forecast, typically 6 h initialized with the previous analysis. This forecast is often referred to as the "first guess" or "background." The job of the analysis algorithm, denoted by the yellow diamond in Fig. 5.4, is to blend the myriad sources of data, which includes radiosonde observations satellite measurements, pilot reports, surface station reports, ship buoy measurements, and more, with the forecast to produce an optimal estimate of the state of the atmosphere on the model grid. The process illustrated in Fig. 5.4 depicts a "3D-Var" system. In 3D-Var, data gathered within an analysis time window, typically 3 h before and after the standard analysis times, is assumed to be synchronous. The analysis then consists of an optimal blending of measurements and forecast background in space.

This optimal blending can be expressed as a "cost function minimization." The cost function is written in matrix form [2, 54, 58]:

$$J(\mathrm{x}) = (\mathrm{x} - \mathrm{x}_b)T\ \mathrm{B}^{-1}(\mathrm{x} - \mathrm{x}_b) \\ + [\mathrm{y} - H(\mathrm{x})]^T\ \mathrm{R}^{-1}[\mathrm{y} - H(\mathrm{x})] + J_b(\mathrm{x})$$

where **x** is a model "state vector," that is *V*, *T*, *q* and possibly other analyzed species on the model's numerical grid (or spectral decomposition). The quantity $\mathbf{x}_b$ is the model state from a forecast, that is, the first guess or background, and the matrix **B** is the background, or forecast, error covariance matrix. This matrix is a key piece of the analysis algorithm, and is estimated by examining differences between forecasts radiosonde observations [59] or more recently by calculating the covariance of different short forecasts, for example, 24 h and 48 h, valid at the same time [60, 61].

The second term in the cost function contains the observation error covariance matrix **R**, the observation vector **y** and a vector *H*(**x**) which is the result of the "observation operator" *H* acting on the model state **x**. In the case of an observation taken by a thermometer placed at a model grid-point, the observation operator would simply select the appropriate element of **x**. However, in the case of remote satellite observations, which directly measure radiances (photons) from the atmosphere, *H* could represent a complex radiative transfer calculation using for example *T* and *q* from the model state to estimate the radiance measured by a particular instrument. The approach of transforming model quantities into a form that is directly comparable with observations, sometimes referred to as "radiance assimilation," led to dramatic increases in the positive impact of satellite measurements on forecasting [62].

The third term in the cost function represents balance constraints on the flow, such as those discussed in the section on "Geostrophically Balanced Flows". Inclusion of such a term at NCEP has eliminated the need for a separate initialization procedure for forecasts [60].

The task of the analysis algorithm is to find the model state $\mathbf{x}_a$ that minimizes the cost function *J*(**x**). For more details on how this solution is actually accomplished the reader is referred to discussion in Chap. 5 of Kalnay (2003) [54]. 3D-var as described here is used by NCEP as well as in slightly modified version by NASA's Global Modeling and Assimilation Office (GMAO) in their GEOS-5 DAS [63]. A somewhat different approach known as 4D-Var is used at ECMWF. This approach takes into account the possibly asynchronous nature of data when formulating the cost function. For more details, see the discussion in Kalnay (2003).

Ensemble Forecasting

The early thinking of researchers in NWP was that the central problem of forecast initialization was to correctly filter out rapid divergent motions, and, that once this was accomplished no fundamental limits on atmospheric predictability existed. A series of seminal papers by Edward N. Lorenz proved this to be incorrect [64–66]. Lorenz showed that simple analogs to atmospheric equations of motion possess a sensitive dependence to initial conditions. In other words, small differences in even well-balanced initial conditions will cause forecasts to become uncorrelated after a finite time.

Modern operational forecasting centers typically perform ensembles of many runs with slightly different initial conditions, as well as a single higher-resolution "deterministic" run to produce forecasts for a given time [54, 67]. The generation of ensemble members is a nontrivial task. Ideally, the members of the ensemble should vary in special directions in phase space that are related to the most rapidly growing instabilities in the flow [54, 68].

Future Directions

Weather forecasts have improved demonstrably during the 50 year history of NWP. This is illustrated in Fig. 5.5, which shows the evolution of r_{500} at 3, 5, and 7 days over the last 30 years in the ECMWF system [69]. Part of the improvement is traceable to the explosion in the amount of satellite data over the last 30 years. However, a large part of the improvement in skill is due to improvements in the forecast and analysis "system," such as increased forecast model resolution, improved analysis algorithms etc. This is nicely demonstrated in the bottom panel of Fig. 5.5, which shows the evolution of skill in retrospective forecasts, using the *current* ECMWF system on the historical data base. The skill of retrospective forecasts is significantly higher, indicating that the improved forecast and analysis system makes a significant, perhaps the dominant, contribution to the overall increase in skill seen in the last 30 years. One aspect of improvement that is clearly due to improved data sources (satellites) is the convergence of skill in southern and northern hemispheres.

Scalable Dynamical Cores

As computing power increases NWP model resolution also continues to increase. In the last several years, the increase in computer power has appeared primarily in the form of massively parallel machines with larger and larger numbers of processors, rather than in the form of faster individual processors. This means that "time-to-solution" has not decreased dramatically in recent years, but the size of feasible calculations has increased dramatically. This has stimulated the development of scalable models. Scalability means that model speed increases more-or-less linearly with the number of processors used. A trivial example of perfectly scalable problem is the addition of 1,000 pairs of numbers a + b = c. If one processor is available then the calculation will take 1,000 CPU time units. If 1,000 processors are available the

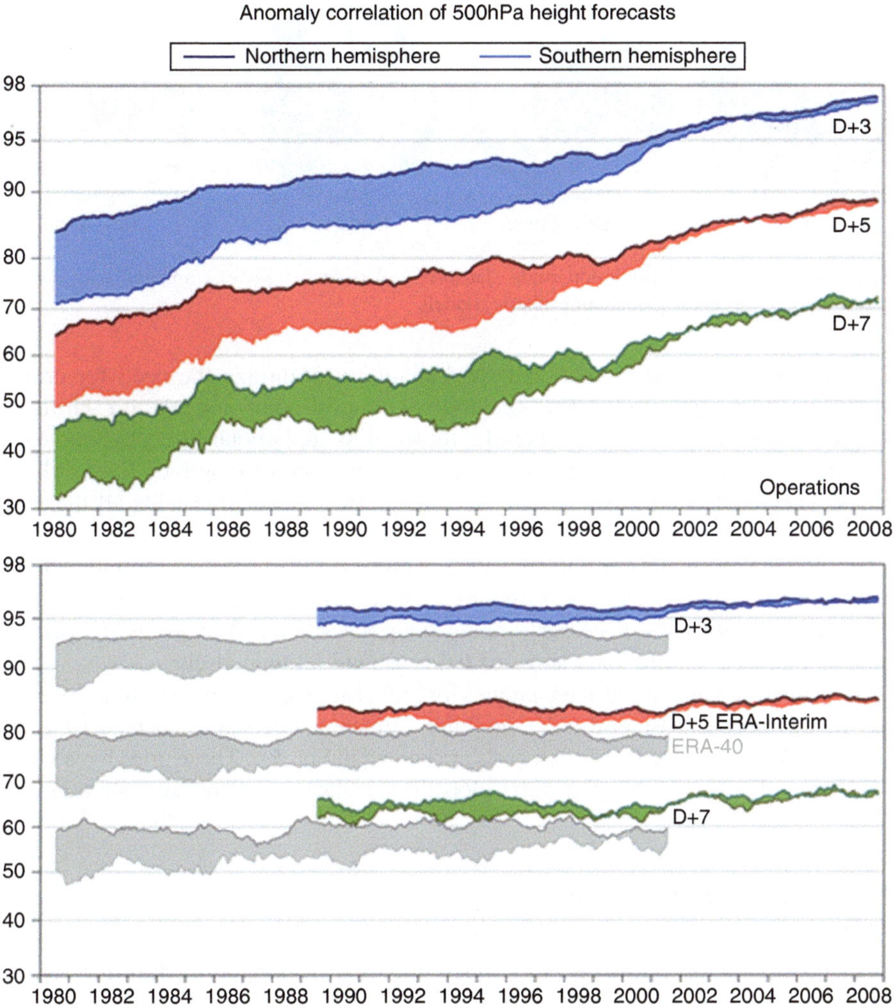

Fig. 5.5 Evolution of forecast skill at ECMWF, adapted and extended from the study of Simmons and Hollingsworth (2002). The *top panel* shows a history of r_{500} at 3, 5, and 7 days from the ECMWF operational forecast beginning in 1980. The *colored* and *shaded areas* are bounded by southern hemisphere skill below and northern hemisphere skill above. The *lower panel* shows retrospective forecasts produced using two versions of current ECMWF analysis and forecast systems. In these cases, the "system" is fixed in time, while data inputs evolve in actual historical fashion

entire calculation can take place in 1 CPU time unit. However, communication between processors also costs time. In any real numerical model of the atmosphere processors eventually need information residing in other processors. This prevents numerical models of the atmosphere from scaling perfectly.

The amount of cross-processor communication required can vary widely depending on model design. Ideally decomposition will maximize the ratio of

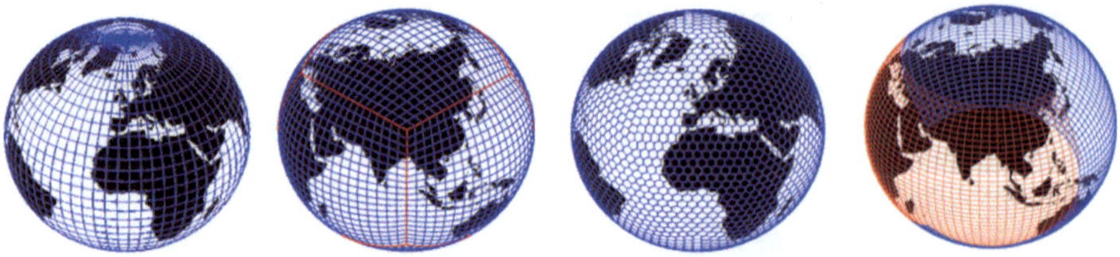

Fig. 5.6 The standard latitude-longitude "lat-lon" grid (*leftmost globe*) compared with newer, non-traditional grids for global atmospheric models

area to perimeter in computational subdomains to minimize the need for cross-processor communication. The scalability of grid-point models on latitude-longitude grids is severely hampered by the need to apply polar filters to overcome numerical instabilities that arise due to the convergence of longitude lines at both poles. These filters typically require knowledge of atmospheric fields all the way around latitude circles. Thus, modelers must either decompose the globe into thin computational domains circling the globe, which will lead to large communication requirements in the north–south direction, or they must pay the cost of frequent "gathers" to obtain the necessary inputs for polar filters. Transforms in spectral models likewise require knowledge of fields around latitude circles.

Recent efforts in numerical techniques for global atmospheric models have focused on the development of grid-point or finite element models on nonstandard grids [70–73]. Several examples of such grids are shown in Fig. 5.6. These grids have fairly uniform grid cell sizes over the entire globe. Polar filters are therefore not required.

Nonhydrostatic Dynamics

Once horizontal resolution becomes much finer than 10 km, nonhydrostatic effects must be taken into account. This will require models based on a different set of equations. One option is to simply use the full Euler or Navier Stokes equations [39, 74] and pay the costs associated with the short time-steps required by the presence of acoustic waves. Another approach is to use an inelastic equations system [75], but this requires solution of an elliptic equation which is an intrinsically nonlocal procedure and again raises cross-processor communication costs.

Seamless Models for Climate and Weather

Modern global NWP and climate models are essentially the same. Both use the same set of dynamical equations (see section on "Primitive Equations"). Both also use the same set of physical parameterization schemes (see section on "Parameterization"). In practice, differences do exist between global models of the atmosphere intended for climate simulation and those intended for forecasting. Operational global NWP models have typically used resolutions that are a factor of 8–16 times higher than those used in long climate simulations. Other differences arise, more-or-less unintentionally, in the tuning process as a consequence of the different metrics used in the evaluation of NWP and climate models (see section on How are NWP models (versus climate models) evaluated?).

Climate researchers understand that atmospheric phenomena such as squall lines and tropical cyclones may play a role in establishing climate on both regional and global scales. Such "mesoscale" features are not resolved in climate models with resolutions of 100 km or coarser. However, the continued increase in computer power, and the recent emphasis on massively parallel architecture, will allow decadal or even century-long simulations at resolutions close to 10 km in the near future. At these resolutions, mesoscale circulations should be well represented. These resolutions will also present novel challenges to the sub-grid parameterizations used in climate models, as assumptions about scale-separation and statistical equilibrium become questionable.

Bibliography

1. Tribbia J, Anthes R (1987) Scientific basis of modern weather prediction. Science 237:493–499
2. Lynch P (2006) The emergence of numerical weather prediction: Richardson's dream. Cambridge University Press, Cambridge, MA
3. Richardson L (1922) Weather prediction by numerical process. Cambridge University Press, Cambridge
4. Gill A (1982) Atmosphere–ocean dynamics. Academic, London
5. Simpson J (1994) Sea breeze and local winds. Cambridge University Press, New York
6. Lau K, Li M (1984) The monsoon of East Asia and its global associations_survey. Bull Amer Meteor Soc 65:114–125
7. Webster P, Tomas R (1998) Monsoons- processes, predictability, and the prospects for prediction. J Geophys Res 103(C7):14451–14510
8. Douglas M, Maddox R, Howard K, Reyes S (1993) The Mexican monsoon. J Clim 6(8):1665–1677
9. Koch S, Handley C (1997) Operational forecasting and detection of mesoscale gravity waves. Weather Forecast 12(2)
10. Moncrieff M, Liu C (1999) Convection initiation by density currents: role of convergence, shear, and dynamical organization. Mon Weather Rev 127(10):2455–2464
11. Lac C, Lafore J, Redelsperger J (2002) Role of gravity waves in triggering deep convection during TOGA COARE. J Atmos Sci 59(8)
12. Houze R Jr (2004) Mesoscale convective systems. Rev Geophys 42. doi:10.1029/2004RG000150
13. Grabowski W, Moncrieff M (2006) Large-scale organization of tropical convection in two-dimensional explicit numerical simulations. Q J Roy Meteor Soc 127(572):445–468
14. Smith R (1979) The influence of mountains on the atmosphere. Adv Geophys 21:87–230

15. Longuet-Higgins M (1964) Planetary waves on a rotating sphere. Proc R Soc Lond A Math Phys Sci 279(1379):446–473
16. Andrews D, McIntyre M (1976) Planetary waves in horizontal and vertical shear: the generalized Eliassen-palm relation and the mean zonal acceleration. J Atmos Sci 33:2031–2048
17. Hoskins B, Ambrizzi T (1993) Rossby wave propagation on a realistic longitudinally varying flow. J Atmos Sci 50:1661–1671
18. Charney J (1947) The dynamics of long waves in a baroclinic westerly current. J Meteor 4(5):135–162
19. Eady E (1949) Long waves and cyclone waves. Tellus 1(3):33–52
20. Kiladis G, Wheeler M, Haertel P, Straub K, Roundy P (2009) Convectively coupled equatorial waves. Rev Geophys 47. doi:10.1029/2008RG000266
21. Burpee R (1972) The origin and structure of easterly waves in the lower troposphere of North Africa. J Atmos Sci 29(1):77–90
22. Landsea C et al (1993) A climatology of intense (or major) Atlantic hurricanes. Mon Weather Rev 121(6):1703–1713
23. Pytharoulis I, Thorncroft C (1999) The low-level structure of African easterly waves in 1995. Mon Weather Rev 127(10)
24. Thorncroft C, Hodges K (2001) African easterly wave variability and its relationship to Atlantic tropical cyclone activity. J Clim 14(6):1166–1179
25. Madden R, Julian P (1994) Observations of the 40–50-day tropical oscillation: a review. Mon Weather Rev 122(5):814–837
26. Zhang C (2005) Madden-Julian oscillation. Rev Geophys 43(2). doi:10.1029/2004RG000158
27. Liebmann B, Hendon H, Glick J (1994) The relationship between tropical cyclones of the western Pacific and Indian Oceans and the Madden–Julian oscillation. J Meteor Soc Japan 72(41):1–412
28. Maloney E, Hartmann D (2000) Modulation of hurricane activity in the Gulf of Mexico by the Madden-Julian oscillation. Science 287(5460):2002
29. Maloney E, Hartmann D (2000) Modulation of eastern North Pacific hurricanes by the Madden-Julian oscillation. J Clim 13(9)
30. Elsner J, Jagger T, Niu X et al (2000) Changes in the rates of North Atlantic major hurricane activity during the 20th century. Geophys Res Lett 27(12):1743–1746
31. Hall J, Matthews A, Karoly D (2001) The modulation of tropical cyclone activity in the Australian region by the Madden–Julian oscillation. Mon Weather Rev 129:12
32. Bond N, Vecchi G (2003) The influence of the Madden–Julian oscillation on precipitation in Oregon and Washington. Weather Forecast 18(4):600–613
33. Jones C, Waliser D, Lau K, Stern W (2004) The Madden–Julian oscillation and its impact on northern hemisphere weather predictability. Mon Weather Rev 132(6)
34. Waliser DE, Lau KM, Stern W, Jones C (2003) Potential predictability of the Madden–Julian oscillation. Bull Amer Meteor Soc 84:33–50
35. Manabe S, Smagorinsky J, Strickler R (1965) Simulated climatology of a general circulation model with a hydrologic cycle. Mon Weather Rev 93(12):769–798
36. Namias J (1980) The early influence of the Bergen School on synoptic meteorology in the United States. Pure App Geophys 119(3):491–500
37. Grenas S (2005) Vilhelm Bjerknes' vision for scientific weather prediction. In: The Nordic seas: an integrated perspective: oceanography, climatology, biogeochemistry, and modeling. American Geophysical Union, Washington, DC, p 357
38. Hinkelmann K (1951) Der Mechanismus des meteorologischen Lärmes. Tellus 3:285–296
39. Laprise R (1992) The Euler equations of motion with hydrostatic pressure as an independent variable. Mon Weather Rev 120(1):197–207
40. Shuman F, Hovermale J (1968) An operational six-layer primitive equation model. J Appl Meteor 7(4):525–547

41. Bengtsson L (2001) The development of medium range forecasts. In: 50th anniversary of numerical weather prediction, commemorative symposium, Potsdam
42. Kasahara A (2000) On the origin of cumulus parameterization for numerical prediction models. Gen Circ Model Dev 70:199–224
43. Tiedtke M (1993) Representation of clouds in large-scale models. Mon Weather Rev 121:3040–3061
44. Del Genio A, Yao M, Kovari W, Lo K (1996) A prognostic cloud water parameterization for global climate models. J Clim 9:270–304
45. Haltiner G, Williams R (1983) Numerical prediction and dynamic meteorology, 2nd edn. Wiley, New York
46. Leith C (1965) Numerical simulation of the earth's atmosphere. In: Methods in computational physics, vol 4. Academic, New York
47. Webster S, Brown A, Cameron D, Jones C (2006) Improvements to the representation of orography in the Met Office Unified Model. Q J Roy Meteor Soc 129(591):1989–2010
48. Clark A, Gallus W Jr, Xue M, Kong F (2009) A comparison of precipitation forecast skill between small convection-allowing and large convection-parameterizing ensembles. Weather Forecast 24(4):1121–1140
49. Bhanu Kumar O, Ramalingeswara Rao S, Muni Krishna K (2009) Role of cumulus parameterization schemes in simulating heavy rainfall episodes off the coast of Maharashtra state during 28 June–4 July 2007. Meteorol Atmos Phys 105(3):167–179
50. Gleckler P, Taylor K, Doutriaux C (2008) Performance metrics for climate models. J Geophys Res 113:D06104
51. Teweles S, Wobus H (1954) Verification of prognostic charts. Bull Amer Meteor Soc 35:455–463
52. Shuman F (1989) History of numerical weather prediction at the National Meteorological Center. Weather Forecast 4:286–296
53. Gandin L, Murphy A (1992) Equitable skill scores for categorical forecasts. Mon Weather Rev 120:361–370
54. Kalnay E (2003) Atmospheric modeling, data assimilation, and predictability. Cambridge University Press, Cambridge, MA
55. Machenhauer B (1977) On the dynamics of gravity oscillations in a shallow water model, with applications to normal mode initialization. Beitr Phys Atmos 50:253–271
56. Baer F, Tribbia J (1977) On complete filtering of gravity modes through nonlinear initialization. Mon Weather Rev 105(12):1536–1539
57. Tribbia J (1984) A simple scheme for high-order nonlinear normal mode initialization. Mon Weather Rev 112:278–284
58. Daley R (1993) Atmospheric data analysis. Cambridge University Press, New York
59. Thiebaux HJ, Pedder MA (1987) Spatial objective analysis with applications in atmospheric science. Academic, London/Orlando, p 308
60. Parrish D, Derber J (1992) The National Meteorological Center's spectral statistical-interpolation analysis system. Mon Weather Rev 120(8):1747–1763
61. Buehner M, Gauthier P, Liu Z (2005) Evaluation of new estimates of background-and observation-error covariances for variational assimilation. Q J Roy Meteor Soc 131(613):3373–3384
62. Derber J, Wu W (1998) The use of TOVS cloud-cleared radiances in the NCEP SSI analysis system. Mon Weather Rev 126(8):2287–2299
63. Rienecker M, Suarez M, Todling R, Bacmeister J, Takacs L, Liu H, Gu W, Sienkiewicz M, Koster R, Gelaro R et al (2008) The GEOS-5 data assimilation system-documentation of versions 5.0.1, 5.1.0. NASA Technical Memorandum, NASA/TM-2007, 104606
64. Lorenz E (1965) A study of the predictability of a 28-variable atmospheric model. Tellus 17(3):321–333
65. Lorenz E (1969) The predictability of a flow which possesses many scales of motion. Tellus 21(3):289–307

66. Lorenz E (1969) Atmospheric predictability as revealed by naturally occurring analogues. J Atmos Sci 26(4):636–646
67. Lewis J (2005) Roots of ensemble forecasting. Mon Weather Rev 133(7):1865–1885
68. Toth Z, Kalnay E (1993) Ensemble forecasting at NMC: the generation of perturbations. Bull Amer Meteor Soc 74:2317–2330
69. Simmons A, Hollingsworth A (2002) Some aspects of the improvement in skill of numerical weather prediction. Q J Roy Meteor Soc 128(580):647–678
70. Ringler T, Heikes R, Randall D (2000) Modeling the atmospheric general circulation using a spherical geodesic grid: a new class of dynamical cores. Mon Weather Rev 128(7):2471–2490
71. Skamarock W, Klemp J, Ringler T, Thuburn J (2008) A hexagonal C-grid atmospheric core formulation for multiscale simulation on the sphere. AGU fall meeting abstracts, San Francisco, CA, p 01
72. Putman W, Lin S (2009) A finite-volume dynamical core on the cubed-sphere grid. Astron Soc Pac Conf Ser 406:268
73. Taylor M, Fournier A (2010) A compatible and conservative spectral element method on unstructured grids. J Comput Phys 229:5879–5895
74. Klemp J, Skamarock W, Dudhia J (2007) Conservative split-explicit time integration methods for the compressible nonhydrostatic equations. Mon Weather Rev 135(8):2897–2913
75. Grabowski W, Smolarkiewicz P (2002) A multiscale anelastic model for meteorological research. Mon Weather Rev 130(4)

Chapter 6
Atmospheric General Circulation Modeling

Philip J. Rasch

Glossary

Aerosols	The small (solid and liquid) particles that are suspended in the atmosphere. Aerosols have both natural (e.g., sea-salt, dust, and some organic compounds released by vegetation) and anthropogenic origins (e.g., the pollution released by power plants, cars, trucks, agricultural burning, etc.).
Climate	The statistical description of characteristics of our environment over long periods, including properties like the mean and extreme values of value of fields like temperature, winds, and moisture.
Climate sensitivity	Usually used to mean the change in globally averaged surface temperature ΔT that would occur in a model if it were allowed to equilibrate to a forcing ΔF associated with a doubling of CO_2. It is sometimes used in a looser fashion to refer to the change in temperature resulting from a change in forcing.
Feedback	A process in the climate system that can either amplify ("positive feedback") or diminish ("negative feedback") a change in climate forcing.
Lapse rate	A term that refers to the vertical temperature decrease in the atmosphere. When that lapse rate exceeds certain thresholds, convective overturning can occur. Two threshold lapse rates are important. The dry adiabatic lapse rate identifies the rate at which an unsaturated parcel will cool if it is lifted adiabatically. If the environmental lapse rate is larger than the dry adiabatic

This chapter was originally published as part of the Encyclopedia of Sustainability Science and Technology edited by Robert A. Meyers. DOI:10.1007/978-1-4419-0851-3

P.J. Rasch
Pacific Northwest National Laboratory, Richland, WA 99352, USA
e-mail: philip.rasch@pnl.gov

P.J. Rasch (ed.), *Climate Change Modeling Methodology: Selected Entries from the Encyclopedia of Sustainability Science and Technology*, DOI 10.1007/978-1-4614-5767-1_6, © Springer Science+Business Media New York 2012

	lapse rate, a parcel lifted adiabatically will gain buoyancy and convective overturning can occur. When saturated air is lifted adiabatically, it will cool at a temperature-dependent rate as phase change occurs. At warm temperatures, the saturated adiabatic lapse rate is less than the dry adiabatic lapse rate. Since the atmosphere will produce overturning to reduce these instabilities associated with buoyant parcels, the lapse rate and water vapor amount play an important role in convection. The moist and dry adiabatic lapse rates explain much of the vertical temperature gradient to lowest order.
Parameterization	The equations and computer code describing the representation of a particular physical process in a climate model, for example, the representation of convection.
Radiative forcing	A change altering the energy budget of the climate system usually associated with changes in the atmospheric abundance of greenhouse gases and aerosols, or factors like solar variability and volcanic. These changes are expressed in terms of radiative forcing, which is used to compare how a range of human and natural factors drive warming or cooling influences on global climate.
Subgrid scale	The behavior of a process at time and space scales that are smaller than the model can resolve.
Tropopause	A permeable boundary separating two layers of the atmosphere: the stratosphere (a relatively stable region above) and the troposphere (a less stable region below where convective overturning often occurs). The tropopause varies quite smoothly in latitude. It is highest in the tropics (18 km) and decreases toward the poles (to 10 km or so).
Weather	The short-term evolution of our environment.

Definition of the Subject

This entry provides a brief introduction to the computer models of the atmosphere used for climate studies. The concepts of atmospheric forcing and response are developed and used to highlight the importance of clouds and aerosols to the climate system and the many uncertainties associated with their representation. Many processes that are important to the accurate representation of clouds and aerosols for climate are subgrid scale, and present both physical and computational challenges in atmospheric modeling. Other factors contributing to uncertainties in models are discussed, and some remaining challenges in atmospheric models are introduced.

Introduction

This entry provides a brief description of models of the atmosphere used for climate studies. These models can be part of a coupled climate system model or Coupled Climate and Earth System Models, as described by Gent elesewhere in the section *Climate Change Modeling and Methodology*, but they can also be used separately with prescribed values for surface fields, or simpler treatments for surface processes.

The atmospheric component of climate models can vary enormously in complexity. Simple atmospheric models based on energy balance arguments can be run on a laptop to provide rapid estimates of global, annual averaged properties of the atmosphere (e.g., surface temperature, e.g., [22]). The simpler types of atmospheric models (Energy Balance Models and Models of Intermediate Complexity) are discussed in the entry Climate Change Projections: Characterizing Uncertainty Using Climate Models by Sanderson, and Knutti, and Edmonds et al.

Much more elaborate models typically used in Earth system models are capable of simulating the distributions (in space and time) of hundreds of atmospheric fields and processes, the interaction between those fields and processes, and their response to external forcing. In this entry, the focus is on the more complex form of atmospheric models. These models are frequently also called General Circulation Models (GCMs) or Atmospheric General Circulation Models (AGCMs). The term GCM will be used here. More detail is found in the textbooks by Washington and Parkinson [20], Jacobson [5], and McGuffie and Henderson-Sellers [9].

GCMs share a great many features with the weather prediction models described by Bacmeister in this section *Climate Change Modeling and Methodology*. Both use the "equations of motion" (simplified versions of the Navier–Stokes for fluid flow, coupled with thermodynamic and mass conservation equations) to describe the evolution of the atmosphere, and parameterizations, but the way the models are used, and the focus is different (discussed more below).

A nice history of climate science and the development of weather and climate models can be found in Weart [21]. The first incarnation of atmospheric models solved on computers can be traced back to the efforts of a small group of meteorologists and physicists initially lead by John von Neumann and later Jule Charney near Princeton, New Jersey, soon after World War II. That effort started with the solution of simplified versions of the equations of motion on the most powerful computers available at the time (less powerful than the laptops in use today).

The complexity of present-day models has increased enormously, and large communities have grown up around those models. There are perhaps a dozen comprehensive independent GCMs in use today around the world, and many more prototypes used for study and development. Those communities include scientists and computer staff engaged in the development of the basic model, including testing and evaluation, as well as scientists that use the model as a tool for investigating climate science.

An easy example of a community activity focused on a particular GCM is the Community Atmosphere Model (CAM; http://www.cesm.ucar.edu/working_groups/Atmosphere/) project, an activity started in the USA about 30 years ago. That model is part of the larger Community Earth System Modeling (CESM) Project (http://www.cesm.ucar.edu).

CAM is a computer code that is order 400,000 lines of FORTRAN90 code. It is capable of being run in a variety of configurations, each optimized for different purposes, for example:

- Configurations useful for paleoclimate problems. These simulations sometimes require different positions of the continents. Simulations might be performed over millennia to explore climate change in response to orbital changes, or the Sun's luminosity.
- Idealized model configurations that simplify the Earth system by assuming the underlying surface is an "aquaplanet" (a world entirely covered by water).
- Configurations able to simulate the reactive photochemistry important for understanding the evolution of ozone distributions in the middle atmosphere handling hundreds of trace constituents, and many chemical reactions between those constituents.
- Configurations useful for study the reasons for climate change arising from nature and mankind over the last two centuries, and useful for producing projections given various scenarios of change in the future.

There are many other model configurations, and this list is not inclusive.

Detailed technical notes and scientific papers describing these types of applications can be found on the web sites mentioned above. The model is "open source," and it can be downloaded and run by anyone (provided a sufficiently powerful computer is available to perform a desired calculation). Archives of previous simulations are also available for download and examination.

The CESM and CAM projects are perhaps the largest of the GCM modeling activities in existence today, and so it provides an easy example to discuss and describe, but perhaps a dozen other activities around the world have similar capabilities.

The rest of this entry discusses the "generic" characteristics of this kind of climate model.

Global GCMs are generally run at resolutions resolving horizontal features larger than hundreds of kilometers and vertical variations of a few hundred meters. These resolutions are somewhat lower than used by weather models, where more detail is often needed. GCMs frequently include representation of processes and variables that are neglected or treated more simply in weather models (e.g., weather models often neglect details of the evolution of aerosols, or the slow evolution of greenhouse gases that affect the Earth system over longer timescales than are important for weather). GCMs may also include external forcing terms (e.g., variations in solar fluxes, or a historic database of volcanic eruption information) that are neglected by weather models.

Weather prediction models have typically been optimized to provide information about local features of the atmosphere over shorter time periods at higher resolution. Because initial conditions for weather models are constantly being "reset" to observed values, less attention has been paid to processes that affect the simulation over longer timescales (months to years). Climate models, on the other hand, focus on a description of the subtle balances and feedbacks occurring between processes and tend to describe these relationships through statistics of their long-term behavior. For many applications, climate models ignore initial conditions (weather modelers have traditionally viewed their simulations as an initial value problem; climate modelers, as a boundary value problem). A focus on statistical properties necessarily requires "multiple samples" from a distribution, with less attention on initial conditions, and more attention on the processes that control the model equilibrium or produce a transition from one climate regime to another. These points of view are changing, as discussed below.

This divergence in focus between weather and climate has led to differences in model design and configuration. Climate scientists have developed models that allow simulations over centuries or millennia. Weather models provide much higher resolution information, but those simulations are often for periods of only a few days or weeks.

GCMs being used around 2010 divide the atmosphere into columns of about a hundred kilometer on a side with 30–50 layers vertically (see Fig. 6.1). Most of the focus is on the atmosphere within the 40 km nearest the Earth's surface. Weather models may use columns as narrow as 10 km on a size with two to three times the number of vertical layers. Global weather models thus divide the atmosphere into volumes about 200 times smaller than climate models. Computational and accuracy constraints require that model time steps decrease in proportion to the size of the model volume. A reasonable first guess on model cost (the number of floating point operations required to complete a simulation of fixed length) is that it scales as the cube of the model resolution. Figure 6.1 shows a typical type of atmospheric model grid (in this case uniform in latitude and longitude), but other discretizations are possible (see Bacmeister, this volume).

This difference in horizontal and vertical resolution produces significant differences in the way some features important to weather and climate are represented in these models. An example is shown in Fig. 6.2, which displays the surface topography for North and South America at a 200-km horizontal resolution typical of climate models and a 20-km resolution typical of weather models. The very sharp, small-scale topographic features like the Andes have been significantly "smoothed out" at low resolution, differing in altitude by almost a factor of 2 and spread over a much broader horizontal extent, with measurable impacts on the role of the Andes as a barrier to winds, and their role in influencing precipitation patterns. The need to resolve some features important for climate at high resolution while minimizing computational cost is one of the motivations for the development of regional climate models (see Regional Climate Models by Leung), or models with variable resolution meshes to put the resolution where it is needed.

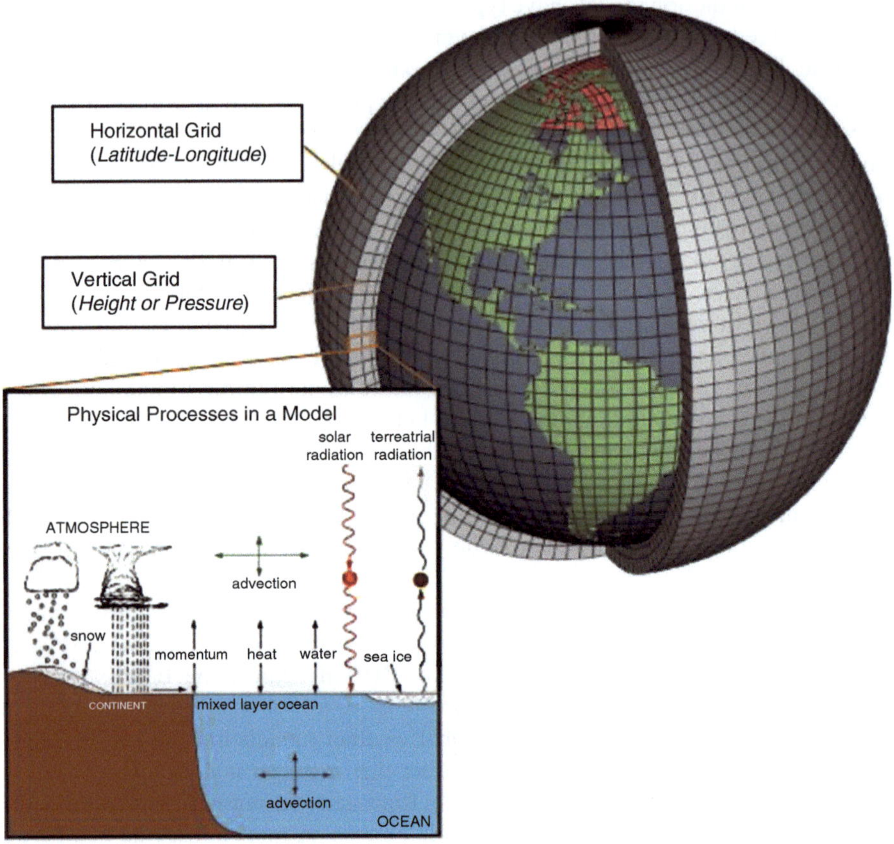

Fig. 6.1 Typical discretization of a GCM or weather model (Figure from http://www.oar.noaa.gov/climate/t_modeling.html)

Figure 6.3 shows a typical layer structure for an atmospheric climate model. Most models in use today rewrite their equations to employ a vertical coordinate that follows the terrain near the surface, with a gradual transition to a fixed height or pressure coordinate at higher altitudes. Model layers are generally concentrated near the Earth's surface to deal with the complexity of processes taking place there due to boundary layer effects, terrain, interactions with surface models, and the fact that mankind lives in that region. Models typically use layers 10–100-m thick near the surface and gradually decrease that resolution to use 1–2-km-thick layers at higher altitudes. Other coordinate systems have also been considered for climate models. The equations of motion are expressed more simply with height- and pressure-based vertical coordinates, but treatment of boundary conditions is more complex. Some modeling groups have explored the use of vertical coordinates that approximately follow a material surface. These formulations result in more complex models with coordinate surfaces that can also intersect the surface of the Earth, introducing additional complexity in the treatment of boundary conditions, but the benefit is a model with much more accurate treatment of vertical transport.

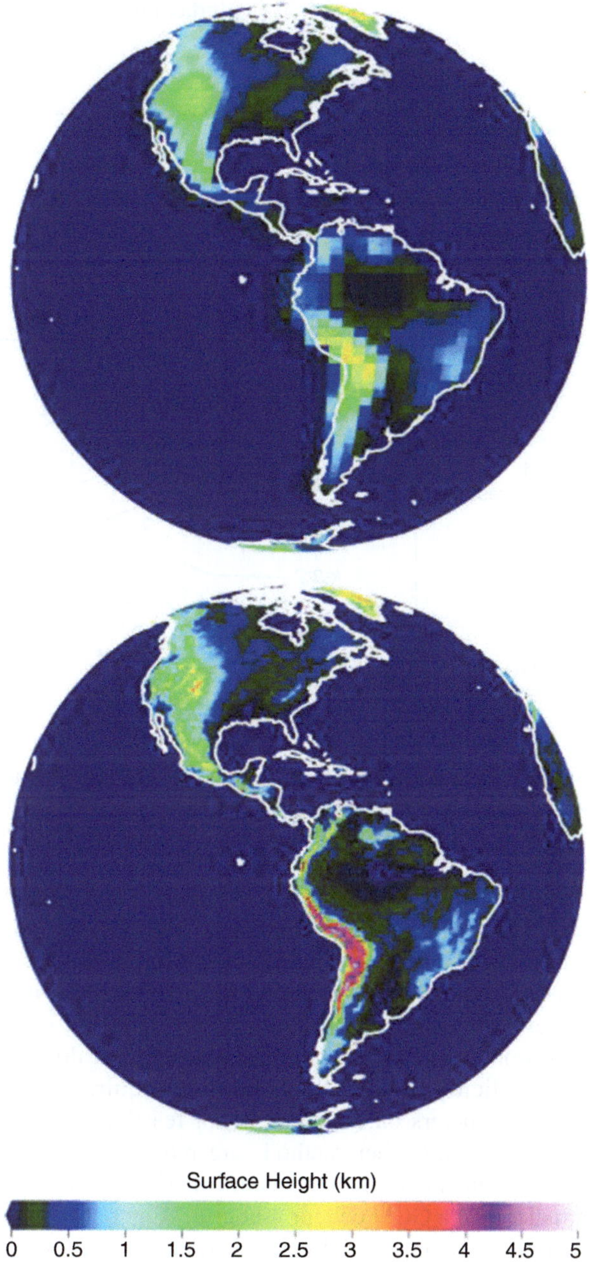

Fig. 6.2 The topography used by a typical "low-resolution" global atmospheric model (approximately 2° horizontal resolution, *upper panel*) and the high-resolution topography more typical of weather models, and next-generation climate models (about 0.25° resolution). Note the factor of 2 difference in height of the Andes and similar differences over the Rockies of N. America

Climate problems require descriptions of physical interactions at multiscale scales, leading to a very demanding challenge in physics and computational mathematics. Small-scale phenomena operating on the scale of molecules (e.g., chemistry, phase change of water, and radiative transfer) influence larger-scale features

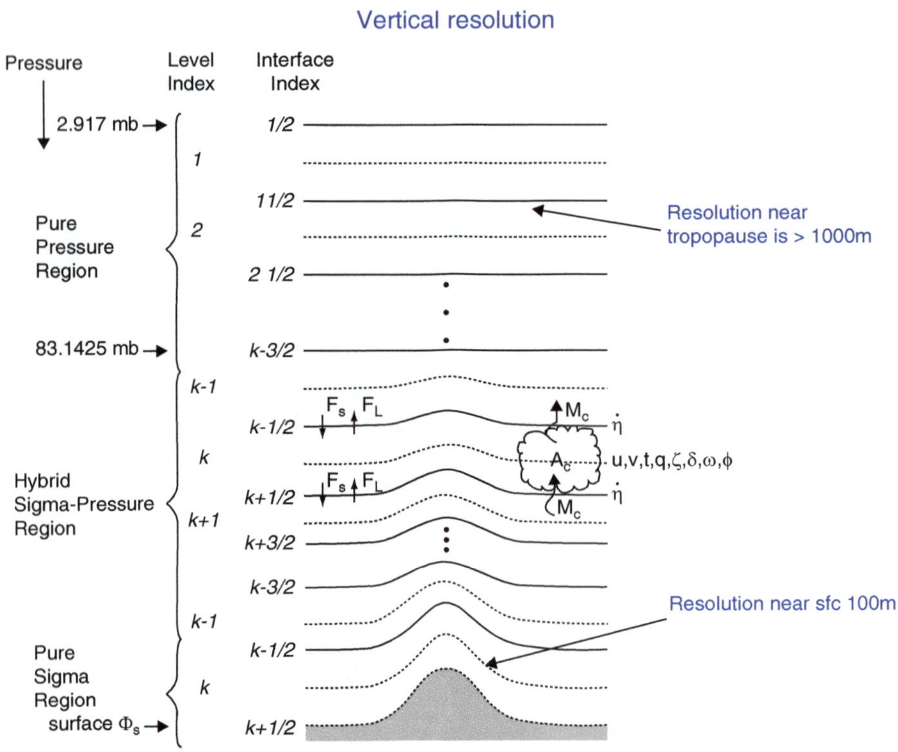

Fig. 6.3 Typical distribution of layers in an atmospheric model. Top model layer will reach around 40 km for models with a focus on the troposphere and much higher for models interested in middle atmosphere problems (Figure from http://www.cesm.ucar.edu/models/atm-cam/docs/description/description.pdf)

and eventually have global impacts. The physics and chemistry occurring at those small scales influence fluid motions through radiative heating and phase change to produce important phenomena like clouds with important features at scales of meters to kilometers, for example, updrafts and downdrafts. A brute force representation to treat fluid motions like up- and downdrafts would require a model that explicitly resolves those features, requiring discretizations with cells as small as a few meters on a side. It is not feasible to represent the whole globe at this resolution, and other methods are required. For this reason, many processes of climate relevance involve treatment of processes and features that are inevitably smaller than a GCM cell, or "subgrid." Figure 6.4 shows a satellite image of a cloud system in the equatorial eastern Atlantic with a typical GCM grid superimposed upon it in the right panel. The cloud features are clearly below the resolution of the model. A zoomed image of a small portion of a grid cell (outlined in red) is presented in the left panel, showing important cloud features at yet finer scales.

Parameterizations have been developed in order to represent processes that are important to the atmosphere, but occur at resolutions much below the scales the

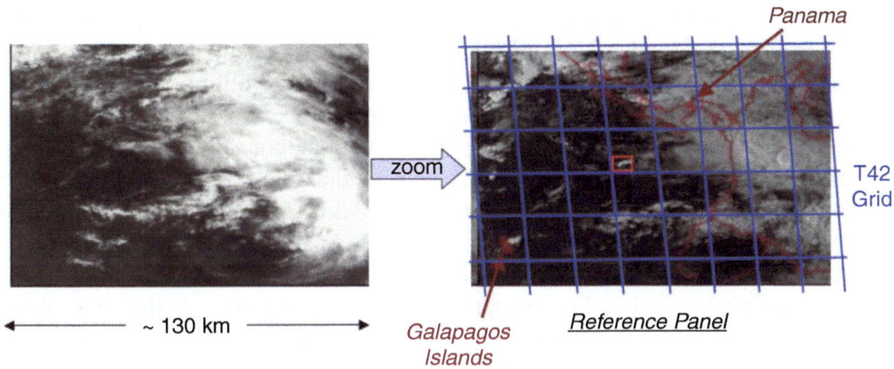

Fig. 6.4 A satellite image (courtesy NASA) of the Eastern Central Pacific showing the cloud features in the context of a typical climate model resolution. The *blue lines* of the *right panel* show a superimposed grid typical of a low-resolution atmospheric model. A zoomed image of the small *red box* shown in the *right panel* appears on the left, showing even more detail

model is able to resolve (see Stensrud [16]). The equations describing the fundamental physics equations are sometimes abandoned, or simplified, through an "abstraction" to approximate that process. Sometimes those simplifications are based upon formal mathematical decompositions, like the turbulence parameterizations that depend upon "Reynolds averaging" of the equations of motion, along with appropriate choices for the constitutive equations, and "closure assumptions." In other parameterizations, the complete equations are simplified to speed the calculation: for example, the equations of radiative heating are often approximated by assuming plane-parallel radiative transfer, and integrated over wavelength intervals to capture the essential absorption and emission for the gases and condensed species present in the atmosphere.

Other parameterizations are more explicitly "empirical," employing process representations which are based upon observed behavior of the atmosphere. For example, some parameterizations for convection [2, 8] attempt to represent the overturning that occurs in the atmosphere when less dense air resides below more dense air by simply adjusting the profiles of temperature and water vapor toward prescribed profiles that agree approximately with observations. Profiles can be defined for shallow (nonprecipitating) and deep precipitating convection based on both observational evidence and theoretical considerations. So, rather than identifying a mechanism through which convection operates to reduce instabilities in the atmosphere, the parameterization makes robust statements about the "end state" of an adjustment process, and introduces empirical tendencies in the evolution equations that adjust the profile to agree with the observed profiles. This type of parameterization is more frequently used in weather prediction than in climate modeling because these empirical parameterizations may not express enough of the physical underpinning to allow inferences to be made about how, why, and where that process is important, or allow extensions to handle additional model needs. Adjustment schemes, for example, would have difficulty handling convective transport of trace species, or adjusting to changes in the fundamental physics that might be occurring as the climate changes (e.g., the response of convective precipitation to pollution).

A third class of parameterizations resort to "process-based models." These parameterizations replace the basic physics with a conceptual model that is assumed to mimic the processes that occur in the real world. An example of a process-based model parameterization can be seen in the use of a "bulk plume model" to represent the role of convective clouds in a model column; this type of model is used in the majority of climate models in use in 2010 (see, e.g., [23]).

In a bulk plume model, the convective overturning occurring in the atmosphere in clouds, like those seen in the right panel of Fig. 6.5, is envisioned to take place through an ensemble of up- and downdrafts. The updrafts are assumed to begin at the "level of free convection" (the level where a parcel lifted from the surface will be both saturated and buoyant with respect to the ambient environment). The updraft is assumed to be driven by heat released during condensation taking place in parcels within the updrafts. The condensation produced in the updraft is assumed to produce rain. The rain falling into surrounding air is assumed to partially evaporate and initiate a saturated downdraft. These up- and downdrafts carry air from one level to another, entraining air from outside the cloud in the lower part of the cloud layers to dilute the updrafts, and detraining air to the environment aloft, to moisten it and redistribute heat. The ensemble of updrafts is represented by a single "bulk" updraft plume that entrains and detrains at multiple levels and a single "bulk" downdraft driven by evaporating rain to produce a conceptual model of convection like that described in the left panel of Fig. 6.5. The details of the representative up- and downdrafts are in turn controlled by specifications of the rate of entrainment and detrainment, how condensation, conversion of condensate to precipitation, and evaporation occurs within those up- and downdrafts, and a "closure assumption" that describes how the buoyancy generation occurring outside of the clouds is reduced by the mass fluxes within the up- and downdrafts. These parameterizations are obviously gross simplifications of the way clouds behave in the real world. The parameterizations introduce many "uncertain parameters" that require tuning to mimic the behavior of clouds in the real world.

The reader will note that the citations chosen here describe convective parameterizations written in, or prior to, the 1990s. Progress has been slow in developing better formulations for convection. Most parameterizations of convection have made progress "around the edges" by incrementally improving some aspect of the parameterization, like "closure assumptions" (the assumptions that deter how buoyancy excesses are removed from a column) or the "microphysical formulations" (controlling the ways that condensation, conversion to precipitation, and evaporation operate within the up- and downdrafts). Cloud parameterizations are viewed by climate scientists as one of the least satisfactory components of a GCM [12]. Convective parameterizations based upon plume models have the advantage over the very simple formulations like the Betts-Miller scheme of providing a physical picture (albeit crude) of how convection works that permits the expression of conservation laws (conservation of energy, enthalpy, momentum,

Fig. 6.5 The conceptual model used to produce a parameterization of a convective cloud like that seen in the figure at right. See text for details

mass, etc.). Plume model parameterizations also allow extensions to represent interactions between aerosols and clouds, for example, or the transport of soluble and insoluble trace constituents through vigorous convection, but they still have many limitations. Recently, a new class of parameterizations has begun to be explored, in which a "nearly cloud-resolving model" is embedded within each column of a GCM (e.g., [6]). These "super-parameterizations" of clouds have their own strengths and weakness: they use equations which are very close to the original equations of motion, but those equations are solved at scales that do not really resolve cloud motions. The parameterizations also increase the cost of the model by at least a factor of 100 over models using more traditional parameterizations so that their behavior for climate problems has not yet been thoroughly explored. Other new frameworks for cloud parameterization have also been suggested [1] that present an interesting approach to extending conventional parameterizations. There has not yet been time to evaluate the approach.

Clouds and Aerosols in Climate Models

The accurate representation of the effect of clouds and aerosols in the atmosphere is one of the most difficult and challenging tasks in climate models for scientists at the time this entry is written.

This task is important because clouds play many roles in the atmosphere: they scatter and absorb radiant energy in both the solar (shortwave) and infrared (longwave) part of the energy spectrum, reflecting sizable amounts of sunlight (shortwave energy) back to space and thus acting to cool the planet, but they also hinder the escape of heat/energy in the longwave, and thus can warm the Earth. Clouds are also reservoirs for heat and water acting to temporarily store energy and water in condensed phases, then return it to the atmosphere at other times and locations; they are sites for important in situ atmospheric chemistry and affect photolysis rates in both clear and cloudy regions of the atmosphere by changing the actinic flux; they are regions responsible for the rapid transport of atmospheric trace constituents from the lower to the upper atmosphere through vigorous convection; and they are also entities responsible for the removal of soluble species (gases and particles) through "wet deposition" processes. And, as discussed in the previous section, they are also incredibly difficult to represent accurately and comprehensively in GCMs.

But clouds are also strongly affected by "aerosols," the small (solid and liquid) particles with sizes less than about 10 um that are suspended in the atmosphere. Aerosols have both natural (e.g., sea-salt, dust, and some organic compounds released by vegetation) and anthropogenic origins (e.g., the pollution released by power plants, cars, trucks, agricultural burning, etc.). Like clouds, aerosols scatter and absorb radiant energy in both the solar and infrared part of the energy spectrum, and thus play a direct role in the energy budget of the planet. Aerosols also affect air quality and can affect ecosystems in a number of ways (e.g., the mobilization of dust particles from deserts, their subsequent transport by winds, followed by deposition; dust deposition is believed to be a source of iron as a nutrient to ocean biota). In addition, some aerosols act as sites that facilitate the phase change of water from vapor to liquid, or ice at far lower vapor pressures than would be needed for the phase change to occur in the absence of the particles. The aerosols that act as sites for water vapor condensation to form liquid cloud drops are known as Cloud Condensation Nuclei (CCN); those that are sites for formation of ice crystals are called Ice Nuclei (IN). Different types of aerosols are more and less effective as CCN and IN, and aerosols "compete" with each other and nearby cloud drops and ice crystals for water vapor, making their interactions extremely complex and hence difficult to model (see, e.g., Seinfeld and Pandis [14] and Lohmann and Feichter [7] for complementary discussions on some of these issues).

The aerosols that become part of cloud drops or ice crystals will eventually be removed from the atmosphere when those drops or crystals get large enough to precipitate out (this is termed nucleation scavenging). Aerosols are also removed as precipitation (raindrops, snow, hail, and graupel) falls and "collects" particles along the way (termed below cloud scavenging). The treatment of aerosols thus depends intimately on the treatment of clouds in GCMs.

Aerosols and clouds thus interact in many ways. It is easy to find examples of situations where aerosols can affect the cellular structure of clouds and their reflectivity. Figure 6.6 shows a dramatic example of the influence of pollution from ship emissions on the brightness of low clouds near the ocean surface. Climate scientists believe that anthropogenic emissions of many aerosol types, from

Fig. 6.6 NASA image (http://eoimages.gsfc.nasa.gov/images/imagerecords/5000/5488/ShipTracks_TMO_2005131_lrg.jpg)

pollution, biomass burning, agriculture, etc., affect both the reflectivity of low clouds with impact on how clouds cool the planet, and the partial opacity of the high ice clouds that hinder the escape of heat from the planet.

The subtle interactions between clouds and aerosols, and their interactions with other components of the climate system, produce some of the largest uncertainties in interpreting the signatures of climate change over the twentieth century and complicate modelers' abilities to produce accurate projections of climate change in the future.

These issues are discussed in great detail in the fourth assessment of the Intergovernmental Panel on Climate Change (AR4, IPCC2007), and it is not possible to provide much detail here. The reader should consult AR4 and the references therein for more detail.

Figure 6.7 shows globally averaged radiative forcing estimates for various forcing agents from IPCC2007. Changes in the atmospheric abundance of greenhouse gases and aerosols, in solar radiation, and in land surface properties alter the energy balance of the climate system. These changes are expressed in terms of a "radiative forcing" (W/m^2), a term used to compare how a range of human and natural factors drive warming or cooling trends on global climate. The three estimates related to aerosols (surface albedo, direct effect, and cloud albedo effect) are particularly noteworthy when compared to other forcing agents.

Surface albedo changes through Black Carbon deposition on snow are estimated to have a relatively small warming effect (positive forcing) on the planet. The forcing is a result of the decrease in reflectivity of the snow that occurs when the dark material is deposited on the snow surface. Since sunlight is more easily absorbed by the darker surface in this situation, the snow melts more quickly,

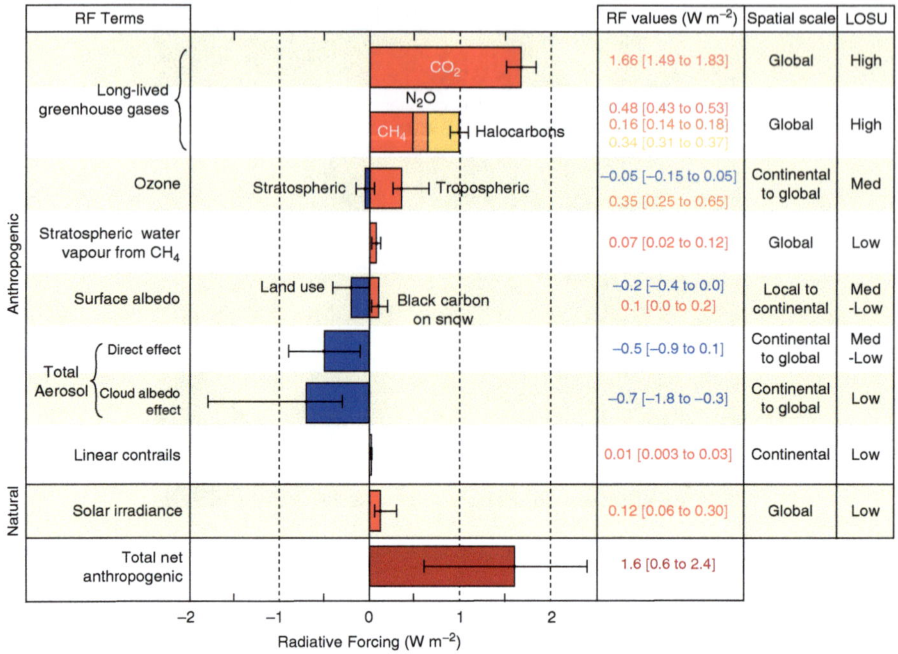

Fig. 6.7 From the summary for policymakers, IPCC AR4 [4] showing the globally averaged radiative forcing estimates for various forcing agents along with uncertainty estimates and level of scientific understanding (LOSU)

revealing other darker surfaces below (vegetation, dirt, sea-ice, etc.), which further increases the warming effect.

The "direct effect" of aerosols refers to the ability of aerosol to reflect or absorb sunlight as it enters the atmosphere. When the aerosols scatter sunlight back to space, the implied forcing is identified as negative (cooling) because less energy is absorbed by the planet. When aerosols absorb sunlight, they reduce the planetary albedo and warm the planet, producing a positive radiative forcing. The net radiative direct effect is estimated to be positive; models estimate that aerosols reflect more energy back to space than they absorb, but the uncertainty indicated by the horizontal whiskers is very large, and the level of scientific understanding (LOSU) is judged to be "low," as indicated in the figure.

The aerosol "indirect effect" refers to the role that aerosols play on clouds. Increasing aerosols (e.g., from pollution) can increase the number of particles available for cloud drops or ice particles to form on by acting as CCN or IN. Those "extra" cloud drops or ice particles introduced by the additional CCN and IN compete with the ambient aerosols for water vapor, and the result is that the cloud drops and ice crystals will be smaller than they would be in the absence of pollution. Smaller drops and particles scatter sunlight more efficiently (as demonstrated by simple physical arguments and the ship tracks seen in Fig. 6.6), and they frequently also precipitate less efficiently, affecting cloud lifetime and areal extent. Models estimate the indirect effect to produce a very large negative forcing, but the whiskers again indicate that the uncertainty is very large, and the LOSU very low. While the processes that produce cloud brightening in the presence of additional CCN and IN are well understood, there are numerous other factors that complicate the response of cloud reflectivity enormously, and scientists know that climate models treat these other factors very crudely, and inaccurately (see Stevens and Feingold [17]). For example, increasing the number of cloud drops and decreasing their size can also cause cloud drops to evaporate more readily, so the cloud reflectivity can actually decrease, and changing the reflectivity of particular regions of a cloud system can induce changes in the cloud dynamics (the intensity and extent of up- and downdrafts that control the precipitation and cloud areal extent; see, e.g., [19]), changing the cloud morphology and thus its radiative forcing.

The take-home message from the figure and discussion above is that aerosols of anthropogenic origin are currently estimated to have offset a substantial fraction of the positive forcing (heating) produced by increasing greenhouse gas concentrations over the twentieth century, but that result is very uncertain, and the level of understanding is low. These uncertainties in the estimates of the role of aerosols, clouds, and their interaction are strongly influenced by the deficiencies in the model representation of these processes, and by remaining deficiencies in our understanding of how these processes act.

Since it is not known how much of the twentieth century climate change (e.g., changes in surface temperature or precipitation) should be assigned to aerosol forcing and how much to the changes in greenhouse gas concentrations, it makes it much more difficult to interpret the system response to that forcing, and using understanding developed from simulations of past climate, provide accurate projections of how climate will change in the future.

Climate Forcing and Response

Climate change can be thought of as the response by the Earth system to the combination of externally imposed natural (e.g., solar variability and volcanic activity that changes the albedo of the planet) and anthropogenic forcings (e.g., greenhouse gases and aerosols), modulated by the internal model processes that allow the system to adjust to the imposed forcing. The response is strongly influenced by internal feedbacks within the Earth system. Negative feedbacks will increase the rate of cooling in the presence of positive external forcing (warming). Positive feedbacks can amplify that warming (see the discussion in the entry Coupled Climate and Earth System Models by Gent). The relative importance of positive and negative feedbacks in the climate system controls the amplitude of climate change produced from a given amount of external forcing.

There are a variety of ways to characterize the ratio of forcing to response. One convenient measure is "climate sensitivity." Climate sensitivity is sometimes expressed in terms of a feedback parameter λ (expressed in W m^{-2} K^{-1}) or its inverse, $1/\lambda$. Oftentimes, the sensitivity is expressed as the change in globally averaged surface temperature ΔT that would occur in a model if it were allowed to equilibrate to a forcing ΔF associated with a doubling of CO_2:

$$\Delta T_{2\times CO_2} = \lambda^* \Delta F_{2\times CO_2}$$

It is generally assumed that λ can be expressed as the sum of a sequence of feedbacks λ_i where i indicates the process responsible for the feedback:

$$\lambda = \sum \lambda_i$$

The equilibrium change in surface temperature is a somewhat arbitrary measure of climate response, and other measurements have also been explored. The oceans have a very large heat capacity, and the rate of transport of heat into the deep oceans is very slow, which means that it would take a very long time (thousands of years) to reach an equilibrium. It is possible, with clever analytic methods, to estimate this equilibrium sensitivity, and other definitions are also used (e.g., the "transient climate sensitivity") to describe the ratio of forcing to response. The different ways are not critical for this discussion, and the equilibrium definition is followed here.

This value is estimated in IPCC2007 (*AR4*) as "*likely to be in the range 2–4.5°C with a best estimate of about 3°C, and is very unlikely to be less than 1.5°C. Values substantially higher than 4.5°C cannot be excluded, but agreement of models with observations is not as good for those values.*"

The first, largest, and perhaps easiest feedback to describe is the so-called Planck Feedback or Planck response, which describes the increase in emission that will occur as temperature increases due to a positive forcing. If one assumes from theory and detailed radiative calculations that the change in forcing F produced by a change in CO_2 from concentration C0 to C (e.g., Myhre [25]) can be written as

$$F = k^* \ln\left(C/C_0\right) \text{ where } k \sim 5\text{Wm}^{-2}$$

then a doubling of CO_2 will result in an increase in the forcing of roughly 4 W/m^2. That positive forcing will tend to increase surface temperature. If one then assumes that (1) the emission temperature of the planet is proportional to the surface temperature, (2) the planet radiates energy in proportion to the Stefan-Boltzman equation (σT^4), and (3) no other atmospheric properties (clouds, water vapor, etc.) change, then model calculations indicate that λ is about -3.2 ± 0.1 W m^{-2} K^{-1} where the sign is chosen negative to indicate that the feedback is negative. And it follows that an increase of 4 W m^{-2} in forcing will result in an approximate change of 1 K in surface temperature.

This feedback estimate is robust, with very little uncertainty, and it is much lower than AGCMs report. The higher values of climate sensitivity are a result of the amplification of the response by other positive feedbacks that exist in the climate system (see, e.g., [4, 15]).

The largest positive feedback is believed to be the water vapor feedback: observations and models indicate that relative humidity (the ratio of ambient water vapor to the saturation value at a given temperature) remains approximately constant as temperature changes, particularly at high altitude (5–20 km). Therefore, an increase in temperature (e.g., from CO_2) will produce an increase in water vapor, which is the strongest of greenhouse gases. That increase in water vapor hinders the escape of energy, and the warming is amplified. IPCC2007 estimated λ_{WV} to be about 1.8 W m^{-2} K^{-1}.

Another important feedback is the "lapse rate feedback" λ_{LR}. When estimating the Planck Feedback, it was assumed that temperature change was constant (in latitude, longitude, and altitude). But it is known that the atmosphere will not produce a uniform change in the presence of a new forcing. The observed atmospheric lapse rate (vertical temperature gradient, see Glossary) roughly follows a "moist adiabatic lapse rate" in the tropics. It is temperature dependent and decreases more rapidly at cool temperatures than warm. So a temperature increase introduced near the surface in the tropics will produce adjustments that approximately follow a moist adiabatic lapse rate, and the perturbation will amplify with altitude. Since emission of infrared radiation varies with temperature, it will be more efficient as temperature increases, producing a negative *lapse rate feedback* that weakens the greenhouse effect. Model studies indicate that λ_{LR} has to be about –.84 W m^{-2} K^{-1}.

It is interesting to note that models suggest that the water vapor feedback and lapse rate feedback are strongly (negatively) correlated, and the agreement by models on the sum of these two feedbacks is much more robust than the individual components: $\lambda_{LR+WV} = 0.95 \pm 0.1$ W m^{-2} K^{-1}.

The surface albedo feedback occurs because an increase in surface temperature due to a positive forcing can melt surface snow and ice. A decrease in ice and snow reduces surface reflectivity, allowing more energy to be absorbed at the surface, producing further warming, and further reducing snow and ice. Soden and Held [15] and IPCC estimate the surface albedo feedback λ_{alb} to be 0.26 $\pm$ 0.08 W m^{-2} K^{-1}.

The last feedback to be discussed is the cloud feedback. A variety of observational, theoretical, and modeling studies suggest that low clouds tend to cool the planet by reflecting sunlight back to space. High ice clouds not only reflect sunlight back to space but also have a "greenhouse effect" and hinder the escape of longwave energy to space. Observational and modeling studies indicate that the net effect of high and low clouds is to cool the Earth (cloud reflection dominates the longwave trapping of energy). But it is by no means clear how cloud radiative forcing will change in the presence of external climate forcing. Clouds are so varied and complex that fewer clear general statements emerge to guide inferences about the sign and amplitude of their feedback processes. This is an area of very active research. Soden and Held found that all the GCMs used for AR4 had a positive cloud feedback (0.68 $\pm$ 0.37 W m^{-2} K^{-1}), even though half the models exhibited a reduction in net radiative forcing in response to a warmer climate. They concluded that change in cloud forcing itself is not a reliable measure of the sign or absolute magnitude of cloud feedback due to noncloud feedbacks on the cloud forcing. Note that the uncertainty in feedback amplitude from clouds is about three times larger than that found for the lapse rate + water vapor feedback, or the albedo feedback.

The combination of positive and negative climate feedbacks produces the likely warming range of 2–4.5°C for a CO_2 doubling cited in IPCC2007. Climate skeptics contend that the planet is unlikely to warm as much as GCMs predict by arguing that negative feedbacks are missing or underestimated or positive feedbacks overestimated in GCMs. These criticisms frequently appear in informal venues (blogs, the popular press, and elsewhere). When they are submitted to peer-reviewed refereed publications, they are taken seriously and scrutinized further. To date, the feedbacks described here have been found to be robust and dominant, and estimates of the range and distribution of climate quite robust.

Calibration (Tuning) and Evaluation of GCMs

There are still aspects of the atmosphere that are poorly characterized, and many processes remained crudely represented in GCMs. Even in situations where the correct physics is known, it is often too expensive to include the knowledge with brute force techniques, and approximations must be employed. Both lack of understanding and process approximation lead to uncertainties, and these uncertainties produce significant variations in model formulations adopted by groups around the world.

Multiple alternative parameterizations may also exist for a particular process (e.g., convection). Even in the event that a certain configuration of parameterizations is selected, there are many "uncertain" parameters within that configuration. The choices adopted for those uncertain parameters can have strong impacts on the behavior of the model.

So, substantial resources in the modeling community are invested in evaluating the behavior of the model in the presence of these uncertainties, and in selecting the parameterizations to be used in the model, or the values of the uncertain parameters to be used in subsequent simulations. The process of choosing the parameter values is known as "calibration" or "tuning." Model tuning has historically been performed in a series of stages, and it is sort of an "Art" that requires a lot of insight by participating scientists, and perhaps multiple repetition of those stages.

One obvious method of tuning is to compare the behavior of a model, or a model component to observations that strongly constrain the process and adjust the parameters until the simulation agrees to some tolerance with observations. This tuning procedure can be considered an optimization problem, and it occurs frequently during model development.

For example, a parameterization of deep convection might first be tuned to make sure that it produces approximately the observed rate of precipitation, and observed tendencies of water vapor, and temperature, at a particular location and time period (the outputs) where deep convection occurs frequently, given a set of measurements of the atmospheric state (the inputs).

After this tuning, parameterization evaluation can begin by looking at its behavior at other locations and times not used for tuning. Some insight is gained during this process about reasonable values for parameter, and parameter sensitivity to variations of tunable parameters, and inputs. Sometimes "single-column model" versions of the GCM are used for this purpose. This is a version of the GCM in which a single-column model is isolated, and all lateral fluxes of information supplied from observations to identify the model behavior in strongly prescribed situations.

The model can also be tuned to optimize relationships after processes equilibrate with each other; the statistical characteristics of the model can be evaluated and compared with statistical properties of the atmosphere. For example, model top of atmosphere energy fluxes could be compared to observed energy fluxes. Small changes in tunable parameters may be performed to assure that the approximate global averages of top of atmosphere fluxes are similar to observed values without significantly degrading the agreement with observations at the process level.

The model can also be compared with observations for fields, processes, or situations that were not part of the calibration process. Figure 6.8 provides one example of this kind of evaluation showing differences between model annually averaged column-integrated water vapor in a long simulation of a climate model (top panel), compared to an estimate of the corresponding observed value (middle panel) and the difference. Column-integrated water is actually a pretty difficult field to observe and observational uncertainty quite high, but a variety of independent methods are available to provide estimates, and the signatures seen in the difference field are quite robust to choices of the observational dataset used. This particular model is quite moist in the tropics.

Another interesting evaluation and calibration method is the use of a climate model in "weather forecast" mode. The climate model is started from initial conditions from archives of meteorological conditions produced from a forecast center and run for a brief (few day) simulation. These kinds of simulations are often called "hindcasts." The evolution of differences between the short simulations and observed fields provides information about how and why errors develop in the model. Individual processes can be examined to help in identifying the role of each process in driving the error growth.

The strategy can also be extended to longer periods (months or seasons) as part of a strategy sometimes called "seamless prediction" [10]. Models are viewed as representing processes that operate over different timescales. Fast processes (e.g., clouds) respond and produce measurable signatures to forcing on timescales of a day or so, systematic changes to heating rates can perturb planetary-wave structures in the atmosphere on the timescale of 10 days, and ocean and land surface react to changes in wave structure on the timescale of 100 days. On timescales of a thousand days and longer, anomalies will produce modifications to the cryosphere and biosphere. Atmospheric models (without coupling to oceans, ice, and the biosphere) can thus react and produce meaningful information

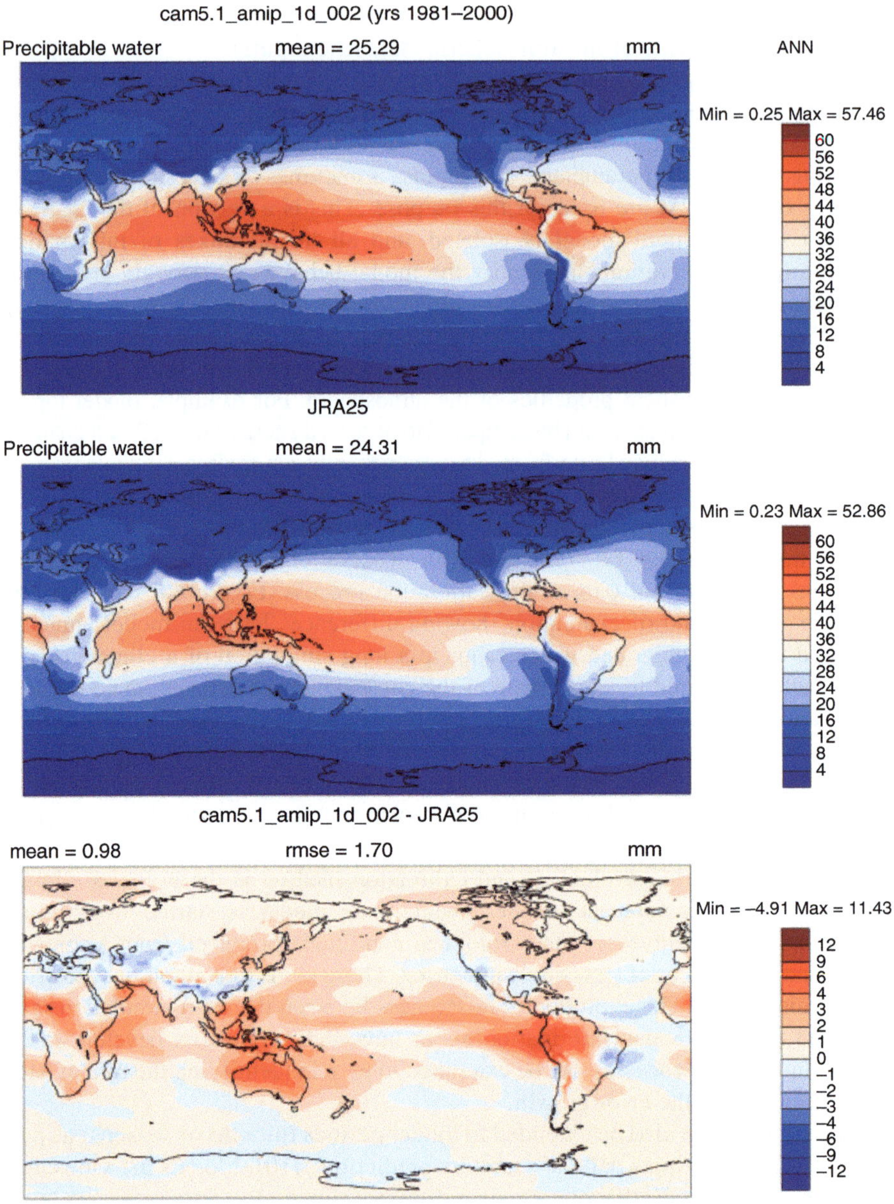

Fig. 6.8 An example of the difference in the vertically integrated amount of water vapor simulated in a climate model run for 20 years with present-day forcing, compared with a reanalysis produce (an estimate produced from observations). The difference is shown in the *lower panel*

on forcing and response through processes like clouds and planetary waves. Longer timescale phenomena need to be evaluated with coupled models. Seamless prediction paradigms ask that models be evaluated and improved on each timescale to provide increasing confidence in the model.

Once an "optimized" or "calibrated" model configuration has been found, it can be compared with other models, or other configurations of the model to quantify its sensitivity to changes in forcing, to identify the reasons that it responds the way it does, and to put that sensitivity in context compared to the behavior of other models. This process is one component of an evolving field of climate change science called "Uncertainty Quantification."

Remaining Challenges in Modeling the Atmosphere

There are many situations where scientists understand the fundamental physics of a physical process much more thoroughly than they can afford to express in a climate model, and scientists know that it is important to represent those processes more accurately. These are, in a sense, "known unknowns" in the parlance of Donald Rumsfeld. Scientists know that the parameterizations of clouds and aerosols used in climate models today can be formulated more accurately, and they know (in principle) how to do it. The challenge is (1) representing fields and processes that are truly important for the relevant climate problem and (2) expressing knowledge about that process in a computationally tractable way. This requires both scientific study and computational work. Many of these improvements can be achieved by clever revisions to the computational representation of critical processes in models. Scientists are rewriting model components to make more efficient use of evolving computer architectures that are providing enormous increments in computational resources, and developing more accurate and efficient computation methods to represent those processes. It may be possible in a few years to "brute force" problems that in the past required clever but potentially inaccurate approximations.

While some efforts to create a global cloud-resolving model have been made, their computation cost is at least a million times higher than models run at current climate resolution (e.g., [13]). It is currently a challenge just to complete a single annual cycle with models at this resolution. Scientists are also rewriting models to support "variable resolution" to allow models to provide higher resolution in areas that require it. One of the immediate challenges in this situation is that many of the approximations currently employed in atmospheric parameterizations are "scale dependent," that is, the quantitative behavior of the parameterization changes behavior as the resolution changes. Current parameterizations are typically "retuned" for different resolutions, and any cloud parameterization that adapts to resolution changes is not yet known, although scale-aware parameterizations of subgrid-scale propagation of waves and wave breaking do exist. It is desirable that parameterizations become "scale aware" in the sense that they adapt to the resolution they are used at. Parameterizations of subgrid-scale processes should recognize that the definition of the phenomena they are being asked to resolve changes as the resolution changes [1]). Scale-aware parameterizations and the model equations that they are embedded in should reduce to "fundamental physics" as model resolution approaches the resolution of the process being represented. This is an area of active research.

There are also challenges for climate models in treating "unknown unknowns," or "missing physics." These are phenomena that scientists believe may be important to represent, but that they do not know about or are very unsure of. An example of this situation is the role of Organic Aerosols on climate. These aerosols are difficult to measure and model. They are semivolatile; the aerosols condense, evaporate, and evolve chemically with atmospheric dilution. Aerosol scientists now believe that these aerosols play a much larger role in the climate system than was originally recognized, but they only have hints as to the complex chain of events that produce these aerosols. Precursor emissions, chemical evolution, and interaction with other aerosols and atmospheric trace species are complex, and they are treated very simply in climate models today.

Finally, there are many ways that climate science can be advanced by improvements in methodology. Ensembles of model simulations can be used to characterize model uncertainty. Model intercomparison activities like AMIP [3] and CMIP5 [18] continue to improve and allow more robust evaluation of the strength and weaknesses of climate models and their ability to represent the real world and their performance as a tool for understanding climate change.

Acknowledgments I would like to thank Sarah Fillmore for her editorial help and my colleagues at the Pacific Northwest National Laboratory and the National Center for Atmospheric Research for their willingness to share their expertise, knowledge, and model results with me over many years.

Bibliography

1. Arakawa A (2011) Toward unification of the multiscale modeling of the atmosphere. Atmos Chem Phys 11:3731–3742. doi:10.5194/acp-11-3731-2011, www.atmos-chem-phys.net/11/3731/2011/
2. Betts AK, Miller MJ (1993) The Betts–Miller scheme. In: Emanuel KA, Raymond DJ (eds) The representation of cumulus convection in numerical models of the atmosphere. American Meteorological Society, Boston
3. Gates WL et al (1999) An overview of the results of the atmospheric model intercomparison project (AMIP I). Bull Atmos Sci 80:29–55
4. IPCC (2007) Summary for policymakers. In: Solomon S, Qin D, Manning M, Chen Z, Marquis M, Averyt KB, Tignor M, Miller HL (eds) Climate change 2007: the physical science basis. contribution of working group I to the fourth assessment report of the intergovernmental panel on climate change. Cambridge University Press, Cambridge, UK/New York
5. Jacobson MZ (2005) Fundamentals of atmospheric modeling, 2nd edn. Cambridge University Press, New York, p 813

6. Khairoutdinov MF, Randall DA, DeMotte C (2005) Simulations of the atmospheric general circulation using a cloud-resolving model as a super-parameterization of physical processes. J Atmos Sci 62:2136–2154
7. Lohmann U, Feichter J (2005) Global indirect aerosol effects: a review. Atmos Chem Phys 5:715–737. doi:1680-7324/acp/2005-5-715, www.atmos-chem-phys.org/acp/5/715/
8. Manabe S, Smagorinsky J, Strickler RF (1965) Simulated climatology of a general circulation model with a hydrologic cycle. Mon Weather Rev 93:769–798
9. McGuffie K, Henderson-Sellers A (2005) A climate modeling primer. Wiley, New York
10. Palmer TN, Doblas-Reyes FJ, Weisheimer A, Rodwell MJ (2008) Towards seamless prediction: calibration of climate-change projections using seasonal forecasts. BAMS 89:459–470
11. Philips TJ et al (2004) Evaluating parameterizations in general circulation models: climate simulation meets weather. Bull Am Meteorol Soc 85:1903–1915
12. Randall D, Khairoutdinov M, Arakawa A, Grabowski W (2003) Breaking the cloud parameterization deadlock. Bull Am Meteorol Soc 84:1547–1564
13. Satoh M, Matsuno T, Tomita H, Miura H, Nasuna T, Iga S (2008) Nonhydrostatic icosahedral atmospheric model (NICAM) for global cloud resolving simulations. J Comput Phys 227: 3486–3514
14. Seinfeld JH, Pandis SN (1997) Atmospheric chemistry and physics. Wiley, New York
15. Soden B, Held I (2006) An assessment of climate feedbacks in coupled ocean–atmosphere models. J Climate 19:3354–3360
16. Stensrud DJ (2007) Parameterization schemes: keys to understanding numerical weather prediction models. Cambridge University Press, Cambridge
17. Stevens B, Feingold G (2009) Untangling aerosol effects on clouds and precipitation in a buffered system. Nature 461:607–613
18. Taylor KE, Stouffer RJ, Meehl GA (2011) An overview of CMIP5 and the experiment design. Bull Am Meteorol Soc. doi:10.1175/BAMS-D-11-00094.1
19. Wang H, Rasch PJ, Feingold G (2011) Manipulating marine stratocumulus cloud amount and albedo: a process-modeling study of aerosol-cloud-precipitation interactions in response to injection of cloud condensation nuclei. Atmos Chem Phys Discuss 11:885–916. doi:10.5194/acpd-11-885-2011
20. Washington W, Parkinson CL (2005) An introduction to three dimensional climate modeling, 2nd edn. University Science, Sausalito, p 354
21. Weart SR (2010) The discovery of global warming. Harvard University Press, Cambridge, MA
22. Wigley TML, Raper SCB (2005) Extended scenarios for glacier melt due to anthropogenic forcing. Geophys Res Lett 32:L05704. doi:10.1029/2004GL021238
23. Zhang GJ, McFarlane NA (1995) Sensitivity of climate simulations to the parameterization of cumulus convection in the Canadian climate centre general circulation model. Atmos Ocean 33(3):407–446
24. Zhu P et al (2005) Intercomparison and interpretation of single-column model simulations of a nocturnal stratocumulus-topped marine boundary layer. Mon Weather Rev 133:2741–2758
25. Myhre G, Highwood EJ, Shine KP, Stordal F (1998) New estimates of radiative forcing due to well mixed greenhouse gases. Geophysical Research Letters 25(14), pp. 2715–2718, doi: 10.1029/98GL01908

Chapter 7
Earth System Model, Modeling the Land Component of

Guo-Yue Niu and Xubin Zeng

Glossary

Aerodynamic resistance	Or drag or aerodynamic drag. The component of force exerted by the air on a liquid or solid object that is parallel and opposite to the direction of flow relative to the object.
Evapotranspiration	A term used to describe the sum of evaporation and plants' transpiration from the Earth's land surface to the atmosphere.
LAI	The ratio of total upper leaf surface area of vegetation divided by the surface area of the land on which the vegetation grows. LAI is a dimensionless value, typically ranging from 0 for bare ground to 6 for a dense forest.
Land surface processes	The various biogeophysical and biogeochemical processes occurring within and over various land surface components and interacting with the atmospheric processes.
Parameterization	A method or process to approximate the effect on the model resolved processes that would be generated by unresolved subgrid-scale processes using formulations containing empirical parameters.

This chapter was originally published as part of the Encyclopedia of Sustainability Science and Technology edited by Robert A. Meyers. DOI:10.1007/978-1-4419-0851-3

G.-Y. Niu (✉)
Biosphere 2, University of Arizona, 32540 S. Biosphere Road, Oracle, AZ 85623, USA
e-mail: niug@email.arizona.edu

X. Zeng
The Department of Atmospheric Sciences, University of Arizona, 1118 E. 4th Street, Tucson, AZ 85721, USA
e-mail: xubin@atmo.arizona.edu

P.J. Rasch (ed.), *Climate Change Modeling Methodology: Selected Entries from the Encyclopedia of Sustainability Science and Technology*,
DOI 10.1007/978-1-4614-5767-1_7, © Springer Science+Business Media New York 2012

Stomatal resistance	The opposition to transport of quantities such as water vapor and carbon dioxide (CO_2) to or from the stomata of plant leaves.
Subgrid heterogeneity	The spatial variation in land surface properties, for example, soil wetness, snow, vegetation, and terrain within an ESM grid cell, which is on the order of ~100 km.
Tile method	A method dealing with subgrid heterogeneity. It gathers similar homogeneous land patches at different geographic locations within an ESM grid-cell into a number of "tiles" or "mosaics"; computes surface fluxes over each "tile" using distinct parameters of each tile while sharing the same atmospheric forcing; and then averages the computed fluxes from the tiles, weighted by the fractional areas of each "tile."
Topographic index	Or "wetness index," = $\ln(a/\tan\beta)$, where a is the specific catchment area, that is, the upstream area above a pixel that drains through the unit contour at the pixel, and $\tan\beta$ is the local surface topographic slope. A greater value corresponds to a pixel of lowland areas that are potentially wetter.

Definition of the Subject and Its Importance

Land covers about 29% of the Earth's surface and represents an important component of the Earth's climate system. The land surface encompasses soil, snow, vegetation, glaciers, urban, lakes and rivers, mountains, etc. Land surface processes refer to various biogeophysical and biogeochemical processes occurring within and over these components and interacting with the atmospheric processes. Land surface models (LSMs) are mathematical descriptions of the land surface processes, with particular attention to momentum, energy, and mass (water, carbon, dust, and other constituents) flux exchanges with its overlaying atmosphere; these fluxes are intimately related to their storages in these components. In the context of an earth system model (ESM), LSMs are designed to provide the atmospheric model with lower boundary conditions over global land areas, describing the interactions of the land surface with the atmosphere. In addition, LSMs provide ocean models with discharges of water, sediments, and solutes. In this chapter, the focus is on descriptions of processes related to momentum, energy, water, and carbon, while processes related to other mass constituents, like dust, sediment, solute transports will not be included. Readers can also further refer to reviews of land surface modeling such as Sellers et al. [1], Pitman [2], and Yang [3].

Most LSMs, if not all, for use in ESMs and operational weather forecasting models are vertically one-dimensional (1D) models solving soil temperature and moisture equations, while parameterizing the effects of lateral flows of surface and subsurface water driven by subgrid topography. Although three-dimensional (3D) hydrological models numerically describing lateral flows of surface and subsurface water and operating at a resolution of meters have existed for years, it is not practical to include them in a climate model due to their expensive computational cost. Vertically, an LSM extends down to a few meters of the soil (2–6 m in different models) and a few meters above the vegetation canopy or the urban canopy. Some LSMs extend their depth to tens of meters deep to consider the long-term effects of permafrost and unconfined aquifers [4].

LSMs describe processes at various spatial scales ranging from millimeter to continental scales. The energy flux exchanges between land surface and the atmosphere involve processes at various scales from scattering of photons by plants and latent and sensible heat transfer by surface layer small eddies to convective planetary boundary layer large eddies. The water flux exchange involves processes at scales from water infiltration and percolation through soil micropores, transpiration through plant stomata, hillslope overland flow to continental-scale hydrological processes. The subgrid, unresolvable small-scale processes are highly parameterized using empirical equations based on a basic understanding of the processes, but not attempting to represent the process by equations representing first principles. This is why LSMs are also referred to as land surface parameterization schemes.

The land surface processes can influence weather and climate at timescales ranging from seconds to millions of years. For instance, the convective atmospheric boundary layer is developed in response to land surface heating on timescales from seconds to hours. Atmospheric temperature and precipitation can be affected through changes in soil water and groundwater, seasonal variations in snow cover and frozen soil, and vegetation phenology on a scale ranging from days to seasons. Temperature and water availability can greatly influence vegetation structure and function from years to centuries [5]. The Earth's climate has been tightly coupled to atmospheric CO_2 levels through the carbonate silicate cycle and/or the organic carbon cycle on geological timescales spanning hundreds of millions of years [6].

Land surface processes are critical for understanding climate dynamics owing to its momentum, energy, and mass exchanges with the overlying atmosphere. Land surface can "memorize" climate dynamics by recording and filtering the signals of weather events through storages of energy, water, and carbon and feed back to the atmosphere through surface fluxes. Pitman [2] reviewed coupled model experiments including sensitivity studies of changing albedo (e.g., [7]), roughness length (e.g., [8]), LAI [9], water-holding capacity [10], and roots [11], etc., deforestation [12], desertification [13], land cover change (e.g., [14]), ecosystem feedback under doubling CO_2 [15], and vegetation-snow-albedo feedback [16]. These sensitivity experiments provided evidence that climate models are sensitive to the land surface processes [2]. In addition, land surface feedbacks may amplify and prolong sea surface temperature (SST) anomaly triggered droughts, playing a critical role in the persistence of droughts and mega-droughts [17, 18]. It is well known that global climate dynamics is greatly affected by ocean–atmosphere interactions. However, land surface processes are of great importance for seasonal, interannual, and decadal climate change over specific land areas. Koster et al. [19] revealed that land and ocean processes have essentially different geographic regions of influence on precipitation variance, that is, the ocean's influence is dominant mainly in the tropical land, while the land's influence is dominant over midlatitude land areas. Koster et al. [20] identified the arid-to-wet transition regions over land as "hot spots" where changes in surface water fluxes associated with soil water storage have greater impacts on precipitation.

The land surface is more heterogeneous than other Earth surface components. Within one grid-cell of an ESM on the order of ~100 km, different land surface types, for example, bare soil, snow, grass, crop, and forests, may coexist. An LSM has to parameterize the integral effects of these subgrid surfaces on surface energy, water, and carbon fluxes over the grid cell. A variety of methods are used, for example, effective parameters, "mosaic" (or tile) methods, probability density function (pdf) methods, or finer grids than the host climate model. LSMs are usually community efforts, because they are inherently interdisciplinary including atmospheric sciences, hydrology, ecology, soil and snow physics, etc., as well as social sciences related to land use and water use managements.

Introduction

LSMs have evolved for three generations [1, 2] since the advent of the first generation global climate models (GCMs). The first LSM was the "bucket model," which describes the land surface as a bucket with a fixed water holding capacity of 15 cm [21]. The bucket can be filled with precipitation and emptied by evaporation. Runoff is generated as the water that exceeds the bucket holding capacity, and evaporation is limited by the water content of the bucket. In the first generation models, the transfer of sensible and latent heat was based on simple aerodynamic bulk transfer formulations, using uniform and prescribed surface parameters, for example, water holding capacity, albedo, and roughness length. These models did not account for ecological processes (e.g., transpiration through plants' stomata) and more detailed hydrological processes (e.g., infiltration and capillary upward flow through soil micropores). Despite their simplifications, the first generation models represent a key step to describe land surface processes in ESMs, and their simulated soil moisture and evapotranspiration (ET) were comparable to more complex models at longer timescales.

From the 1980s to early 1990s, a second generation of models appeared. They explicitly represented vegetation effects and more complex soil hydrology in the calculation of surface energy and water balances [22–24]. Representative second-generation models are the Biosphere-Atmosphere Transfer Scheme (BATS) [23] and the Simple Biosphere Model (SiB) [24]. During this period, two dozen second-generation models were developed with emphasis on various specific processes (e.g., [25–28]). These models vary in detail, but they have many common features. They usually have a number of soil layers, a single canopy layer (or an additional understory canopy layer like in SiB), and a bulk snow layer. Micrometeorological, hydrological, and ecological processes are more explicitly represented in these models. Instead of the fixed surface albedo in the first-generation models, more explicit radiation transfer through the vegetation canopy differentiating visible and near-infrared and beam and diffusive lights was taken into account. Turbulent transfer of heat from multiple surface sources, for example, canopy leaves and the ground surface was considered, and the under-canopy and leaf boundary layer turbulences were parameterized based on atmospheric boundary layer similarity theories. Transpiration through plants' stomata were considered using the Jarvis-type stomatal conductance scheme [29], a simple empirical formulation as a function of light (photosynthetically active radiation; PAR), water (leaf water potential), and environmental conditions (temperature and humidity). Whether the soil moisture Richards equation is solved or not, vertical gravity and capillary flows of water were taken into account. Runoff was parameterized as various functions of soil water storage or excess above infiltration, neglecting topographic effects. Accumulation and ablation of snow on the ground were represented with a bulk layer of snow, neglecting snowpack internal processes.

Since the early 1990s, third-generation models have been developed. The third-generation models were mainly characterized by the capability of simulating carbon uptake by plants and vegetation dynamics, or "greening" of the land surface [1, 2]. Other aspects of the model were substantially improved at the same time. Some models also began to incorporate treatments of nutrient dynamics and biogeography, so that vegetation systems can change location in response to climate shifts. Collatz et al. [30] and Sellers et al. [31] began to integrate stomatal conductance and photosynthesis into LSMs based on the work of Farquhar et al. [32]. A semiempirical model of leaf conductance was proposed [30] based on the understandings of the limitations on carbon assimilation by leaf and maximum use of water by plant [33].

During the same time period, LSMs were substantially improved in other aspects with regard to representations of hydrological processes in ESMs. LSMs started to account for subgrid variability in a more explicit way, for example, the "tile" approach to represent vegetation patchiness [34, 35] and the statistical approach to represent subgrid distribution of soil moisture and runoff generation [36], spatial variability of infiltration [37], topographic control on subgrid soil moisture distribution and runoff generation [38, 39], and subgrid snow distributions [40, 41]. LSMs started to include more physically based, multilayer snow submodels to accommodate more internal processes by parameterizing growth of grain size and liquid water retention and percolation within snowpack [42–44]. Ice content within soil (frozen soil) was explicitly resolved as a new prognostic variable, and its impacts on runoff and infiltration were investigated [45–47]. Most recently, groundwater processes were implemented in LSMs [48–52], and LSMs entered a new era involving multiple disciplinary sciences.

Major Processes Represented in an LSM

Subgrid Heterogeneity and Surface Data

Within an ESM grid on the order of $\sim$100 km, land surface properties, for example, soil wetness, snow, vegetation, and terrain, vary at the scale of a few meters. Because of the strongly nonlinear dependence of surface fluxes on surface properties and states, for example, the dependence of ET on soil wetness, simple area-averaging approaches used to aggregate surface properties and states is problematic. The use of LSM grids whose resolution is higher than ESMs may partially improve the accuracy of grid cell averaged fluxes [53] but it is still very difficult to explicitly resolve the variability. Therefore, representing subgrid variability within ESMs will always remain a challenge no matter how fine the model resolution is [36, 54]. There are three major approaches to representing the subgrid heterogeneity, that is, effective parameters, "mosaic" or "tile" approach, and statistical approach.

Earlier LSMs used effective parameters to obtain the correct fluxes averaged over patches of a grid cell. Some effective parameters can be obtained by simply averaging the parameters for various patches, e.g., albedo, because of the linear relationship between radiative fluxes and surface albedo. But other effective parameters, which have a nonlinear relationship with surface fluxes, for example, the effective roughness length, cannot be obtained by simply averaging parameter values of individual patches and need to be derived in such a way that correct area-averaged fluxes [55–57] can be ensured.

The statistical approach uses pdf of input variables to derive grid cell averaged runoff and ET. Assuming point precipitation intensity and soil wetness are spatially distributed over a large ESM grid cell according to exponential and gamma distributions respectively, Entakhabi and Eagleson [36] derived grid cell mean surface runoff ratio (to grid cell mean precipitation), bare soil evaporation efficiency (ratio of actual to potential), and transpiration efficiency (ratio of actual to potential evaporation). Liang et al. [37] implemented spatially variability of infiltration capacity in the Variable Infiltration Capacity (VIC) model. Niu et al. [39] used an exponential distribution of topographic index (or wetness index) to parameterize fractional saturated area and saturation excess runoff.

The tile approach [34, 35] gathers similar homogeneous land patches at different geographic locations within an ESM grid cell into a number of "tiles" or "mosaics," computes surface fluxes over each "tile" using distinct parameters of each tile while sharing the same atmospheric forcing, and then averages the computed fluxes from the tiles, weighted by the fractional areas of each "tile." The performance of the tile approach has been evaluated in numerous offline studies (e.g., [35]), boundary layer or mesoscale atmospheric models (e.g., [34]), and coupled ESMs [26, 58]. The tile can be also generated using different variables, for example, topographic index [38], soil wetness [54], or soil ice content [59]. The tile approach may overlap too much canopy shadows and thus induce errors in computing net solar radiation absorbed by the ground, especially over high latitude, boreal forest regions [60].

However, this approach is widely used in numerous second- and third-generation LSMs, because of its readiness to make full use of high-resolution satellite remotely sensed land surface data to generate the "tiles."

As an example, the National Center for Atmospheric Research (NCAR) Community Land Model version 4.0 (CLM4) [61] adopted the tile approach and parameterized some key processes using the statistical approach. CLM4 accounts for the subgrid heterogeneity as a nested subgrid hierarchy of landunits, soil/snow columns, and plant functional types (PFTs). Each grid cell can have a number of different landunits, for example, glacier, wetland, lake, urban, natural vegetation, and managed vegetation (crops). Each landunit can have a number of different columns (e.g., irrigated and rainfed soil), and each column can have multiple PFTs. CLM4 adopted the statistical method of Niu et al. [39] to parameterize surface saturated area and associated runoff. Surface datasets to feed CLM4 include percentages of glacier, lake, wetland, urban, and various PFTs to determine the fractional area of these surface tiles. PFTs are based on MODIS satellite data at 30 s ($\sim$1 km) resolution [62]. Prescribed PFT LAI was derived from the MODIS satellite data of Myneni et al. [63]. Prescribed PFT stem area index (SAI) is derived from PFT LAI phenology combined with the method of Zeng et al. [64]. Prescribed canopy top and bottom heights are from Bonan et al. [65]. Soil color dataset (affecting surface albedo) is based on Lawrence and Chase [62]. The IGBP soil dataset (Global Soil Data Task 2000) of 4,931 soil mapping units and their sand and clay content for each soil layer is used to create a mineral soil texture dataset [65] and an organic matter density dataset [66] that vary with depth.

Surface Energy Balance

Over an ESM grid cell, the net radiation (R_n) absorbed by the land surface is balanced by latent (LE), sensible (H), and ground heat fluxes (G):

$$R_n = LE + H + G \tag{7.1}$$

Note that the chemical energy stored during photosynthesis and released by respiration is not included in the above equation as it amounts to less than 1% of the absorbed insolation [2]. R_n includes net shortwave and net longwave radiation fluxes. LSMs that do not differentiate the vegetation canopy and the ground, for example, Noah [67] use the above equation and compute a single temperature for the land surface. More complex models that have a separate canopy layer from the ground, like versions of SiB and CLM, use two energy balance equations, one for the vegetation canopy and another for the ground surface, to solve their temperatures. The partitioning of available energy (R_n) into latent and sensible heat fluxes represented by Eq. 7.1 is greatly controlled by the land surface hydrological conditions. For an LSM using the tile subgrid approach, Eq. 7.1 is applied over each tile.

Net Shortwave Radiation

LSMs compute net shortwave radiation over various land surfaces:

$$S_a = S(1 - \alpha_s) \tag{7.2}$$

where S is the incoming solar radiation from the host ESM, and α_s is the surface albedo, varying with solar zenith angle (SZA), vegetation types, snow cover, and soil moisture conditions. In earlier LSMs, α_s was prescribed for various land surface types. More complex second- and third-generation models used Beer's Law and two-stream approximation scheme [68, 69] to compute radiative transfer through the vegetation canopy, net radiation absorbed by the vegetation canopy and the ground, and the surface albedo.

Beer's law assumes that irradiance decreases exponentially with path length through an absorbing medium. Thus, the radiation reaching the ground, $S_g = S\ e^{(-\kappa LAI)}$, where κ is a constant attenuation coefficient, and LAI is the effective leaf area index of the canopy. However, Beer's law does not account for the scattering and multiple reflections by the canopy leaves [70].

The two-stream approximation [68, 69] accounts for beam and two main streams (vertical upward and downward) of diffusive radiation over two wavebands (visible and near-infrared). However, it assumes that the canopy leaves are evenly distributed within a grid cell. On the basis of complex radiation transfer models [71], Yang and Friedl [72] proposed a modified two-stream scheme to account for aggregation of the evenly distributed leaves represented by a two-stream scheme into canopy crowns with between-canopy and within-canopy gaps, which vary with radius and thickness of the canopy, tree density (the distance between trunks), and SZA. The scheme needs prescribed optical parameters, for example, reflectivity and absorptivity of the vegetation and ground, for various land surface types. The two-stream approximation also computes the fraction of sunlit and shaded leaves, which are crucial to photosynthesis processes.

Net Longwave Radiation

The net longwave radiation (or infrared) is computed as the residual of the incoming ($L_\downarrow$) and emitted by the surface:

$$L_a = \epsilon(L_\downarrow - \sigma T_s^4) \tag{7.3}$$

The upward longwave radiative flux emitted by the surface is based on the Stefan-Boltzmann Law. In the above equation σ is the Stefan-Boltzmann constant, ε is the emissivity of the land surface, and T_s is the surface skin temperature. For a model that has a canopy layer, computation of L_a becomes more complicated due to the interactions between the ground surface and its overlaying canopy [61].

Sensible and Latent Heat Fluxes

The turbulent transfer of latent and sensible heat fluxes in most LSMs is based on Monin-Obukhov similarity theory [73, 74]. LSMs compute surface sensible and latent heat fluxes using the bulk transfer scheme:

$$H = \rho C_p C_H u (T_0 - T_a) \tag{7.4}$$

$$LE = \frac{\rho C_p}{\gamma} C_W u (e_0 - e_a) \tag{7.5}$$

where ρ is air density, C_p is the specific heat of dry air, γ is the psychrometric constant, C_H and C_W are the turbulent exchange coefficients for heat and water vapor, respectively, T_a and u are the air temperature and wind speed at the lowest layer of the host ESM, and T_0 and e_0 are the temperature and water vapor pressure at the height of the surface roughness length, z_0. The first-generation models prescribed C_H and C_W with different values for various land surfaces. The second- and third-generation models calculate C_H and C_W according to Monin-Obukhov similarity theory, accounting for surface stability conditions. For simplicity, $C_H = C_W$ and is assumed to be equivalent to that for momentum C_D in some LSMs. More elaborate schemes differentiate C_H (=C_W) from C_D using roughness lengths different for heat z_{0H} and momentum z_0; and usually $z_0 > z_{0H}$, because the transfer of momentum is more efficient (through pressure and diffusion) than those of heat and mass (through diffusion only). The value of $\ln(z_0/z_{0H})$ is about 2.0 over a range of vegetation surfaces, and in practical applications, $z_0/z_{0H} = 10$ [75]. In analogy with Ohm's law in electricity, the drag coefficient can be transformed to an aerodynamic resistance:

$$r_{a,H} = r_{a,W} = (C_H u)^{-1} \tag{7.6}$$

Shuttleworth [75] provided a review on formulations of the aerodynamic resistances for estimating surface ET and stability correction.

Complex models compute sensible and latent heat fluxes from multiple sources of the land surface such as soil, snow, plants' leaves, and stomata using equations similar to Eqs. 7.4 and 7.5. Over the soil surface, because of the strong temperature gradient between the soil skin (at 0 m) and z_0 as well as the difficulty in computing the temperature at z_0, LSMs use the skin temperature T_s to compute sensible and latent heat fluxes for convenience. The skin temperature over these surfaces can be computed iteratively using Eq. 7.1 till the surface energy is balanced. Because of the use of T_s rather than T_0 in Eqs. 7.4 and 7.5, an extra resistance, r_{soil}, is needed [76]:

$$H = \rho C_p \frac{(T_s - T_a)}{r_{a,H} + r_{soil}} \tag{7.7}$$

$$LE = \frac{\rho C_p}{\gamma} \frac{(e^*(T_s) - e_a)}{r_{a,W} + r_{soil}} \tag{7.8}$$

where $e^*(T_s)$ is the saturated vapor pressure at T_s, and r_{soil} accounts for the resistance to water vapor and heat transfer from the surface soil pore space to z_{0H} (roughness sub-layer) through molecular diffusion. Some LSMs also added a relative humidity term before $e^*(T_s)$ to estimate the water vapor pressure within surface soil pore space. Sakaguchi and Zeng [77] reviewed numerous soil resistance schemes. This expression for soil surface evaporation is similar to that for transpiration through leaf stomata

$$LE = \frac{\rho C_p}{\gamma} \frac{(e^*(T_v) - e_a)}{r_{a,v} + r_c} \tag{7.9}$$

where T_v is the vegetation canopy temperature, $r_{a,v}$ is the leaf boundary layer resistance per unit LAI [73], and r_c is the canopy-scale stomatal resistance that is scaled up from the stomatal-scale stomatal resistance, r_{st}, using LAI: $r_c = r_{st}/LAI$. Most second-generation models implemented the Jarvis-type stomatal resistance [29], which relates r_{st} to multiplicative stress factors:

$$1/r_{st} = g_0 f_S(PAR) f_D(\Delta e) f_T(T_a) f_W(\psi_l) \tag{7.10}$$

where g_0 is a vegetation type-dependent maximum stomatal conductance and f_S, f_D, f_T, and f_W, all of which range from 0 to 1, are environmental stress factors associated with PAR, water vapor deficit of the surrounding air (Δe), air temperature (T_a), and leaf water potential that relates to root-zone soil moisture, respectively.

Ground Heat Flux

The ground heat flux is controlled by the soil thermal diffusion (or conduction) process and can be expressed as

$$G = -\lambda \frac{T_1 - T_s}{\Delta z_1 / 2} \tag{7.11}$$

where λ is the soil thermal conductivity, T_1 is the temperature of the first soil layer, and Δz_1 is the layer thickness of the first soil layer. By definition ($G = -\lambda (\partial T / \partial z)_{z=0}$), the smaller Δz_1, the more accurate G is. In case of soil covered by snow, LSMs that have a bulk snow layer replace $\Delta z_1/2$ in the above equation with ($\Delta z_1/2 + h_{snow}$), where h_{snow} is the snow depth. Thus, when the snowpack becomes thicker, G becomes more inaccurate. This is one of the major reasons that complex models use a multilayer structure for the snowpack. G is then used as the upper boundary condition of soil temperature equation.

Heat Storage in the Soil and Snow

The soil thermal conduction (or diffusion) equation is written as

$$C\frac{\partial T}{\partial t} = \frac{\partial}{\partial z}(\lambda \frac{\partial T}{\partial z}) \tag{7.12}$$

where C and λ are the volumetric heat capacity and thermal conductivity, respectively. LSMs solve the above equation using various methods, for example, the more efficient force-restore method [22, 78] as used in BATS and SiB and the finite difference method as used in the Noah and CLM4 models.

The force-restore method is based on an analytical solution of the above equation under two assumptions that the surface forcing is periodic and the soil thermal properties are uniform. It is named after the two terms controlling surface temperature, that is, the forcing term (G in Eq. 7.11) and a restoring term due to heat diffusion from surface to a deeper soil layer. Recognizing that the two assumptions are not always valid, Dickinson [78] took into consideration the nonuniform thermal properties induced by snow and soil moisture and generalized the force-restore method. However, the surface temperature solved through the force-restore method is not so responsive to variations in surface forcing with a frequency higher than diurnal (e.g., cloud effects on solar radiation) as the iterative method solving the surface energy balance equation (Eq. 7.1). Therefore, it may cause difficulties in resolving the available energy for snowmelt, which occurs at a sub-daily scale [60].

Current LSMs explicitly solve Eq. 7.12 using finite difference methods for layered soil and snow. The upper boundary condition is the surface forcing (G), and lower (bottom) boundary condition is either specified as a zero flux as used in CLM or as the long-term averaged surface air temperature at a deep soil layer (e.g., 8 m) as used in Noah [67]. The zero flux condition is more readily satisfied when the bottom layer is deeper. The accuracy of the solution increases as the number of layers increases. Due to the computational cost, the number of layers cannot be very large. CLM has ten soil layers, while many other LSMs have three to five layers. To account for the thermal inertia of deep ground, the number of soil layers of CLM4 is changed from 10 to 15 layers, extending the total depth of the soil column down to 42 m [4].

When covered by snow, the soil temperature is solved together with the temperature of snow of a varying number of layers [43, 60]. The layering of snowpack and the number of layers vary with the total snow depth and is more complex than layering soil (e.g., [43, 61]). For frozen soil, a source/sink term due to phase change, $\rho_{ice} L_f \partial\theta_{ice}/\partial t$, where ρ_{ice}, L_f, θ_{ice}, and t are density of ice, heat for fusion, volumetric ice content, and time, respectively, is added to the right-hand side of Eq. 7.12. This term accounts for the heat released by freezing liquid water or heat consumed to melt ice. It is usually combined with heat conductivity C and named as "apparent heat capacity," C_a,

$$C_a = C - \rho_{ice} L_f \frac{d\theta_{ice}}{dT} \tag{7.13}$$

Surface Water Balance

The temporal variation of the water storage (S) in a grid cell of an ESM is controlled by the water incident on land surface, that is, precipitation (P) and the water leaving the grid cell, that is, ET and runoff (R):

$$\frac{dS}{dt} = P - ET - R \tag{7.14}$$

The above equation neglects the lateral surface and subsurface flows that transport water to adjacent grid cells. The storage term in the above equation includes water (ice and liquid water) intercepted by the canopy, snow on the ground, water in the soil and aquifers. ET includes ground evaporation, canopy evaporation, and transpiration, which can be derived from their corresponding latent heat fluxes by dividing the heat of vaporization, L. For most LSMs, R includes surface runoff and subsurface runoff, which will be discussed later.

Canopy-Intercepted Water

Although the amount of the canopy-intercepted water is relatively small compared to other water storages, it plays an important role in contributing to the total ET and altering the canopy reflectance by intercepting snowfall. Most LSMs takes into consideration liquid water interception by the vegetation canopy, while few LSMs explicitly consider additional processes of loading and unloading of snowfall [79, 80]. The temporal variation of the liquid water stored on the vegetation canopy, $W_{c,liq}$, is determined by the rates of interception (R_{intr}), evaporation/dew (E_c), and drip (R_{drip}):

$$\frac{dW_{c,liq}}{dt} = R_{\text{int}\,r} - E_c - R_{drip} \tag{7.15}$$

The interception rate is parameterized as a function of LAI and SAI. The interception capacity of the vegetation canopy is also related to the subgrid distribution of vegetation and precipitation. Considering only a fraction of the vegetation canopy within the large area of an ESM grid cell receives precipitation, CLM4 introduced a fractional area of precipitation as a limiting factor (0.25) to its interception rate [61]. However, this factor should vary with the distribution of precipitation that relates to the ratio of convective precipitation to total precipitation [39] and should be further investigated using high-resolution precipitation data.

The interception capacity for snow is estimated to be about 50 times greater than for rain [80]. Interception of snowfall by the vegetation canopy significantly reduces snow mass on the forest floor. About 30–40% of the annual snowfall over complete coniferous canopies sublimates from the canopy and thus never reaches the ground [80]. Depending on meteorological conditions, the intercepted snow may fall to the ground, sublimate, melt, or refreeze. Implementing a snowfall interception model and properly representing the sublimation of the canopy-intercepted snow can significantly improve the simulation of snow on the ground [81]. More importantly, properly representing interception of snow can significantly improve the simulation of surface albedo. However, most LSMs do not explicitly distinguish between solid and liquid phases of water on the canopy surface, and they use the same interception capacity for both snowfall and rainfall.

Based on the water stored on the vegetation canopy, a wet fraction of the canopy can be computed. Many LSMs (e.g., BATS and CLM) calculate the wet fraction as the 2/3 power of the ratio of canopy water to its maximum value, following Deardorff [22], while others (e.g., SiB) compute it as the ratio of the canopy water storage to its maximum value when the saturation vapor pressure at leaf temperature is less than the vapor pressure of the surrounding air. The maximum value of the canopy water is parameterized as a function of LAI.

Snow Water

Snow cover is an important part of the climate system because of its high albedo, low thermal conductivity, and ability to absorb heat when melting. In addition, the snowmelt water is the primary source of stream flow and groundwater recharge in cold and alpine regions. The impacts of snow-cover processes on global and regional climate have been investigated by numerous researchers (e.g., [82–84]) using earlier GCMs. In addition, numerous sensitivity studies using GCMs found that the removal of all forests north of 45°N would lead to cooling and delay snowmelt because of the increased surface albedo (e.g., [12, 85]), indicating the important role of forest canopies in modifying the surface energy budget and climate over snow covered regions.

Snow covered area exhibits spatial and temporal fluctuations ranging from 7% to 40% seasonally in the Northern Hemisphere [86]. Snow cover extent in the Northern Hemisphere has been decreasing since the mid-1980s in response to a global warming trend [87, 88]. On the other hand, Arctic warming may be accelerated by decreases in snow cover due to the positive snow-albedo feedback. Chapin et al. [89] reported that Arctic summer warming is highly related to an increase in snow-free days and the transition from tundra (short to be easily buried by snow) to forest.

LSMs compute snow water equivalent (SWE), snow depth, and fractional area covered by snow. The representation of snow processes in the first- and second-generation LSMs was relatively simple using a bulk snow layer (e.g., [26, 90, 91]). Third-generation models tend to have multiple snow layers to resolve the internal processes of snowpack such as densification, liquid water flow, and multiphase changes of water [42–44, 92]. The multilayer structure also has a thin surface layer designed to produce a more accurate ground heat flux and snow skin temperature [60].

Bulk-layer models combine the snowpack with the top soil layer as a bulk layer. These models, for instance, BATS and SiB, use the force-restore approach to solve snow temperature. However, the force-restore approach cannot accurately resolve high-frequency variations in snow surface forcing and temperature, and thus induce errors in snowmelt. Most of the multilayer snow models were simplified from schemes with detailed internal processes such as grain-size growth and gravitational flows of liquid water within a snowpack (e.g., [93, 94]). In a multilayer model like CLM4, the snow mass balance includes ice and liquid water content, but neglects the vapor phase. Densification processes include destructive or equi-temperature metamorphism, compaction due to the weight of the overlying layers of snow mass, and melt metamorphism, following Anderson [93]. Excessive liquid water above the holding capacity (ratio of liquid water volume to the free space of a snow layer) flow down to next layer. The melting (or freezing) energy for a layer is assessed as the energy excess (or deficit) needed to change the temperature of the layer to the freezing point [60].

Researchers also recognized that snowpack modeling for hydrological applications and climate studies needs to account for the radiative effects of the vegetation canopy (e.g., [95]) and the impacts of the interception of snow by the vegetation canopy [80, 96]. In addition, the turbulent transfer of sensible heat flux below the vegetation canopy under a stable condition during melting season significantly affects melting snow on the ground [81]. However, these studies were not fully introduced into ESMs.

Snow depth varies greatly at subgrid-scales because of heterogeneities in land cover, terrain, snow deposition, snowmelt, and meteorological conditions [40]. In LSMs, subgrid snow distributions are represented as snow cover fraction (SCF), that is, the fraction of a grid cell covered by snow, through the relationship between SCF and snow depth. At an ESM grid cell scale, one of the largest uncertainties in modeling snow and its interactions with the atmosphere comes from SCF formulations and their associated parameters. Various SCF formulations as summarized in Liston [40] result in a wide spread of SCFs between models. ESMs vary significantly in simulating SCF, and most of them underestimate SCF [97] because of unrealistic SCF formulations. Most SCF formulations are parameterized as a function of grid cell mean snow depth and the ground roughness length [40, 41, 81, 90]. Some of them also considered the impact of subgrid topography variations on SCF [91]. Using gridded snow depth and SWE reanalysis and the advanced very high resolution radiometer (AVHRR) SCF [87], Niu and Yang [41] found the SCF–snow depth relationship varies with seasons and introduced snow density into an existing SCF formulation [90] to approximate the seasonal variation in the SCF–snow depth relationship.

Soil Water

Although soil water accounts for a very small portion ($\sim$1%) of the terrestrial water storage, it plays an important role in the climate system through its controlling role on ET. Soil moisture anomalies (i.e., deviation from the mean) can persist from weeks to seasons [98, 99]. Numerous researchers have investigated its impact on climate predictability using GCMs (e.g., [100]). Koster et al. [19] revealed that the land's influence is dominant over midlatitude land areas. In analogy to the oceanic hot spot, for example, the eastern equatorial Pacific, playing a key role in the El Niño-La Niña cycle, Koster et al. [20] found the terrestrial "hot spots," that is, the arid-to-wet transition regions, where precipitation is affected by soil moisture anomalies during boreal summer. These hot spots were further confirmed by Zeng et al. [101] using an alternative index accounting for the land-precipitation coupling strength. For this reason, a proper initialization of soil moisture can enhance precipitation prediction skill over these hot spots [102–104]. Also, upgrades of LSMs with more elaborate representations of soil moisture can improve climate prediction skill especially during ENSO neutral years when the SST signal is relatively weak [104].

Temporal variation in the total soil moisture of a soil column within a certain depth (e.g., 3.4 m in CLM4) is controlled by infiltration, that is, the residual of precipitation minus runoff and evaporation at the soil surface, root uptake through transpiration, and the water flux at the bottom. The vertical distribution of soil water is controlled by vertical root distribution and soil property that determine gravitational and capillary water fluxes. Evolving from the first-generation "bucket" models, which neglects the capillary water flux, some second- and third-generation models directly solve the 1D, mass-based (θ) Richards equation to calculate the vertical distribution of soil moisture [61, 67]. Modern LSMs also tend to increase the number of soil layers to improve the accuracy of the solution. Various model intercomparison projects indicated that LSMs have a much better ability to simulate the soil moisture anomalies than to simulate the absolute value of soil moisture (e.g., [105, 106]).

The mass-based form (θ) of the Richards equation cannot resolve the soil water distribution at both unsaturated zone and saturated zone (groundwater) as continuously as a pressure-based (ψ) Richards equation. However, when solved through finite-difference numerical scheme, it conserves mass better than the pressure-based form. Zeng and Decker [107] demonstrated that a previous version of CLM cannot maintain the hydrostatic equilibrium of soil moisture distribution, and this problem cannot be resolved by increasing the vertical resolution of the soil column. In CLM4, they proposed a revised form of the Richard equation by explicitly subtracting the hydrostatic equilibrium pressure profile, $\psi_E(z)$, from the soil water pressure profile, $\psi(z)$.

Frozen Soil

Frozen soil has a tremendous impact on soil thermal and hydrological properties, ecosystem diversity and productivity, and greenhouse gas release from the soil. Freezing of soil water delays the winter cooling of the land surface; thawing of the frozen soil delays the summer warming of the land surface. Frozen soil also affects the snowmelt runoff and soil hydrology by reducing the soil permeability. Runoff from the Arctic river systems is about 50% of the net flux of freshwater to the Arctic Ocean [108]. This is a large percentage when compared to the freshwater inputs to the tropical oceans, where freshwater input is dominated by precipitation. Runoff and the degree of freshening can affect ocean salinity, sea ice conditions, and thermohaline circulation.

Earlier LSMs did not explicitly solve soil ice content. These LSMs showed a much larger scatter in the simulated soil moisture and runoff in spring than in other seasons due to different representations of the effects of frozen soil on infiltration [109]. Some of them switched off infiltration for subfreezing temperatures, for example, in simplified SiB [25] and BATS [110]. This treatment failed to produce the spring peaks of soil moisture due to the underestimated infiltration [111, 112]. Xue et al. [112] improved SSiB's ability to simulate the spring peaks of soil moisture by gradually decreasing the hydraulic conductivity at a rate of 10% per degree for subfreezing temperatures following SiB2 [113]. Pitman et al. [45] implemented an explicit representation of the hydrological and thermal effects of soil ice in their LSM but found that the representation degraded runoff simulation in a large-scale river basin.

Field studies showed that the effect of frozen soil on infiltration depends on scale, surface conditions of snow and vegetation, and soil structures. Shanley and Chalmers [114] showed that the effects of frozen soil on runoff are scale-dependent. There was no significant correlation between seasonal runoff ratios and ground frost depth for the 15 year of record from the Sleepers River watershed, USA, with an area of 111 km^2, while the increased runoff due to frozen ground was observed occasionally in its 0.59-km^2 agricultural subcatchment. Lindstrom et al. [115] also concluded that there were no clear effects of frozen soil on the timing and magnitude of runoff from an analysis of 16-year data in a 0.5-km^2 watershed in northern Sweden. Researchers have demonstrated that soil structure, air-filled porosity, ice content, and the number of freezing and thawing cycles are the governing factors affecting the infiltration capacity of frozen soil. Even at very local scales, recent laboratory and field studies using dye tracer techniques (e.g., [116]) revealed that water can infiltrate into deeper soil through preferential pathways where air-filled macropores exist at the time of freezing.

One-dimensional numerical models using the fully coupled heat and mass balance equations [46, 59, 117–119] used a variety of ways to parameterize the hydraulic properties of frozen soil. Some of the models [117, 119] assume that the freezing–thawing process is similar to the drying–wetting process with regard to the dependence of the soil matric potential on the liquid water content. This assumption leads to a very low infiltration rate or even upward water movements resulting in ice heave in surface layers. However, other modelers [46, 47, 59, 118] proposed various schemes to compute hydraulic properties as a means to produce greater infiltration rates. Stähli et al. [118] proposed two separate domains for the water infiltration into frozen soil: the low-flow domain where water flows through the liquid water film absorbed by the soil particles and the high-flow domain where water flows through the air-filled macropores. Koren et al. [46] assumed that frozen soil is permeable due to soil structural aggregates, cracks, dead root passages, and worm holes and reduced the effects of frozen soil on runoff production. Cherkauer and Lettenmaier [59] assumed that surface water tends to find areas of higher infiltration capacity as it flows across a frozen surface. They split their model domain into 10 bins with each having different ice content, which was derived from the observed spatial distribution of soil temperature, to increase the infiltration rate in VIC model. Niu and Yang [47] introduced fractional permeable area, which partitions an ESM grid cell into an impermeable fraction and a permeable fraction, and used the total soil moisture (liquid water and ice) to calculate soil matric potential and hydraulic conductivity over the permeable fraction.

When soil water freezes, the water closest to soil particles remains in liquid form due to the absorptive and capillary forces exerted by soil particles. For this reason, there is a certain amount of liquid water coexisting with ice over a wide range of temperatures below 0°C. This supercooled soil water at subfreezing temperatures can be computed with the freezing point depression equation [46, 47] or parameterized with observational data [120]. Most LSMs, for example, CLM4, consider ice and liquid water content but neglect the water vapor phase. The energy available for phase change between ice and liquid water is assessed in the same way as for snow except that only the excess above the supercooled liquid water can be frozen in the soil. Li et al. [121] provided a thorough summary of frozen soil models with different levels of complexity.

Groundwater

Groundwater storage constitutes about 30% of the terrestrial water storage. Groundwater storage shows as large variations as that of soil water at monthly or longer timescales in Illinois [122]. The groundwater level also shows a strong diurnal cycle in aquifers where the water table depth is less than 2 m because of the water uptake by the roots of the aboveground plants [51]. The rise and fall of the water table directly interacts with soil moisture in the vegetation root zone and thus acts as a source term for ET. Groundwater dynamics also control runoff generation, which can further affect the computation of soil moisture and ET in a climate model. Observational data show that runoff is much more related to the water table depth than to precipitation at monthly timescale in Illinois [50].

Earlier LSMs did not explicitly account for groundwater storage and its interactions with soil moisture. When solving the soil moisture equation, these LSMs assumed zero flux or free drainage at the bottom of the soil column as the lower boundary condition. For instance, an earlier version of CLM (version 3.0) used gravitational free drainage as its lower boundary condition at a depth of 3.4 m. Water drained from the soil bottom under gravity during wet seasons or wet years (parameterized as subsurface runoff) would never come back to the soil column in dry seasons or dry years. In reality, when the soil is drying, the soil can draw water through capillary suction from its underlying aquifer (or saturated layers), which has a longer memory of the past precipitation events than the soil. LSMs implicitly include such a process when the water table is shallow enough to be within the model soil layers. However, when the water table is below the model's soil bottom, LSMs neglect this upward water flow through capillary forces while taking into consideration gravitational free drainage at the model bottom, resulting in unrealistically dry deep soil.

During the past decade, the need for a groundwater component in LSMs has received increasing attention. As a result, a number of researchers have incorporated a groundwater component into LSMs [48–52, 123]. These models, for use in climate studies, can be divided in two categories: 3D models [51, 123] and 1D models [48–50, 52]. Because the horizontal transport of groundwater is more pronounced at a smaller scale, 3D models are more applicable to regional climate and water resource studies, while 1D models are often used in ESMs.

Researchers [48–50] have proposed more realistic solutions to solve the continuum of soil and relatively shallow aquifers at the expense of adding more integration elements, that is, 100 nodes, 50 layers, and 10 layers, respectively, in the above-referenced studies. Even with these additional layers, the models are only applicable for relatively shallow aquifers (<5 m) because of the prescribed depth of the model. Niu et al. [52] added a single integration element below the soil of CLM and parameterized groundwater recharge rate according to Darcy's law to account for the interactions between soil water and groundwater. Zeng and Decker [107] proposed another simple way to represent the groundwater recharge rate using the deviation of soil water pressure from its hydrostatic equilibrium at the soil bottom. All these developments represent a step forward from the current LSMs using gravitational free drainage or zero flux as their lower boundary condition. Further development and evaluation are needed to more realistically represent groundwater in ESMs.

Runoff

Runoff is one of the major components of the global water cycle and accounts for about 40% of the precipitation that falls on land. It plays an important role in the global climate system by affecting ET and freshwater inputs to the oceans, which in turn affects the ocean thermohaline circulation. LSMs' runoff formulation directly affects the partitioning of precipitation into runoff and ET and soil moisture.

Runoff is conceptually difficult to represent in ESMs. The environmental factors that control runoff, precipitation, soil moisture, and topography, often vary considerably on local scales. The plethora of conceptualizations for runoff schemes reflects the uncertainties in understanding and representing fundamental runoff processes. Land model intercomparison projects (e.g., [106, 124]) summarized various implementations of runoff schemes ranging from simple bucket models to more sophisticated topography-based runoff models. The partitioning of precipitation into evapotranspiration, surface runoff and subsurface runoff (baseflow) varies widely among these LSMs. Climate models have been adjusted so that the global, multiyear average runoff production is about one third of the average precipitation. Runoff in BATS is divided approximately equally between surface and subsurface runoff to match the early observational estimates [110].

Most LSMs parameterize runoff processes as a sum of surface runoff (fast component) and subsurface runoff or base flow (slow component). The surface runoff is generated by two major mechanisms: Horton runoff (infiltration excess) and Dunne runoff (saturation excess). The former is assessed as the excess of precipitation intensity over soil infiltration capability, while the latter is estimated as the precipitation over saturated and impermeable surfaces. Horton runoff is more difficult to represent in ESMs because the precipitation intensity produced by ESMs is unrealistically small ("drizzle" problem) and thus precipitation needs to be distributed in both space and time to mimic storms. Hydrologists introduced the concept of fractional saturated area as the dominant control on surface runoff. Surface runoff is largely dominated by the fractional saturated area, that is, the saturated fraction of a grid cell of ESMs, where rainfall or snowmelt water immediately flows into local river systems [125–127].

The fractional saturated area is conceptually correlated with near-surface soil moisture as it is represented in BATS [110]. More recent implementations [38, 39, 127–129] relate the fractional saturated area to the water table depth (or water deficit depth) and the subgrid characteristics of the topography following TOPMODEL [126, 130]. The latter is described by statistics of the "topographic index" (or "wetness index," $\lambda = \ln(a/\tan\beta)$, where a is the specific catchment area, i.e., the upstream area above a pixel that drains through the unit contour at the pixel, and $\tan\beta$ is the local surface topographic slope) computed from the high-resolution digital elevation model (DEM) [131]. Famiglietti and Wood [38] proposed a discretized framework in which the distribution of the topographic index was disaggregated into a number of bands, each representing a fraction of the watershed with similar water table depth and soil moisture, to parameterize the subgrid variability in soil moisture and runoff. However, its structural conflicts with climate models and its high computation costs impeded its application to ESMs. Most applications (e.g., [127–130]) used a three-parameter gamma distribution function to represent the pdf of the topographic index. By fitting the pdf of the topographic index with an exponential function, Niu et al. [39] simply expressed the saturated fraction as an exponential function of the water table depth.

The slow component was formulated as drainage under gravity in BATS [110] or a downslope lateral drainage in SiB [24]. TOPMODEL formulated its base flow as an exponential function of the soil water storage deficit [126] or the water table depth [130]. The TOPMODEL-based runoff schemes for ESMs also parameterized base flow as an exponential function of the catchment water storage deficit [128] or the water table depth [38, 39, 127, 129]. The exponential relationship between runoff (surface runoff and base flow) and the water table depth was demonstrated by observations [50].

Carbon Budgets and Vegetation Dynamics

The terrestrial ecosystem stores very large amounts of carbon, which is estimated at about 500 Pg of carbon (1 Pg C = 10^{15} g C) in plant biomass and 2,000 Pg of carbon in soil organic matter (top 1 m) [132]. The atmosphere, with a CO_2 concentration of 370 ppmv, contains about 785 Pg C as CO_2 [132]. Terrestrial ecosystem absorbs atmospheric CO_2 through photosynthesis at a rate of about 120 Pg C per year (gross primary productivity; GPP) and releases about half of that back to the atmosphere as CO_2 by respiration, resulting in net primary production (NPP) at about 60 Pg C per year. At the same time, heterotrophic respiration from soil and fire return about 60 Pg C per year back to atmospheric CO_2, closing the loop. The carbon fluxes between terrestrial ecosystem and the atmosphere have been quite stable for millennia. However, losses of C from land use change have been steadily increasing over the last one-and-half centuries, approaching rates of about 2 Pg C per year, mostly from tropical deforestation. Net cumulative CO_2 emissions from land use change may be close to 200 Pg C during the human history [132], about 25% of the atmospheric CO_2 at the level of 370 ppmv. Therefore, ESMs simulating the evolution of climate for centuries of the past and the future should represent these processes more realistically in their LSMs. ESMs also need to realistically simulate the relative role of land versus ocean in taking up the CO_2 released by anthropogenic sources.

Representing carbon budgets and vegetation dynamics is a major feature of the third-generation LSMs. The third-generation models try to fully represent plant physiology and phenology. Most of these LSMs use a semiempirical model of leaf conductance [1, 30, 33] to link the stomatal resistance with photosynthesis processes:

$$1/r_{st} = m \frac{A_n}{c_a} \frac{e_a}{e_{sat}} P + g_0 \qquad (7.16)$$

where m is an empirical coefficient (~9 for most C_3 plants and ~4 for C_4 plants; a larger m means the leaf consumes more water to assimilate the same carbon), A_n is the net carbon assimilation rate, c_a and e_a are the CO_2 concentration and vapor pressure at the leaf surface, respectively, e_{sat} is the saturated vapor pressure at the leaf temperature, P is the atmospheric pressure, and g_0 is the minimum stomatal conductance.

The net carbon assimilation is calculated as the minimum of three limiting factors: the Rubisco-limited, light-limited, and export-limited rates of carboxylation rate for C_3 plants and PEP-carboxylase limitation for C_4 grasses based on Collatz et al. [30, 133]. Both Rubisco-limited and export-limited rates of carboxylation are related to the maximum rate of carboxylation, which varies with temperature, foliage nitrogen, and soil water:

$$V_{\max} = V_{\max 25} a_{v\max}^{\frac{T_v - 25}{10}} f(N) f(T_v) \beta \tag{7.17}$$

where $V_{\max 25}$ is maximum carboxylation rate at 25°C depending on plant functional types (PFTs) and $\alpha_{v\max}$is the Q_{10} parameter (i.e., for an increase in temperature by 10°C, $V_{\max}$ would increase by $\alpha_{v\max}$), $f(T_v)$ is a function that mimics thermal breakdown of metabolic processes [30, 32], $f(N)$ is a foliage nitrogen factor and, the β factor is the soil moisture stress factor.

Most third-generation models, for example, SiB2 [113], IBIS [5], MOSES [117], and CLM4 [61] represent plant physiology using the above formulations. To simulate the net carbon budgets from the terrestrial ecosystem, the models also included representations of autotrophic respiration and heterotrophic respiration. For instance, CLM4 represents autotrophic respiration as the sum of maintenance respiration and growth respiration. Maintenance respiration is a function of temperature and tissue nitrogen concentration for live biomass [134], while growth respiration is assumed to be about 0.3 of the total carbon in new growth on a given time step based on construction cost for a range of woody and nonwoody tissues. CLM4 also represents decomposition, mineralization, and immobilization processes of carbon and nitrogen stored in three litter pools, three soil organic matter pools, and a coarse woody debris pool, under the soil moisture and temperature controls [134].

These models also describe plant phenology. For instance, CLM4 predicts LAI, SAI, and vegetation heights, including three distinct phenological types: evergreen, seasonal-deciduous, and stress-deciduous. The seasonal-deciduous phenology algorithm is based on the parameterizations for leaf onset and offset for temperate deciduous broadleaf forest [135]. Initiation of leaf onset is triggered when a common degree-day summation exceeds a critical value, and leaf litterfall is initiated when daylength is shorter than a critical value. The stress-deciduous phenology algorithm is based in part on the grass phenology model proposed by White et al. [135]. The algorithm handles phenology for vegetation types such as grasses and tropical drought-deciduous trees that respond to both cold and drought stresses, and that can have multiple growing seasons per year. The overall reasonable performance of CLM4 has been documented in Lawrence et al. [4].

Future Directions

LSMs have been developed for more than 3 decades, focusing more on representations of biogeophysical processes, that is, the energy and water processes occurring in various terrestrial components, and are now at their third generation. They are becoming more complex and more realistically process based than earlier generations. The third-generation models already include terrestrial ecosystem dynamics and associated physiological effects on terrestrial carbon budgets.

However, processes affecting greenhouse gas (GHG) emissions from various natural and anthropogenic sources over land are still not fully represented in these models. GHG emissions from various land surfaces are intimately coupled with various terrestrial biogeochemical processes. As suggested by Pitman [2], the need for an LSM to capture biogeochemical processes, which are tightly coupled to hydrological processes, is increasing and this provides focuses for future developments.

Numerous researchers have investigated the feedback of terrestrial ecosystem to climate change for the next century using the third-generation models (e.g., [15, 136]). Despite differences in details, these results demonstrate an important but uncertain role of the future ecosystems in the climate system. In these simulations, the terrestrial ecosystem functions as a sink through about 2050 and then turns into a source due to the collapse of the soil-carbon sink, resulting in an acceleration of global warming (1.5 K higher than without the carbon feedback). However, these results are questionable, because the terrestrial ecosystem and related biogeochemical processes are more complex than what are currently represented in these LSMs. For instance, these models do not fully represent key processes for stabilizing soil carbon, the hysteresis relationship between soil respiration and temperature, and the isoprene protective effects on photosynthesis under high temperature and light stress, etc. As an example, CLM4 includes the controlling role of soil temperature and water potential in the decomposition of leaf litter and soil organic carbon, but it lacks a representation of key processes such as biogeochemical weathering, heteroaggregate formation, and microbial activities. To represent more sensitive links between climate, microbial physiology, and C transformations, the response of microbial enzymes and active fungi to bacteria ratios to climatic variation and plant cover should be included [137, 138]. CLM4 considered emissions of biogenic volatile organic carbon (BVOC) including isoprene and monoterpenes [139] based on the emission model of Guenther et al. [140]. However, CLM4 neglects the allocation of the assimilated carbon through photosynthesis to BVOC emissions versus respiratory CO_2 emissions and CO_2 inhibitive effects on isoprene emissions. As recently reviewed by Vickers et al. [141], a rich literature exists demonstrating the role of isoprene in protecting photosynthesis under high temperature and light stresses. Therefore, third-generation models have recently started to implement photosynthesis-based isoprene emission algorithms [142]. In addition, a growing body of evidence suggests a temperature-respiration relationship that exhibits a diurnal hysteresis pattern [143]. A representative model (based on a Q_{10} factor) that does not consider hysteresis would overestimate soil CO_2 efflux for the entire growing season [143]. It has also been found that the hysteresis is caused by the imbalance of production and diffusion of CO_2 within soil, and soil moisture controls the transition from the imbalanced hysteresis pattern under wet soil conditions to a balanced pattern without hysteresis under dry soil conditions. As such, Riveros-Iregui et al. [143] suggested that the role of soil moisture in controlling the hysteresis temperature–respiration relationship should be considered in models. In addition, autotrophic respiration from roots and more elaborate representation of vertical root distribution should also receive more attention.

A major indication of the anthropogenic effects on the climate system is the large-scale conversion of natural ecosystems to cropland throughout human history. Associated with this conversion are changes in GHG emissions and related water use. Global agriculture is a major contributor to increasing the GHGs, with ~18% of the total anthropogenic emissions, including 9% of the total CO_2 emissions due to land use (e.g., deforestation), 40% of the total CH_4 emissions due to rice cultivation, and 62% of the total N_2O emissions due to fertilizer application for year 2000 [144]. For irrigation, global cropland consumes about 2,500 km^3 of water each year, which represents almost 70% of the global fresh water withdrawal from surface reservoirs and aquifers [145]. LSMs have started to include crop models and irrigation schemes (e.g., CLM4). Fully representing agricultural practices, for example, sowing, dates, cultivars, irrigation, and fertilization and associated biogeochemical processes that affect GHG emissions and water use represents a direction in future LSM development. Crop models developed by the agriculture community and measurements from the Free-air Controlled Enhancement (FACE) [146] over various natural and agricultural ecosystems should be helpful for developing models and test models' response to enhanced temperature and CO_2 concentration.

Almost all LSMs for use in ESMs are 1D, solving the governing equations in the vertical direction. However, water flows also in lateral directions at surface and in subsurface driven by gravity; the lateral water flow controls the formation and persistence of wetlands and governs the dynamics of freshwater inputs to the oceans. There is an increasing need to implement 3D hydrologic model into LSMs for use in ESMs. Methane emission, which is the second largest contributor to the anthropogenic GHG, is highly related to wetland dynamics. Of all the natural and anthropogenic sources of methane, methane emission from wetlands is known to be the single largest [147]. However, most LSMs lack a representation of wetland dynamics, prescribing wetland area as a fixed value. To more realistically represent wetland area, Fan et al. [148] developed an efficient hydrologic framework to simulate wetlands for climate models, involving large-scale groundwater convergence that feeds wetland. The spatial resolution of ESMs is increasing (~10 km) with recent computational advances. Implementing a quasi 3D hydrological model, which describes lateral surface and saturated subsurface (groundwater) flows and operates at a high spatial resolution (1 km), to LSMs will become possible in the near future. The hydrology community recently proposed hyper-resolution (1 km) global land surface modeling for monitoring Earth's terrestrial water resources, studies of biogeochemical processes, and agricultural applications, etc. [149]. Lessons learned from developing such a hyper-resolution model will certainly be beneficial for developing advanced 3D LSMs for use in ESMs.

Acknowledgments This review was partially supported by DOE (DE-SC0006773), NSF (AGS-0944101, and NASA (NNX09A021G).

Bibliography

Primary Literature

1. Sellers PJ, Dickinson RE, Randall DA, Betts AK, Hall FG, Berry JA, Collatz GJ, Denning AS, Mooney HA, Nobre CA, Sato N, Field CB, Henderson-Sellers A (1997) Modelling the exchanges of energy, water and carbon between continents and the atmosphere. Science 275:502–509
2. Pitman AJ (2003) The evolution of, and revolution in, land surface schemes designed for climate models. Int J Clim 23:479–510
3. Yang Z-L (2004) Modeling land surface processes in short-term weather and climate studies. In: Zhu X (ed) Observations, theory, and modeling of atmospheric variability, vol 3, World scientific series on meteorology of East Asia. World Scientific, Singapore, pp 288–313
4. Lawrence DM, Oleson KW, Flanner M, Thornton PE, Swenson SS, Lawrence PJ, Zeng X, Yang L, Levis S, Sakaguchi K, Bonan GB, Slater AG (2010) Parameterization improvements and functional and structural advances in version 4 of the community land model. J Adv Model Earth Syst 3:M03001, 27p. doi:10.1029/2011MS000045
5. Foley JA, Prentice IC, Ramankutty N, Levis S, Pollard D, Sitch S, Haxeltine A (1996) An integrated biosphere model of land surface processes, terrestrial carbon balance, and vegetation dynamics. Global Biogeochem Cycles 10:603–628
6. Pielke RA, Avissar R, Raupach M, Dolman AJ, Zeng X, Denning AS (1998) Interactions between the atmosphere and terrestrial ecosystems: influence on weather and climate. Glob Chang Biol 4:461–475
7. Charney J, Quirk WJ, Chow S-H, Kornfield J (1977) A comparative study of the effects of albedo change on drought in semi arid regions. J Atmos Sci 34:1366–1385
8. Sud YC, Shukla J, Mintz Y (1988) Influence of land surface roughness on atmospheric circulation and precipitation: a sensitivity study with a general circulation model. J Appl Meteorol 27:1036–1054
9. Chase TN, Pielke RA, Kittel TGF, Nemani R, Running SW (1996) Sensitivity of a general circulation model to global changes in leaf area index. J Geophys Res 101:7393–7408
10. Milly PCD (1997) Sensitivity of greenhouse summer dryness to changes in plant rooting characteristics. Geophys Res Lett 24:269–271
11. Zeng X, Dai J, Dickinson RE, Shaikh M (1998) The role of root distribution for climate simulation over land. Geophys Res Lett 25:4533–4536
12. Bonan GB, Pollard D, Thompson SL (1992) Effects of boreal forest vegetation on global climate. Nature 359:716–718
13. Xue Y (1997) Biosphere feedback on regional climate in tropical North Africa. Q J R Meteorol Soc 123:1483–1515
14. Betts RA (2000) Offset of the potential carbon sink from boreal forestation by decreases in surface albedo. Nature 408:187–190
15. Cox PM, Betts RA, Jones CD, Spall SA, Totterdell IJ (2000) Acceleration of global warming due to carbon-cycle feedbacks in a coupled climate model. Nature 408:184–187
16. Foley JA, Kutzback JE, Coe MT, Levis S (1994) Feedbacks between climate and boreal forests during the Holocene epoch. Nature 371:52–54
17. Hong SY, Kalnay E (2000) Role of sea surface temperature and soil-moisture feedback in the 1998 Oklahoma–Texas drought. Nature 408(6814):842–844
18. Schubert SD, Suarez MJ, Pegion PJ et al (2004) On the cause of the 1930s dust bowl. Science 303(5665):1855–1859
19. Koster RD, Suarez MJ, Heiser M (2000) Variance and predictability of precipitation at seasonal-to-interannual timescales. J Hydrometeorol 1(1):26–46
20. Koster RD et al (2004) Regions of strong coupling between soil moisture and precipitation. Science 305:1138–1140. doi:10.1126/science.1100217

21. Manabe S (1969) Climate and the ocean circulation: 1. The atmospheric circulation and the hydrology of the Earth's surface. Mon Weather Rev 97:739–805
22. Deardorff JW (1978) Efficient prediction of ground surface temperature and moisture with inclusion of a layer of vegetation. J Geophys Res 83:1889–1903
23. Dickinson RE (1984) Modelling evapotranspiration for three dimensional global climate models. In: Hansen JE, Takahasi T (eds) Climate processes and climate sensitivity, vol 29, Geophysical monograph. American Geophysical Union, Washington, DC, pp 58–72
24. Sellers PJ, Mintz Y, Sud YC, Dalcher A (1986) A simple biosphere model (SiB) for use within general circulation models. J Atmos Sci 43:505–531
25. Xue Y, Sellers PJ, Kinter JL, Shukla J (1991) A simplified biosphere model for climate studies. J Climate 4:345–364
26. Verseghy DL, McFarlane NA, Lazare M (1993) CLASS – a Canadian land surface scheme for GCMs. II: vegetation model and coupled runs. Int J Climatol 13:347–370
27. Thompson SL, Pollard D (1995) A global climate model (GENESIS) with a land-surface transfer scheme (LSX). Part 1: present climate simulation. J Climate 8:732–761
28. Desborough CE, Pitman AJ (1998) The BASE land surface model. Global Planet Change 19:3–18
29. Jarvis PG (1976) The interpretation of the variations in leaf water potential and stomatal conductance found in canopies in the field. Philos Trans R Soc Lond A Ser B 273:593–610
30. Collatz GJ, Ball JT, Grivet C, Berry JA (1991) Physiological and environmental regulation of stomatal conductance, photosynthesis and transpiration: a model that includes a laminar boundary layer. Agr Forest Meteorol 54:107–136
31. Sellers PJ, Berry JA, Collatz GJ, Field CB, Hall FG (1992) Canopy reflectance, photosynthesis and transpiration. III. A reanalysis using improved leaf models and a new canopy integration scheme. Remote Sens Environ 42:187–216
32. Farquhar GD, von Caemmerer S, Berry JA (1980) A biochemical model of photosynthetic CO_2 assimilation in leaves of C3 species. Planta 149:78–90
33. Ball JT (1988) An analysis of stomatal conductance. PhD thesis, Stanford University, Stanford
34. Avissar R, Pielke RA (1989) A parameterization of heterogeneous land surfaces for atmospheric numerical models and its impact on regional meteorology. Mon Weather Rev 117:2113–2136
35. Koster RD, Suarez MJ (1992) Modelling the land surface boundary in climate models as a composite of independent vegetation stands. J Geophys Res 97:2697–2715
36. Entekhabi D, Eagleson PS (1989) Land surface hydrology parameterization for atmospheric general circulation models including subgrid scale spatial variability. J Climate 2:816–831
37. Liang X, Lettenmaier DP, Wood EF, Burges SJ (1994) A simple hydrologically based model of land surface water and energy fluxes for general circulation models. J Geophys Res 99:415–428
38. Famiglietti JS, Wood EF (1994) Multi-scale modeling of spatially variable water and energy balance processes. Water Resour Res 30(11):3061–3078
39. Niu G-Y, Yang Z-L, Dickinson RE, Gulden LE (2005) A simple TOPMODEL-based runoff parameterization (SIMTOP) for use in global climate models. J Geophys Res 110:D21106. doi:10.1029/2005JD006111
40. Liston GE (2004) Representing subgrid snow cover heterogeneities in regional and global models. J Climate 17:1381–1397
41. Niu G-Y, Yang Z-L (2007) An observation-based formulation of snow cover fraction and its evaluation over large North American river basins. J Geophys Res 112:D21101. doi:10.1029/2007JD008674
42. Jin JM, Gao X, Yang ZL, Bales RC, Sorooshian S, Dickinson RE (1999) Comparative analyses of physically based snowmelt models for climate simulations. J Climate 12:2643–2657
43. Dai Y et al (2003) The common land model (CLM). Bull Am Meteorol Soc 84(8):1013–1023

44. Yang Z-L, Niu G-Y (2003) The versatile integrator of surface and atmosphere processes (VISA) part 1: model description, global planet. Change 38:175–189
45. Pitman AJ, Slater AG, Desborough CE, Zhao M (1999) Uncertainty in the simulation of runoff due to the parameterization of frozen soil moisture using the GSWP methodology. J Geophys Res 104:16 879–16 888
46. Koren V, Schaake JC, Mitchell KE, Duan Q-Y, Chen F, Baker JM (1999) A parameterization of snowpack and frozen ground intended for NCEP weather and climate models. J Geophys Res 104:19,569–19,585. doi:10.1029/1999JD900232
47. Niu G-Y, Yang Z-L (2006) Effects of frozen soil on snowmelt runoff and soil water storage at a continental scale. J Hydrometeorol 7:937–952. doi:10.1175/JHM538.1
48. Liang X, Xie Z, Huang M (2003) A new parameterization for surface and groundwater interactions and its impact on water budgets with the variable infiltration capacity (VIC) land surface model. J Geophys Res 108(D16):8613. doi:10.1029/2002JD003090
49. Maxwell RM, Miller NL (2005) On the development of a coupled land surface and groundwater model. J Hydrometeorol 6:233–247
50. Yeh PJF, Eltahir EAB (2005) Representation of water table dynamics in a land surface scheme, part I: model development. J Climate 18(12):1861–1880
51. Fan Y, Miguez-Macho G, Weaver CP, Walko R, Robock A (2007) Incorporating water table dynamics in climate modeling: 1. Water table observation and equilibrium water table simulations. J Geophys Res 112:D10125. doi:10.1029/2006JD008111
52. Niu G-Y, Yang Z-L, Dickinson RE, Gulden LE, Su H (2007) Development of a simple groundwater model for use in climate models and evaluation with gravity recovery and climate experiment data. J Geophys Res 112:D07103. doi:10.1029/2006JD007522
53. Hahmann AN, Dickinson RE (2001) A fine-mesh land approach for general circulation models and its impact on regional climate. J Climate 14:1634–1646
54. Sellers PJ, Fennessy MJ, Dickinson RE (2007) A numerical approach to calculating soil wetness and evapotranspiration over large grid areas. J Geophys Res 112:D18106. doi:10.1029/2007JD008781
55. Mahrt L (1987) Grid-averaged surface fluxes. Mon Weather Rev 115:1550–1560
56. Mason PJ (1988) The formation of areally-averaged roughness lengths. Q J R Meteorol Soc 114:399–420
57. Garratt JR (1992) The atmospheric boundary layer. Cambridge University Press, Cambridge
58. Essery RLH, Besst MJ, Betts RA, Cox PM (2003) Explicit representation of subgrid heterogeneity in a GCM land surface scheme. J Hydrol 4:530–543
59. Cherkauer K, Lettenmaier DP (2003) Simulation of spatial varibility in snow and frozen soil. J Geophys Res 108:8858. doi:10.1029/2003JD003575
60. Niu G-Y et al (2011) The community Noah land surface model with multi-parameterization options (Noah-MP): 1. Model description and evaluation with local-scale measurements. J Geophys Res 116:D12109. doi:10.1029/2010JD015139
61. Oleson KW et al (2010) Technical description of version 4.0 of the community land model. NCAR Technical Note NCAR/TN-478+STR. National Center for Atmospheric Research, Boulder, 257 pp
62. Lawrence PJ, Chase TN (2007) Representing a MODIS consistent land surface in the community land model (CLM 3.0). J Geophys Res 112:G01023. doi:10.1029/2006JG000168
63. Myneni RB et al (2002) Global products of vegetation leaf area and fraction absorbed PAR from year one of MODIS data. Remote Sens Environ 83:214–231
64. Zeng X, Shaikh M, Dai Y, Dickinson RE, Myneni R (2002) Coupling of the common land model to the NCAR community climate model. J Climate 15:1832–1854
65. Bonan GB, Levis S, Kergoat L, Oleson KW (2002) Landscapes as patches of plant functional types: an integrating concept for climate and ecosystem models. Global Biogeochem Cycles 16:5.1–5.23
66. Lawrence DM, Slater AG (2008) Incorporating organic soil into a global climate model. Clim Dyn 30. doi:10.1007/s00382-007-0278-1

67. Ek MB, Mitchell KE, Lin Y, Rogers E, Grunmann P, Koren V, Gayno G, Tarpley JD (2003) Implementation of Noah land surface model advancements in the national centers for environmental prediction operational mesoscale Eta model. J Geophys Res 108(D22):8851. doi:10.1029/2002JD003296
68. Dickinson RE (1983) Land surface processes and climate-surface albedos and energy balance. In: Saltzman B (ed) Theory of climate, vol 25, Advances in geophysics. Academic, San Diego, pp 305–353
69. Sellers PJ (1985) Canopy reflectance, photosynthesis and transpiration. Int J Remote Sens 6:1335–1372. doi:10.1080/01431168508948283
70. Nijssen B, Lettenmaier D (1999) A simplified approach for predicting shortwave radiation transfer through boreal forest canopies. J Geophys Res 104(D22):27,859–27,868
71. Li X, Strahler AH, Woodcock CR (1995) A hybrid geometric optical and radiative transfer approach for modeling albedo and directional reflectance of discontinuous canopies. IEEE Trans Geosci Remote Sens 33(2):466–480
72. Yang R, Friedl M, Ni W (2001) Parameterization of shortwave radiation fluxes for nonuniform vegetation canopies in land surface models. J Geophys Res 106:14,275–14,286
73. Brutsaert WA (1982) Evaporation into the atmosphere. Reidel, Dordrecht, 299 pp
74. Garratt JR, Francey RJ (1978) Bulk characteristics of heat transfer on the instable baroclinic atmospheric boundary layer. Bound -Lay Meteorol 15:399–421
75. Shuttleworth WJ (2007) Putting the 'vap' into evaporation. Hydrol Earth Syst Sci 11(1):210–244
76. Zeng X, Dickinson RE (1998) Effect of surface sublayer on surface skin temperature and fluxes. J Climate 11:537–550
77. Sakaguchi K, Zeng X (2009) Effects of soil wetness, plant litter, and under-canopy atmospheric stability on ground evaporation in the community land model (CLM3.5). J Geophys Res 114:D01107. doi:10.1029/2008JD010834
78. Dickinson RE (1988) The force-restore model for surface temperatures and its generalizations. J Climate 1:1086–1097
79. Roesch A, Wild M, Gilgen H, Ohmura A (2001) A new snow cover fraction parameterization for ECHAM4 GCM. Clim Dyn 17:933–946
80. Hedstrom NR, Pomeroy JW (1998) Measurements and modeling of snow interception in the boreal forest. Hydrol Proc 12:1611–1625
81. Niu G-Y, Yang Z-L (2004) Effects of vegetation canopy processes on snow surface energy and mass balances. J Geophys Res 109:D23111. doi:10.1029/2004JD004884
82. Yeh T-C, Wetherald RT, Manabe S (1983) A model study of the short-term climate and hydrologic effects of sudden snow-cover removal. Mon Weather Rev 111:1013–1024
83. Walsh JE, Ross B (1988) Sensitivity of 30-day dynamical forecast to continental snow cover. J Climate 1:739–754
84. Cess RD et al (1991) Intercomparison of snow-feedback as produced by general circulation models. Science 253:888–892
85. Douville H, Royer JF (1997) Influence of the temperate and boreal forests on the northern hemisphere climate in the meteo-France climate model. Clim Dyn 13:57–74
86. Hall DK (1988) Assessment of polar climate change using satellite technology. Rev Geophys 26:26–39
87. Robinson DA, Frei A (2000) Seasonal variability of northern hemisphere snow extent using visible satellite data. Prof Geogr 52(2):307–315
88. Brown RD (2000) North hemisphere snow cover variability and change, 1915–97. J Climate 13:2,339–2,355
89. Chapin FS, Sturm M, Serreze MC et al (2005) Role of land-surface changes in Arctic summer warming. Science 310:657–660
90. Yang Z-L, Dickinson RE, Robock A, Vinnikov KY (1997) Validation of the snow sub-model of the biosphere-atmosphere transfer scheme with Russian snow cover and meteorological observational data. J Climate 10:353–373

91. Douville H, Royer J-H, Mahfouf J-F (1995) A new snow parameterization for the meteo-France climate model, part I: validation in standalone experiment. Clim Dyn 12:21–35
92. Lynch-Stieglitz M (1994) The development and validation of a simple snow model for the GISS GCM. J Climate 7:1842–1855
93. Anderson EA (1976) A point energy and mass balance model of a snow cover. NOAA Technical Report NWS 19. Office of Hydrology, National Weather Service, Silver Spring, 150 pp
94. Jordan R (1991) A one-dimensional temperature model for a snow cover. Special Report 91–16. Cold Regions Research and Engineering Laboratory, U.S. Army Corps of Engineers, Hanover
95. Davis RE, Hardy JP, Ni W, Woodcock C, McKenzie JC, Jordan R, Li X (1997) Variation of snow cover ablation in the boreal forest: a sensitivity study on the effects of conifer canopy. J Geophys Res 102:29,389–29,385
96. Essery R, Pomeroy J, Parviainen J, Storck P (2003) Sublimation of snow from coniferous forests in a climate model. J Climate 16(11):1855–1864
97. Frei A, Gong G (2005) Decadal to century scale trends in North America snow extent in coupled atmosphere–ocean general circulation models. Geophys Res Lett 32:L18502. doi:10.1029/2005GL023394
98. Pielke RA, Liston GE, Eastman JL, Lu L, Coughenour M (1999) Seasonal weather prediction as an initial value problem. J Geophys Res 104:19,463–19,479. doi:10.1029/1999JD900231
99. Schlosser CA, Milly PCD (2002) A model-based investigation of soil moisture predictability and associated climate predictability. J Hydrometeorol 3:483–501. doi:10.1175/1525-7541(2002) 003<0483: AMBIOS>2.0.CO;2
100. Shukla J, Mintz Y (1982) The influence of land-surface evapotranspiration on the Earth's climate. Science 215(4539):1498–1501
101. Zeng X, Barlage M, Castro C, Fling K (2010) Comparison of land-precipitation coupling strength using observations and models. J Hydrometeorol 11:980–995
102. Koster RD et al (2004) Realistic initialization of land surface states: impacts on subseasonal forecast skill. J Hydrometeorol 5:1049–1063. doi:10.1175/JHM-387.1
103. Koster RD et al (2010) Contribution of land surface initialization to subseasonal forecast skill: first results from a multi-model experiment. Geophys Res Lett 37:L02402. doi:10.1029/2009GL041677
104. Yang RQ, Mitchell K, Meng J, Ek M (2011) Summer-season forecast experiments with the NCEP climate forecast system using different land models and different initial land states. J Climate 24:2319–2334. doi:10.1175/2010JCLI3797.1
105. Entin JK, Robock A, Vinnikov KY, Zabelin V, Liu S, Namkhai A, Adyasuren T (1999) Evaluation of global soil wetness project soil moisture simulations. J Meteorol Soc Jpn 77:183–198
106. Boone A et al (2004) The Rhone-aggregation land surface scheme intercomparison project: an overview. J Climate 17:187–208
107. Zeng X, Decker M (2009) Improving the numerical solution of soil moisture-based Richards equation for land models with a deep or shallow water table. J Hydrometeorol 10:308–319. doi:10.1175/2008JHM1011.1
108. Barry RG, Serreze MC (2000) Atmospheric components of the Arctic Ocean freshwater balance and their interannual variability. In: Lewis EL et al (eds) The freshwater budget of the Arctic Ocean. Kluwer Academic Publishers, New York, pp 45–56
109. Luo LF et al (2003) Effects of frozen soil on soil temperature, spring infiltration, and runoff: results from the PILPS 2(d) experiment at Valdai, Russia. J Hydrometeorol 4:334–351
110. Dickinson RE, Henderson-Sellers A, Kennedy PJ (1993) Biosphere- atmosphere transfer scheme (BATS) version 1e as coupled to the NCAR community climate model. NCAR Technical Note, NCAR/TN- 387 + STR. National Center for Atmospheric Research, Boulder
111. Robock A, Vinnikov KY, Schlosser CA, Speranskaya NA, Xue Y (1995) Use of midlatitude soil moisture and meteorological observations to validate soil moisture simulations with biospere and bucket models. J Climate 8:15–35

112. Xue Y, Zeng FJ, Schlosser CA (1996) SSiB and its sensitivity to soil properties: a case study using HAPEX-mobilhy data. Global Planet Change 13:183–194
113. Sellers PJ, Randall DA, Collatz GJ, Berry JA, Field CB, Dazlich DA, Zhang C, Collelo GD, Bounoua L (1996) A revised land surface parameterization (SiB2) for atmospheric GCMs: part I. Modelformulation. J Climate 9:676–705. doi:10.1175/1520-0442(1996)009<0676:ARLSPF>2.0.CO;2
114. Shanley JB, Chalmers A (1999) The effect of frozen soil on snowmelt runoff at Sleepers River, Vermont. Hydrol Process 13:1843–1857
115. Lindstrom G, Bishop K, Lofvenius MO (2002) Soil frost and runoff at Svartberget, northern Sweden – measurements and model analysis. Hydrol Process 16:3379–3392
116. Stadler D, Stähli HM, Aeby P, Flühler H (2000) Dye tracing and image analysis for quantifying water infiltration into frozen soils. Soil Sci Soc Am J 64:505–516
117. Cox PM, Betts RA, Bunton CB, Essery RLH, Rowntree PR, Smith J (1999) The impact of new land surface physics on the GCM simulation of climate and climate sensitivity. Clim Dyn 15:183–203
118. Stähli M, Nyberg L, Mellander P-E, Jansson P-E, Bishop KH (2001) Soil frost effects on soil water and runoff dynamics along a boreal forest transact: 2. Simulations. Hydrol Process 15:927–941
119. Hansson K, Simunek J, Mizoguchi M, Lundin L-C, van Genuchten MT (2004) Water flow and heat transport in frozen soil: numerical solution and freeze-thaw applications. Vadose Zone J 3:693–704
120. Decker M, Zeng X (2006) An empirical formulation of soil ice fraction based on in situ observations. Geophys Res Lett 33:L05402. doi:10.1029/2005GL024914
121. Li Q, Sun S, Xue Y (2010) Analyses and development of a hierarchy of frozen soil models for cold region study. J Geophys Res 115:D03107. doi:10.1029/2009JD012530
122. Rodell M, Famiglietti JS (2001) An analysis of terrestrial water storage variations in Illinois with implications for the gravity recovery and climate experiment (GRACE). Water Resour Res 37(5):1327–1339
123. Kollet SJ, Maxwell RM (2008) Capturing the influence of groundwater dynamics on land surface processes using an integrated, distributed watershed model. Water Resour Res 44: W02402. doi:10.1029/2007WR006004
124. Bowling LC et al (2003) Simulation of high latitude hydrological processes in the torne-kalix basin: PILPS phase 2e. 1 – Experimental description and summary intercomparisons. Global Planet Change 38:1–30
125. Dunne T, Black RD (1970) Partial area contributions to storm runoff in a small New England watershed. Water Resour Res 6(5):1296–1311
126. Beven KJ, Kirkby MJ (1979) A physically based, variable contributing model of basin hydrology. Hydrol Sci Bull 24:43–69
127. Stieglitz M, Rind D, Famiglietti J, Rosenzweig C (1997) An efficient approach to modeling the topographic control of surface hydrology for regional and global modeling. J Climate 10:118–137
128. Ducharne A, Koster RD, Suarez MJ, Stieglitz M, Kumar P (2000) A catchment-based approach to modeling land surface processes in a general circulation model: 2. Parameter estimation and model demonstration. J Geophys Res 105(D20):24,823–24,838
129. Chen J, Kumar P (2001) Topographic influence on the seasonal and inter-annual variation of water and energy balance of basins in North America. J Climate 14:1989–2014
130. Sivapalan M, Beven K, Wood EF (1987) On hydrologic similarity: 2. A scaled model of storm runoff production. Water Resour Res 23:2266–2278
131. Quinn PF, Beven J, Lamb R (1995) The ln(a/tan b) index: how to calculate it and how to use it within the TOPMODEL framework. Hydrol Process 9:161–182
132. Janzen HH (2004) Carbon cycling in earth systems – a soil science perspective. Agric Ecosyst Environ 104(3):399–417. doi:10.1016/j.agee.2004.01.040

133. Collatz GJ, Ribas-Carbo M, Berry JA (1992) A coupled photosynthesis–stomatal conductance model for leaves of C4 plants. Aust J Plant Physiol 19:519–538
134. Thornton PE, Rosenbloom NA (2005) Ecosystem model spin-up: estimating steady state conditions in a coupled terrestrial carbon and nitrogen cycle model. Ecol Model 189:25–48
135. White MA, Thornton PE, Running SW (1997) A continental phenology model for monitoring vegetation responses to interannual climatic variability. Global Biogeochem Cycle 11:217–234
136. Cramer W, Bondeau A, Woodward FI, Prentice IC, Betts RA, Brovkin V, Cox PM, Fisher V, Foley JA, Friend AD, Kucharik C, Lomas MR, Ramankutty N, Sitch S, Smith B, White A, Young-Molling C (2001) Global response of terrestrial ecosystem structure and function to CO_2 and climate change: results from six dynamic global vegetation models. Glob Chang Biol 7:357–373
137. Schimel JP, Weintraub MN (2003) The implications of exoenzyme activity on microbial carbon and nitrogen limitation in soil: a theoretical model. Soil Biol Biochem 35:549–563
138. Allison SD, Wallenstein MD, Bradford MA (2010) Soil-carbon response to warming dependent on microbial physiology. Nat Geosci. doi:10.1038/NGEO846 (Advance online publication)
139. Levis S, Wiedinmyer C, Bonan GB, Guenther A (2003) Simulating biogenic volatile organic compound emissions in the community climate system model. J Geophys Res 108(D21):4659. doi:10.1029/2002JD003203
140. Guenther A, Karl T, Harley P, Wiedinmyer C, Palmer PI, Geron C (2006) Estimates of global terrestrial isoprene emissions using MEGAN (model of emissions of gases and aerosols from nature). Atmos Chem Phys 6:3181–3210
141. Vickers CE, Gershenzon J, Lerdau MT, Loreto F (2009) A unified mechanism of action for volatile isoprenoids in plant abiotic stress. Nat Chem Biol 5:283–291
142. Pacifico F, Harrison SP, Jones CD, Arneth A, Sitch S, Weedon GP, Barkley MP, Palmer PI, Serca D, Potosnak M, Fu TM, Goldstein A, Bai J, Schurgers G (2011) Evaluation of a photosynthesis-based biogenic isoprene emission scheme in JULES and simulation of isoprene emissions under present-day climate conditions. Atmos Chem Phys 11:4371–4389
143. Riveros-Iregui DA, Emanuel RE, Muth DJ, McGlynn BL, Epstein HE, Welsch DL, Pacific VJ, Wraith JM (2007) Diurnal hysteresis between soil CO_2 and soil temperature is controlled by soil water content. Geophys Res Lett 34:L17404. doi:10.1029/2007GL030938
144. Olivier JGJ, Berdowski JJM (2001) Global emissions sources and sinks. In: Berdowski J, Guicherit R, Heij BJ (eds) The climate system. A.A. Balkema/Swets & Zeitlinger, Lisse, pp 33–78. ISBN 90 5809 255 0
145. Rost S, Gerten D, Bondeau A, Luncht W, Rohwer J, Schaphoff S (2008) Agricultural green and blue water consumption and its influence on the global water system. Water Resour Res 44:W09405. doi:10.1029/2007WR006331
146. Kimball BA (2011) Lessons from FACE: CO_2 effects and interactions with water, nitrogen, and temperature. In: Hillel D, Rosenzweig C (eds) Handbook of climate change and agroecosystems: impacts, adaptation, and mitigation. World Scientific, Singapore, pp 81–107
147. Houghton JT et al (eds) (2001) Projections of future climate change, climate change 2001: the scientific basis, contribution of working group I to the third assessment report of the intergovernmental panel on climate change. Cambridge University Press, New York, 881 pp
148. Fan Y, Miguez-Macho G (2011) A simple hydrologic framework for simulating wetlands in climate and earth system models. Clim Dyn 37:253–278. doi:10.1007/s00382-010-0829-8
149. Wood EF et al (2011) Hyperresolution global land surface modeling: meeting a grand challenge for monitoring Earth's terrestrial water. Water Resour Res 47:W05301. doi:10.1029/2010WR010090

Chapter 8
Integrated Assessment Modeling

James A. Edmonds, Katherine V. Calvin, Leon E. Clarke, Anthony C. Janetos, Son H. Kim, Marshall A. Wise, and Haewon C. McJeon

Glossary

Climate policy (greenhouse gas mitigation policy) A climate policy refers to a policy scheme designed to deliberately limit the magnitude of climate change, often involving mitigation of greenhouse gases. Integrated assessment models (IAMs) represent climate policies in abstract forms. The most commonly modeled climate policy is attaching a universal price on emissions of carbon dioxide (or carbon dioxide equivalent of other greenhouse gases). Such policy represents a universal carbon tax or an economy-wide cap-and-trade policy. Other forms of climate policies, such as differential carbon price by sector or renewable portfolio standards, have also been used in IAMs.

Cost of greenhouse gas mitigation (economic cost) Integrated assessment models (IAMs) employ varies metrics for estimating the economic cost of mitigation policy. One common approach estimates reduction in GDP, a proxy for slowdown in economic activity due to increased price of energy and agricultural products. Another approach estimates the (gross) loss in social welfare due to a policy by measuring the area under the marginal abatement cost

This chapter was originally published as part of the Encyclopedia of Sustainability Science and Technology edited by Robert A. Meyers. DOI:10.1007/978-1-4419-0851-3

J.A. Edmonds (✉) • K.V. Calvin • L.E. Clarke • A.C. Janetos • S.H. Kim • M.A. Wise • H.C. McJeon
Joint Global Change Research Institute (JGCRI),
Pacific Northwest National Laboratory (PNNL), College Park, MD, USA
e-mail: jae@pnnl.gov; katherine.calvin@pnnl.gov; leon.clarke@pnnl.gov; anthony.janetos@pnnl.gov; skim@pnnl.gov; marshal.wise@pnnl.gov; hmcjeon@pnnl.gov

P.J. Rasch (ed.), *Climate Change Modeling Methodology: Selected Entries from the Encyclopedia of Sustainability Science and Technology*,
DOI 10.1007/978-1-4614-5767-1_8, © Springer Science+Business Media New York 2012

curve. Other metrics include foregone consumption, compensated variation, and equivalent variation.

Integrated assessment model (IAM) — Integrated assessment model (IAM) in climate change research is a model which simulates the interactions of human decision-making about energy systems and land use with biogeochemistry and the natural Earth system. IAMs can be divided into two categories.

Higher resolution IAMs focus on explicitly representing processes and process interactions among human and natural Earth systems.

Highly aggregated IAMs use highly reduced-form representations of the link between human activities, impacts from climate change, and the cost of emissions mitigation.

Integrated earth system model (iESM) — Integrated Earth System Models (iESMs) are a class of models under development by collaboration between integrated assessment modeling community and climate modeling community. By fully integrating the human dimension from an IAM and the natural dimension from a climate model, iESM allows simultaneously estimating human system impacts on climate change and climate change impacts on human systems, as well as examining the effects of feedbacks between the components.

Land use (land-use emissions) — Land use is one of the largest anthropogenic sources of emissions of greenhouse gases, aerosols, and short-lived species. Emissions, as well as sequestration of emissions, may occur from land-use practices, changes in land cover, or changes in forested area or the density. On the other hand, land-use patterns are affected by the changes in the climate. As such, modeling land use has been an important component of the integrated assessment modeling of climate change.

Representative concentration pathways (RCPs) — The Representative Concentration Pathways (RCPs) are the most recent set of emission scenarios generated by integrated assessment models. Four scenarios explicitly considering emission mitigation efforts that were sufficiently differentiated in terms of radiative forcing at the end of the century were selected from published literature. RCPs are designed to facilitate the interactions with climate models by including geospatially resolved emissions and land-use data.

Definition of the Subject

This entry discusses the role of integrated assessment models (IAMs) in climate change research. IAMs are an interdisciplinary research platform, which constitutes a consistent scientific framework in which the large-scale interactions between human and natural Earth systems can be examined. In so doing, IAMs provide insights that would otherwise be unavailable from traditional single-discipline research. By providing a broader view of the issue, IAMs constitute an important tool for decision support. IAMs are also a home of human Earth system research and provide natural Earth system scientists information about the nature of human intervention in global biogeophysical and geochemical processes.

Introduction

Integrated assessment models (IAMs) are a class of models which simulate the interactions of human decision-making about energy systems and land use with biogeochemistry and the natural Earth system (see Fig. 8.1). In so doing, IAMs provide insights that would otherwise be unavailable from investigating either

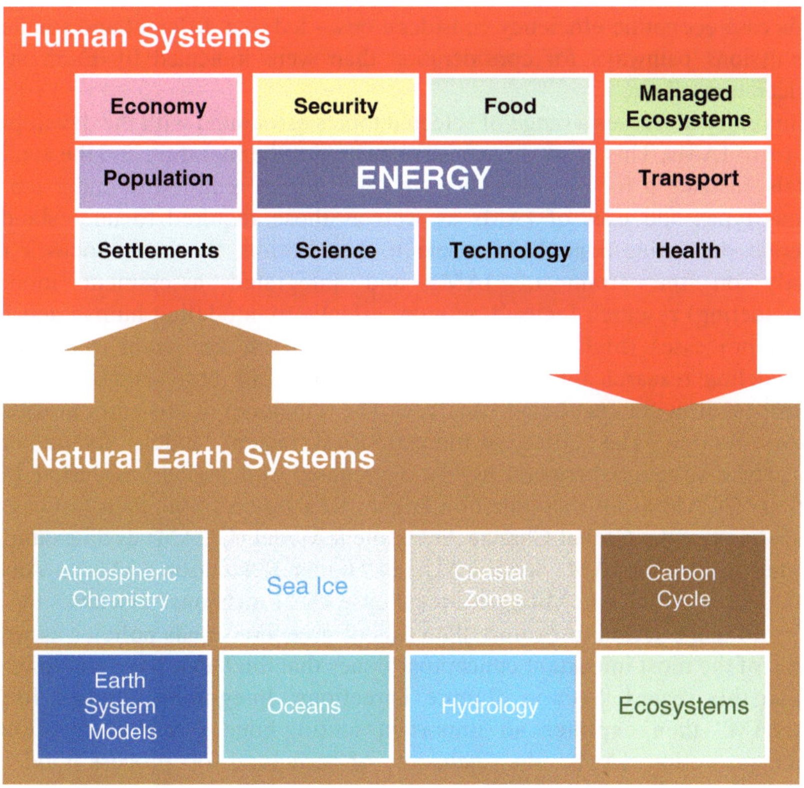

Fig. 8.1 Integrated assessment models integrate human and physical Earth system climate science (Source: Janetos et al. [72])

natural systems or human systems, or their various components, alone. By their nature, IAMs capture interactions between complex and highly nonlinear systems. IAMs serve multiple purposes. One purpose is to provide natural science researchers with information about human systems such as greenhouse gas emissions, land use, and land cover. Another purpose of IAMs is to help human system researchers – such as social scientists – better understand the nature of the human impacts on the natural Earth systems.

Traditionally, researchers have relied on models that are each built on the foundations of a single discipline – such as economics, geography, meteorology, etc. By integrating research methods from various disciplines that characterize both the human and natural Earth systems, IAMs produce insights that would not otherwise be available from disciplinary research. The work of Wigley, Richels, and Edmonds [1] provides a classic example of the nature of insights that are available from the explicit linking of human and Earth systems. Wigley et al. showed that the consideration of economic efficiency in the context of the physical carbon cycle carried important implications for the timing of emissions and emissions mitigation in a world seeking to stabilize the concentration of atmospheric CO_2. In other words, the imposition of human system considerations – in this case economic efficiency considerations – led to a different and smaller set of emissions pathways for consideration than were indicated by Earth system considerations alone.

This entry discusses a range of selected topics associated with the development and use of IAMs. This is not an extensive survey of the literature and the available models. Instead, it focuses on a selected set of topics required to understand the various types and uses of IAMs as well as those required to understand the direction of cutting-edge IAM research. In addition, the entry focuses more heavily on the strain of IAMs and integrated assessment modeling (IA modeling) research focused on more effectively modeling human and Earth system processes (higher resolution IAMs) than on the strain of IAMs and IA modeling research that focuses on more aggregate representations of these systems to allow for cost-benefit analysis. The remainder of this entry proceeds as follows. Section "The Variety of Integrated Assessment Models" focuses on the emerging distinction between highly aggregated and higher resolution IAMs. Section "GCAM as an Example of a Higher Resolution IAM" then follows with a discussion of the Global Change Assessment Model (GCAM) as an example of a higher resolution IAM. Section "Using Higher Resolution IAMs to Analyze the Impact of Policies to Mitigate Greenhouse Gas Emissions" discusses the long history of using IAMs to explore the costs of greenhouse gas policies as well as several of the most important conceptual issues that the IAMs have had to wrestle with in this regard. Section "Future Directions: Integrating Climate Impacts with IAM" then explores an important cutting-edge research direction for higher resolution IAMs: the inclusion of structural or process models of climate impacts.

The Variety of Integrated Assessment Models

There are many approaches that have been used to develop and use IAMs. Indeed, every IAM is different. One of the most important ways that IAMs are distinguished from one another is the level of resolution at which they model the underlying human and natural Earth system process. At one end of the spectrum are *highly aggregated* IAMs. Highly aggregated IAMs use highly reduced-form representations of the link between human activities, impacts from climate change, and the cost of emissions mitigation. At the other end of the spectrum are *higher resolution* IAMs. Higher resolution IAMs focus on explicitly representing processes and process interactions among human and natural Earth systems. The following two subsections provide background on each of these two classes of IAMs.

Highly Aggregated IAMs

The highly aggregated class of IAMs was developed to be able to explore the general shape of optimal climate policy, taking into account both the economic costs of mitigation and the economic damages from a changing climate. Highly aggregated IAMs typically frame the climate change mitigation problem in a cost-benefit framework, choosing emission pathways by explicitly weighing the economic costs of mitigation with the economic benefits of reduced impacts. For this reason, highly aggregated IAMs often focus on issues such as the social cost of carbon or optimal tradeoffs over time between mitigation and impacts. Simplicity and parsimony are main virtues of highly aggregated IAMs.

The oldest of the highly aggregated IAMs is the DICE (Dynamic Integrated model of Climate and the Economy) model, whose antecedents have roots in the work of Nordhaus and Yohe [2]. The original DICE model [3] was utilized to explore the integration of human and natural Earth systems as part of a cost-benefit calculation. Originally developed as a one-region global model, DICE was soon followed by a multiregional version, RICE (Regional dynamic Integrated model of Climate and the Economy) [4]. Other such models also emerged building on the Nordhaus-Yohe and DICE paradigm of combining economic costs and benefits in a single framework. These models include, among others, ICAM (the Integrated Climate Assessment Model) [5], PAGE (Policy Analysis of the Greenhouse Effect) [6], and FUND (Climate Framework for Uncertainty, Negotiation and Distribution) [7] (Weyant et al. [8] and Parson and Fisher-Vanden [9] provide good sources of information on pioneering IAMs).

Highly aggregated IAMs are generally composed of three parts: emissions and mitigation, atmosphere and climate, and climate impacts. Mitigation cost and climate change damages are typically monetized (i.e., expressed in dollars or another currency) to allow comparison between mitigation and impacts on a common basis. Highly aggregated IAMs do not attempt to describe in detail either the energy system or the land-use systems that generate emissions. Similarly, detailed descriptions of the physical process links between climate change and emissions are generally beyond their scope. Instead these models use emissions mitigation supply schedules and climate damage functions. The former maps the relationship between the degree of emissions mitigation and associated cost, while the latter represents the relationship between a measure of climate change and the economic value of damages including both damages from market and nonmarket activities. The strength of these reduced-form representations is that they allow highly aggregated IAMs to weigh costs and benefits explicitly. The drawback is that they cannot provide insight into the actual processes that lead to these costs and benefits.

The technical structure of highly aggregated models is simple, but the equations and associated parameterizations are carefully estimated to capture the behavior of more complex systems. These functions are parameterized by either approximating the behavior of more complex process models, or by fitting simple equations to highly aggregated variables. Analyses using FUND, for example, often produce simple equations that capture the behavior of systems that are represented in more complex models and data. Some models use a simpler approach, in which the economic damages from a prescribed level of climate change are first estimated – for example, a 2°C global mean surface temperature change (GMST) relative to preindustrial level – and a simple function that passes through the estimate – for example, a power function – is assigned to represent the relationship between GMST and total economic damages.

A principle role of highly aggregated IAMs is to integrate and to compare in a common metric, both mitigation effort and climate change impact – each estimated from different disciplines – in order to determine the optimal pathway of emissions reductions or the social cost of carbon. Valuation of damages provides substantial conceptual challenges for highly aggregated IAMs. For example, they must put a value on the loss of human lives as well as nonmarket damages. Another difficult challenge faced by highly aggregated IAMs concerns the relative valuation of impacts that occur at different points in time. See Box 8.1 for details.

Other issues that arise within the highly aggregated IAM paradigm include the problem of interactive effects, that is, the state of one system directly affects the state of another. For example, emissions mitigation may have large-scale effects on land use, which in turn affect the climate, or the climate system may change as a consequence of land-use policy. A challenge for highly aggregated IAMs is to represent such complex interactions in a simple model structure.

Another challenge for highly aggregated IAMs is to determine how to treat impacts occurring outside of the country undertaking the valuation. Early work with highly aggregated IAMs looked at the problem of climate change from the perspective of a single, global, infinitely lived decision maker. But, more recent work has shifted from the perspective of the globe (e.g., [3, 11]) to the perspective of a single country, for example, the United States [14].

Box 8.1. Valuation over Time and Across Generations

Climate change is an issue that is inherently long term as well as global. The nature of carbon cycle processes and their associated time scales create a cumulative relationship between CO_2 emissions and concentrations in the atmosphere (and ocean). Thus, unlike traditional atmospheric pollution problems, control of emissions to a level is insufficient to control the concentration of greenhouses in the atmosphere. In other words, CO_2 and other greenhouse gases are stock pollutants.

One of the most important determinants of the social cost of carbon is the rate at which future events are discounted back to the present. Nordhaus [10] argues that the order of magnitude difference between his estimate of the social value of carbon, derived using DICE, and the value estimated in the Stern Report [11], derived using PAGE, is predominantly a result of the differences in valuing the present relative to the future.

The problem is that there is no consensus on precisely how to approach discounting over periods of time long enough to connect multiple generations. The issues are laid out in Portney and Weyant [12], where the editors note in their overview chapter that "those looking for guidance on the choice of a discount rate could find justification for a rate at or near zero, as high as 20% and any and all values in between" ([12], p. 4). The range of estimates for the appropriate discount rate is generally nonnegative, though even that generalization has its exceptions, for example [13].

Methods for determining the appropriate method for discounting the future can be grouped into two general categories – those which are *prescriptive* and those which are *descriptive*. The prescriptive approach appeals to ethical and moral grounds for choosing a discount rate, while the descriptive approach appeals to observed rates of return on assets in economic markets. It is frequently observed that prescriptive approaches tend to generate lower discount rates than descriptive approaches.

The Higher Resolution IAMs

The higher resolution IAMs have roots in the same era as the highly aggregated IAMs. However, they were developed along different lines to serve different purposes. The higher resolution IAMs were developed to provide detailed information about human and natural Earth system processes and the interactions between these processes. The initial focus of these models was the determinants of anthropogenic carbon emissions. To address this problem, IAMs developed detailed representations of the key features determining long-term energy production, transformation, and end use. The higher

Table 8.1 Some higher resolution integrated assessment models

Some higher resolution integrated assessment models with interdisciplinary research teams		
Model	Home institution	Web link
AIM Asia-Pacific integrated model	National Institutes for Environmental Studies, Tsukuba, Japan	http://www-iam.nies.go.jp/aim/
GCAM Global change assessment model	Joint Global Change Research Institute, PNNL, College Park, MD	http://www.globalchange.umd.edu/models/gcam/
IGSM Integrated global system model	Joint Program on the Science and Policy of Global Change, MIT, Cambridge, MA	http://globalchange.mit.edu/igsm/
IMAGE The integrated model to assess the global environment	PBL Netherlands Environmental Assessment Agency, Bildhoven, The Netherlands	http://themasites.pbl.nl/en/themasites/image/
MERGE Model for evaluating the regional and global effects of GHG reduction policies	Electric Power Research Institute, Palo Alto, CA	http://www.stanford.edu/group/MERGE/
MESSAGE Model for energy supply strategy alternatives and their general environmental impact	International Institute for Applied Systems Analysis; Laxenburg, Austria	http://www.iiasa.ac.at/Research/ENE/model/message.html
ReMIND Refined model of investments and technological development	Potsdam Institute for Climate Impact Research; Potsdam, Germany	http://www.pik-potsdam.de/research/sustainable-solutions/models/remind/

resolution models distinguished different forms of energy, their supplies, demands, and their transformation from primary energy to fuels and electricity for use in end-use sectors such as buildings, transportation, and industry. Examples of higher resolution IAMs are provided in Table 8.1.

Over time these models have grown in complexity. The models have added increasing detail to their representations of both the energy system and the economy. They also broadened their scope, adding natural Earth system processes such as carbon cycle. The current generations of higher resolution IAMs also typically contain representations of agriculture, land use, land cover, and terrestrial carbon cycle processes in addition to representations of atmosphere and climate processes. While all of the higher resolution IAMs model both human and natural Earth system processes, each model was developed independently and each IAM development path emphasized different features of the climate change problem. Some emphasized the development of detailed atmosphere and climate system models. Some focused on detailed representations of technology. Others focused on regional differences in emission patterns and energy systems data. The complex nature of the models requires interdisciplinary research and modeling teams, some of which are listed in Table 8.1.

Because the higher resolution IAMs have grown in their complexity over time, describing the structure of each model in detail is beyond the scope of this entry. For a reference, comparison of three IAMs – IGCM, MERGE, and MiniCAM (the direct ancestor of GCAM) – can be found in [14]. Here, we present the summary comparison table from the report in Table 8.2. All three of these modeling systems have evolved considerably in the subsequent years.

GCAM as an Example of a Higher Resolution IAM

Introduction to GCAM

Rather than try to describe and compare the set of higher resolution IAMs, we have chosen to describe here the Global Change Assessment Model (GCAM) as an example of the higher resolution IAM genre. GCAM is the oldest of the higher resolution IAMs. It traces its roots to work initiated in the late 1970s. The model's first applications were completed in the early 1980s by Edmonds and Reilly [15–18]. Over time the model has developed and evolved through a series of advances documented in a variety of papers including [19–22]. Documentation for GCAM under its previous name, MiniCAM, can be found at http://www.globalchange.umd.edu/models/MiniCAM.pdf/. Other higher resolution IAMs, such as IMAGE and MESSAGE, also use MAGICC to represent atmosphere and climate processes.

At the top level the GCAM model is broken into two interacting system, human Earth system and natural Earth systems. Each of these systems in turn is made up of subsystems. This is the basic structure of all IAMs. GCAM and the other higher resolution IAMs are distinguished from the highly aggregated IAMs in the degree of detail that is incorporated in describing human and natural Earth systems.

All higher resolution IAMs emphasize the representation of human activities and their connection to the sources of greenhouse gas emissions. However, each modeling team has taken a different approach. For example, the IGSM employs a computable general equilibrium (CGE) model of the economy [23]. CGE models emphasize the structure of the economy and the interaction of economic sectors with each other and with labor and capital markets. The MERGE model also employs a highly aggregated CGE model in combination with more highly disaggregated energy sector models all embedded in an intertemporal optimization framework [24, 25]. The Asia-Pacific Integrated Model (AIM) employs a set of models that are used in combination [26]. The GCAM model uses a partial equilibrium framework, rather than a CGE framework. Partial equilibrium models delve into more detail in sectors that are directly related to the analysis in question (e.g., energy supply and demand, agricultural production, land use, and land-use change), and treat other sectors of the economy in aggregate.

Table 8.2 Characteristics of the three integrated assessment models

Feature	IGSM (with EPPA economics component)	MERGE	MiniCAM
Regions	16	9	14
Time horizon, time steps	2100, 5-year steps	2200, 10-year steps	2095, 15-year steps
Model structure	General equilibrium	General equilibrium	Partial equilibrium
Solution	Recursive dynamic	Inter-temporal optimization	Recursive dynamic
Final energy demand sectors in each region	Households, private transportation, commercial transportation, service sector, agriculture, energy-intensive industries, and other industry	A single, nonenergy production sector	Buildings, transportation, and industry (including agriculture)
Capital turnover	Five vintages of capital with a depreciation rate	A putty clay approach wherein the input-output coefficients for each cohort are optimally adjusted to the future trajectory of prices at the time of investment	Vintages with constant depreciation rate for all electricity-sector capital; capital structure not explicitly modeled in other sectors
Goods in international trade	All energy and nonenergy goods as well as emissions permits	Energy, energy-intensive industry goods, emissions permits, and representative tradable goods	Oil, coal, natural gas, biomass, agricultural goods, and emissions permits
Emissions	CO_2, CH_4, N_2O, HFCs, PFCs, SF_6, CO, NOx, SOx, NMVOCs, BC, OC, NH_3	CO_2, CH_4, N_2O, long-lived F-gases, short-lived F-gases, and SOx	CO_2, CH_4, N_2O, CO, NOx, SO_2, NMVOCs, BC, OC, HFC245fa, HFC134a, HFC125, HFC143a, SF_6, C_2F_6, and CF_4
Land use	Agriculture (crops, livestock, and forests), biomass land use, and land use for wind and/or solar energy	Reduced-form emissions from land use; no explicit land-use sector; assume no net terrestrial emissions of CO_2	Agriculture (crops, pasture, and forests) as well as biomass land use and unmanaged land; the agriculture-land-use module directly determines land-use change emissions and terrestrial carbon stocks

(continued)

Table 8.2 (continued)

Feature	IGSM (with EPPA economics component)	MERGE	MiniCAM
Population	Exogenous	Exogenous	Exogenous
GDP growth	Exogenous productivity growth assumptions for labor, energy, and land; exogenous labor force growth determined from population growth; endogenous capital growth through savings and investment	Exogenous productivity growth assumptions for labor and energy; exogenous labor force growth determined from population growth; endogenous capital growth through savings and investment	Exogenous productivity growth assumptions for labor; exogenous labor force growth based on population demographics
Energy efficiency change	Exogenous	Proportional to the rate of GDP growth in each region	Exogenous
Energy resources	Oil (including tar sands), shale oil, gas, coal, wind and/or solar, land (biomass), hydro, and nuclear fuel	Conventional oil, unconventional oil (coal-based synthetics, tar sands, and shale oil), gas, coal, wind, solar, biomass, hydro, and nuclear fuel	Conventional oil, unconventional oil (including tar sands and shale oil), gas, coal, wind, solar, biomass (waste and/or residues and crops), hydro, and nuclear fuel (uranium and thorium); includes a full representation of the nuclear fuel cycle
Electricity technologies	Conventional fossil (coal, gas, and oil), nuclear, hydro, natural gas combined cycle (NGCC) with and without capture, integrated coal gasification with capture, and wind and/or solar, biomass	Conventional fossil (coal, gas, and oil), nuclear, hydro, new coal and gas with and without CCS, other renewables	Conventional fossil (coal, gas, and oil) with and without capture; IGCCs with and without capture; NGCC with and without capture; Gen II, III, and IV reactors and associated fuel cycles; hydro, wind, solar, and biomass (traditional and modern commercial)

(continued)

Table 8.2 (continued)

Feature	IGSM (with EPPA economics component)	MERGE	MiniCAM
Conversion technologies	Oil refining, coal gasification, and bio-liquids	Oil refining, coal gasification and liquefaction, bio-liquids, and electrolysis	Oil refining, natural gas processing, natural gas to liquids conversion, coal, and biomass conversion to synthetic liquids and gases; hydrogen production using liquids, natural gas, coal, biomass; and electrolysis, including direct production from wind and solar, and nuclear thermal conversion
Atmosphere – ocean	2-dimensional atmosphere with a 3-dimensional ocean general circulation model, resolved at 20 minute time steps, 4° latitude, 4 surface types, and 12 vertical layers in the atmosphere	Parameterized ocean thermal lag	Global multi-box energy balance model with upwelling-diffusion ocean heat transport
Carbon cycle	Biogeochemical models of terrestrial and ocean processes; depends on climate and/or atmospheric conditions with 35 terrestrial ecosystem types	Convolution ocean carbon cycle model assuming a neutral biosphere	Globally balanced carbon cycle with separate ocean and terrestrial components, with terrestrial response to land-use changes
Natural emissions	CH_4, N_2O, and weather and/or climate dependent as part of biogeochemical process models	Fixed natural emissions over time	Fixed natural emissions over time
Atmospheric fate of GHGs, pollutants	Process models of atmospheric chemistry resolved for urban and background conditions	Single box models with fixed decay rates. No consideration of reactive gases	Reduced-form models for reactive gases and their interactions
Radiation code	Radiation code accounting for all significant GHGs and aerosols	Reduced form, top-of-the-atmosphere forcing	Reduced form and top-of-the-atmosphere forcing; including indirect forcing effects

Source: Clarke et al. [14]

The GCAM model drives the scale of human activities for each of its 14 geopolitical regions utilizing assumptions about future labor force – determined by working-age population, labor participation, and unemployment rate assumptions – along with the assumptions about labor productivity growth. The highly disaggregated energy, agriculture, and land-use components of GCAM are driven by the scale of human activity. The GCAM geopolitical regions are explicitly linked through international trade in energy commodities, agricultural and forest products, and other goods such as emissions permits.

The human dimension of the Earth system as shown in Fig. 8.2 integrates the energy system and the agriculture and land-use system, as well as the economic system that drives the activity in both systems. An important feature of the GCAM architecture is that the GCAM terrestrial carbon cycle model is embedded within the agriculture-land-use system model; that is, the agriculture-land-use system model explicitly calculates net land-use-change emissions from changes in land-use patterns over time. The energy system model produces and transforms energy for use in three end-use sectors: buildings, industry, and transport. The global human Earth systems are modeled for 14 geopolitical regions.

GCAM is a dynamic-recursive market equilibrium model. In each period of time the model's solution algorithm reconciles the supplies and demands for goods and services in all markets by finding a set of market-clearing prices. That market solution establishes the foundation from which the model steps forward to the next time period. Other IAMs, such as MERGE and MESSAGE, are built on an intertemporal optimization framework. These models solve all periods simultaneously so that expectations about the future are consistent with the model's future realizations in each time period. In contrast, GCAM, and other dynamic-recursive models, do not assume such intertemporal optimization takes place. Decisions taken in one period contain only expectations about future market conditions. These expectations will not necessarily be realized in the future. In other words, the economic agents in GCAM make decisions based on a less-than-perfect foresight, and the agents' only recourse in the subsequent period is to make another set of decisions, which can also be suboptimal.

The GCAM's time step is variable, but in general is set to 5 years, which is relatively common among integrated assessment models. GCAM tracks 16 different greenhouse gases, aerosols, and short-lived species. The GCAM physical atmosphere and climate are represented by the Model for the Assessment of Greenhouse-Gas Induced Climate Change (MAGICC) [27–29].

In the remainder of this section, we discuss in more detail two of the most important model components in GCAM: the representation of the energy system and the representation of agriculture and land use more generally.

The Energy System in GCAM

In GCAM, the energy system represents processes of energy resource extraction, transformation, and delivery, ultimately producing services demanded by end users

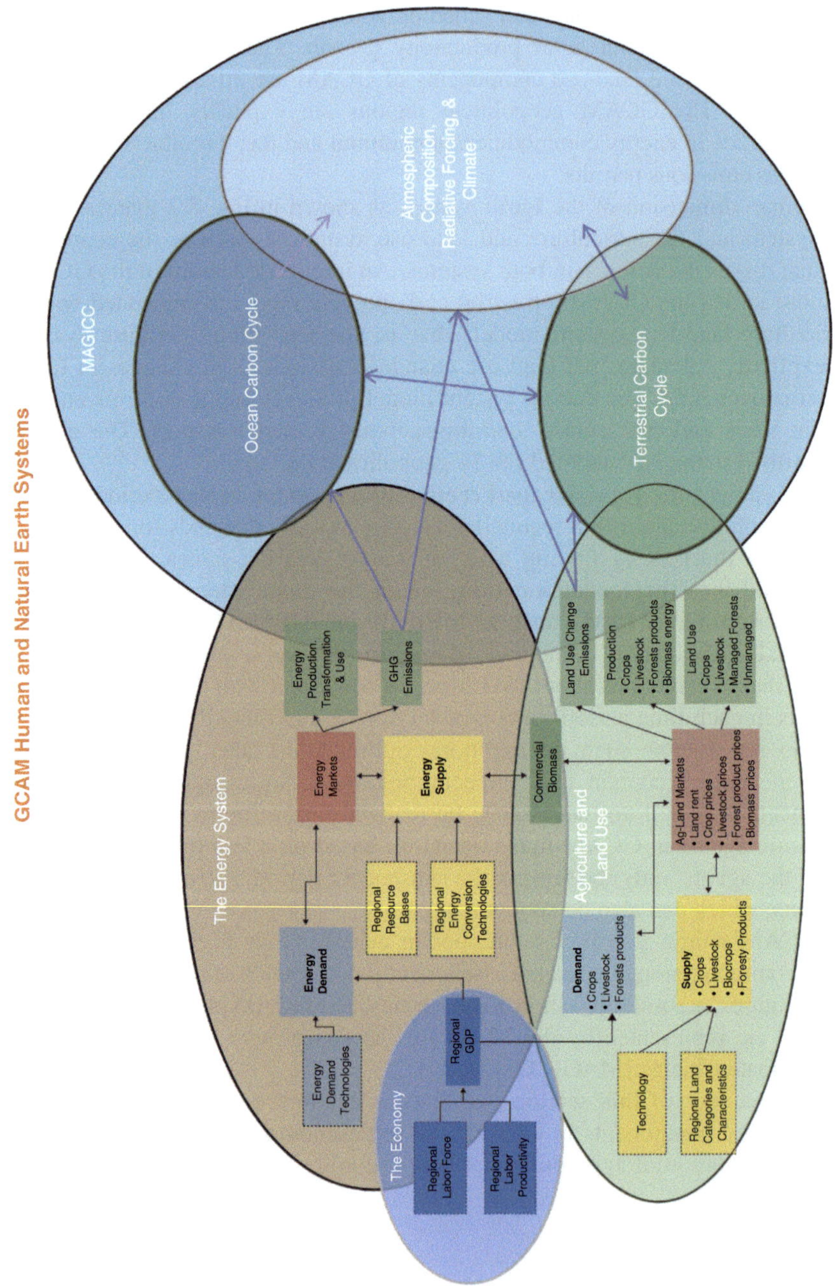

Fig. 8.2 Human and natural Earth systems of the global change assessment model

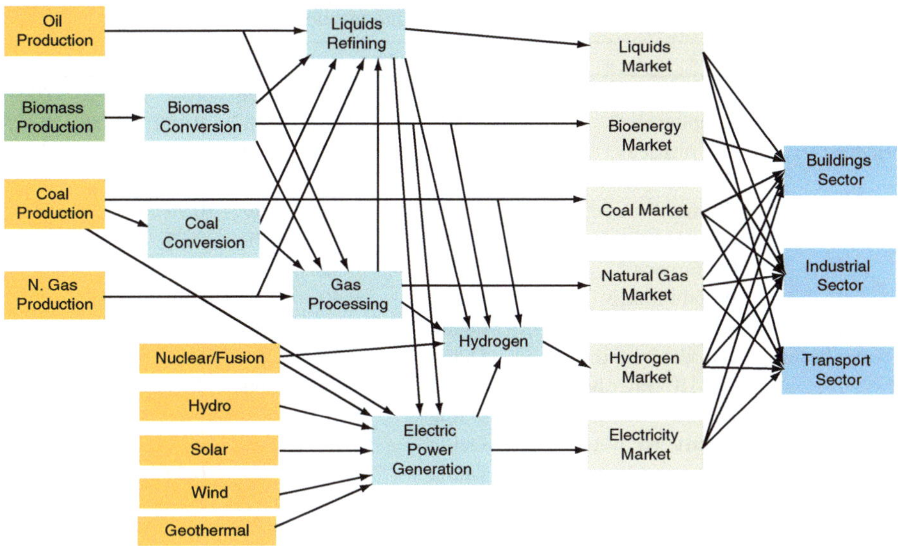

Fig. 8.3 The energy system in GCAM

(Fig. 8.3). In each time period, the market prices of all goods and services, including primary energy resources, land, agricultural goods, and other products, are determined by the market equilibrium.

Primary energy production is limited by regional resource availability. Fossil fuel and uranium resources are finite, graded, and depletable. Wind, solar, hydro, and geothermal resources are also finite and graded, but renewable. Bioenergy is also renewable, but is treated as an explicit product of the agriculture-land-use portion of the model. Extraction costs for graded resources rise as the resource consumption increases, but can fall with improvement in extraction technologies, and can rise or fall depending on other environmental costs.

Primary energy forms can be transformed into six final energy products:

- Refined liquid energy products (oil and oil substitutes)
- Processed gas products (natural gas and other artificially gasified fuels)
- Coal
- Bioenergy solids (various forms of biomass)
- Electricity
- Hydrogen

Energy transformation sectors convert resources initially into fuels, which may be consumed by either other energy transformation sectors or ultimately into goods and services consumed by end users. In each energy sector, multiple technologies compete for market share; shares are allocated among competing technologies using a logit choice formulation [30–32]. The cost of a technology in any period is determined by two key exogenous input parameters – the nonenergy cost and the efficiency of energy transformation – as well as the prices of the fuels it consumes. The nonenergy cost

Table 8.3 Residential sector efficiencies by service and technology (Source: Kyle et al. [79])

Residential			Reference			Advanced	
Service	Technology	unit	2005	2050	2095	2050	2095
	Building shell	W/m^2	0.232	0.182	0.150	0.163	0.125
Heating	Gas furnace	Out/in	0.82	0.90	0.97	Same as Ref	
	Gas heat pump	Out/in	n/a	n/a	n/a	1.75	2.45
	Electric furnace	Out/in	0.98	0.99	0.99	Same as Ref	
	Electric heat pump	Out/in	2.14	2.49	2.79	2.94	4.12
	Oil furnace	Out/in	0.82	0.86	0.93	Same as Ref	
	Wood furnace	Out/in	0.40	0.42	0.44	Same as Ref	
Cooling	Air conditioning	Out/in	2.81	3.90	4.88	4.59	7.19
Water heating	Gas water heater	Out/in	0.56	0.61	0.64	0.79	0.88
	Gas HP water heater	Out/in	0.89	1.09	1.22	1.75	2.45
	Electric water heater	Out/in	0.88	0.93	0.97	Same as Ref	
	Electric HP water heater	Out/in	n/a	2.46	2.75	2.75	3.45
	Oil water heater	Out/in	0.55	0.56	0.59	Same as Ref	
Lighting	Incandescent lighting	Lumens/W	14	15	16	Same as Ref	
	Fluorescent lighting	Lumens/W	60	75	94	Same as Ref	
	Solid-state lighting	Lumens/W	100	112	125	156	245
Appliances	Gas appliances	Indexed	1.00	1.12	1.25	Same as Ref	
	Electric appliances	Indexed	1.00	1.23	1.38	1.44	2.01
Other	Other gas	Indexed	1.00	1.12	1.25	Same as Ref	
	Other electric	Indexed	1.00	1.08	1.21	1.40	1.96
	Other oil	Indexed	1.00	1.12	1.25	Same as Ref	

represents all fixed and variable costs incurred over the lifetime of the equipment (except for fuel costs), amortized into a unit cost of output. For example, a coal-fired electricity plant incurs a range of costs associated with construction (a capital cost) and annual operations and maintenance. The efficiency of a technology determines the amount of fuel required to produce each unit of output (e.g., the fuel efficiency of a vehicle in passenger-km per GJ, or the electricity generation efficiency of a coal-fired power plant). The prices of different fuels are calculated endogenously in each time period based on supplies, demands, and resource depletion.

The representation of energy technologies in GCAM is highly disaggregated. Table 8.3 shows, for example, the set of technologies with accompanying assumptions of technology change over time, for the detailed US representation of residential buildings in GCAM.

Other energy sectors in GCAM have similar, high degrees of technology disaggregation. There are, for example, multiple technology options for generating electric power which include a variety of technologies utilizing solar energy as well as technology options to capture, transport, and store CO_2 in geologic repositories (CCS). The deployment of CCS technology in conjunction with bioenergy is of special interest in the consideration of very low long-term limits on CO_2 concentrations in that this combination potentially allows the production of energy with negative net CO_2 emissions. We discuss this particular technology combination in greater detail in a subsequent section of this entry.

Agriculture and Land Use in GCAM

Overview of the Agriculture and Land-Use Model in GCAM

Land use is one of the largest anthropogenic sources of emissions of greenhouse gases, aerosols, and short-lived species. The conversion of grasslands and forests to agricultural land results in a net emission of CO_2 to the atmosphere. In the nineteenth century, the conversion of forests to agricultural land was the largest source of anthropogenic carbon emissions. In the future, biomass energy crops could compete for agricultural land with traditional agricultural crops, providing a crucial linkage between land use and the energy system. Efforts to sequester carbon in terrestrial reservoirs, such as forests, may limit deforestation activities, and potentially lead to afforestation or reforestation activities. Interactions with crop prices may also prove important. Since land is limited, increasing the demand for land either to protect forests or to plant bioenergy crops could put upward pressure on crop prices that would not otherwise occur [33].

Many higher resolution IAMs include representations of agriculture, land use, and land cover. For some models, such as IGSM or IMAGE, a separate ecosystem model is used to represent terrestrial systems, which is then loosely coupled to the other elements of the IAM. These models represent land use, land cover, and the terrestrial carbon cycle. The IGSM model employs the Terrestrial Ecosystems Model [23], while IMAGE employs their terrestrial environment system submodel [34]. Since these models represent terrestrial processes at fine geographic scales – ½ degree by ½ degree gridded maps, for example – land use is determined by coupling an aggregated model of agriculture with a downscaling algorithm.

GCAM uses a model of land use and land cover, which allocates land area within each of its 14 global geopolitical regions among different land uses and tracks production from these uses and corresponding carbon flows into and out of terrestrial reservoirs. The GCAM agriculture, land use, land cover, terrestrial carbon cycle module determines the demands for and production of agricultural products, the prices of these products, the allocation of land to competing ends, and the carbon stocks and flows associated with land use.

Land is allocated between alternative uses based on expected profitability, which in turn depends on the productivity of the land-based product (e.g., mass of harvestable product per hectare), product price, and non-land costs of production (labor, fertilizer, etc.). The allocation of land types takes place in the model through global and regional markets for agricultural products. These markets include those for raw agricultural products as well as those for intermediate products such as poultry and beef. Demands for most agricultural products, with the exception of biomass products, are driven primarily by income and population. Land allocations evolve over time through the operation of these markets, in response to changes in income, population, technology, and prices.

The boundary between managed and unmanaged ecosystems is assumed to be elastic in GCAM. The area of land under cultivation expands and contracts as crops

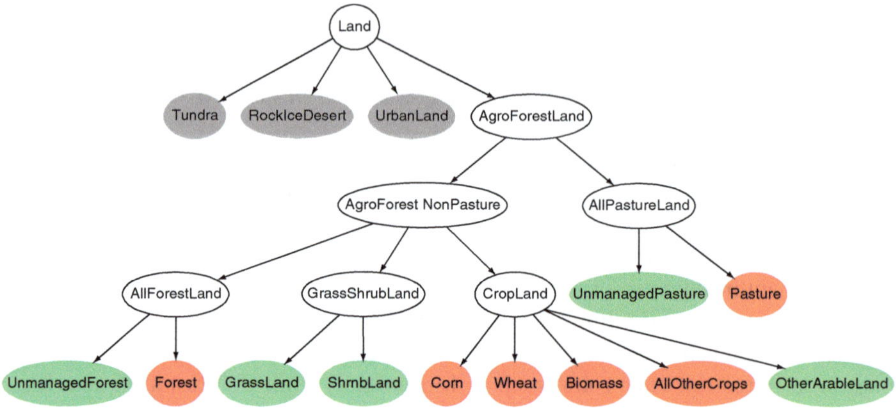

Fig. 8.4 *Competition for land in GCAM. Gray* exogenous in future periods, *Green* unmanaged land use, *Red* managed land use. AgLU tracks carbon content in different land uses. Changes in land use result in carbon flux to the atmosphere. Land owners compare economic returns across crops, biomass, pasture, and (future) forest, based on underlying probability distribution of yields per hectare

become more or less profitable. Thus, increased demands for land result in higher cropland profitability and expansion into unmanaged ecosystems and vice versa. Competition between alternative land uses in the GCAM is modeled using a nested logit architecture [30–32] as depicted in Fig. 8.4.

The costs of supplying agricultural products are based on regional characteristics, such as the productivity of land and the variable costs of producing the crop. The productivity of land-based products is subject to change over time based on future estimates of crop productivity change. It has been shown that the rate of crop yield improvement is a critical determinant of land-use change emissions [33, 35–37].

Bioenergy in GCAM's Agriculture and Land-Use Model

Bioenergy supply is determined by the agriculture-land-use component (AgLU) of GCAM, while bioenergy demand is determined in the energy component of the model. For example, the larger the value of carbon, the more valuable biomass is as an energy source and hence the greater the price the energy markets will be willing to pay for biomass. Conversely, as populations grow and incomes increase, competing demands for land may drive down the amount of land that would be available for biomass production at a given price.

There are three types of bioenergy produced in GCAM: traditional bioenergy production and use, bioenergy from waste products, and purpose-grown bioenergy. Traditional bioenergy consists of straw, dung, fuel wood, and other energy forms that are utilized in an unrefined state in the traditional sector of an economy. Traditional bioenergy use, although significant in developing nations, is a relatively small component of global energy. Traditional bioenergy is modeled as a function of regional income levels with its use diminishing as per capita incomes rise.

Other two types of bioenergy products are fuels that are consumed in the modernized sectors of the economy. Bioenergy from waste products are by-products of another activity. Examples in the model include forestry and milling by-products, crop residues in agriculture, and municipal solid waste. The availability of byproduct energy feedstocks is determined by the underlying production of primary products and the cost of collection. The total potential agricultural waste available is calculated as the total mass of the crop less the portion that is harvested for food, grains, and fibers, and the amount of bioenergy needed to prevent soil erosion and nutrient loss and sustain the land productivity. The amount of potential waste that is converted to bioenergy is based on the price of bioenergy.

The third category of bioenergy is purpose-grown energy crops. Purpose-grown bioenergy refers to crops whose primary purpose is the provision of energy. These would include, for example, switchgrass and woody poplar. The profitability of purpose-grown bioenergy depends on the expected profitability of growing and selling that crop relative to other land-use options in GCAM. This in turn depends on numerous other model factors: in the agricultural sector, bioenergy crop productivity (which in turn depends on the character of available land as well as crop type and technology) and nonenergy costs of crop production, and in the fuel processing sector, cost and efficiency of transformation of purpose-grown bioenergy crops to final energy forms (including liquids, gases, solids, electricity, and hydrogen), cost of transportation to the refinery, and the price of final energy forms. Furthermore, the price of final energy forms is determined endogenously as a consequence of competition between alternative energy resources, transformation technologies, and end-use energy service delivery technologies. In other words, prices are determined so as to simultaneously match demand and supplies in all energy markets as well as all land-use markets.

A variety of crops could potentially be grown as bioenergy feedstocks. The productivity of those crops will depend on where they are grown – which soils they are grown in, climate characteristics and their variability, whether or not they are fertilized or irrigated, the availability of nitrogen and other minerals, ambient CO_2 concentrations, and their latitude. GCAM typically include a generic bioenergy crop, with its characteristics similar to switchgrass that is assumed to be grown in all regions. Productivity is based on region-specific climate and soil characterizes and varies by a factor of three across the GCAM regions. GCAM allows for the possibility that bioenergy could be used in the production of electric power and in combination with technologies to provide CO_2 emissions captured and stored in geological reservoirs (CCS). This particular technology combination is of interest because bioenergy obtains its carbon from the atmosphere and if that carbon were to be captured and isolated permanently from the atmosphere the net effect of the two technologies would be to produce energy with negative CO_2 emissions. See, for example, [33, 38].

Pricing Carbon in Terrestrial Systems

Efficient climate policies are those that apply an identical price to greenhouse gas emissions wherever they occur. Hence, an efficient policy is one that applies identical prices to land-use change emissions and fossil and industrial emissions. This efficient approach is used as the default for emissions mitigation scenarios, though other policy options have also been modeled (A change in atmospheric CO_2 concentration has the same impact on climate change no matter what the source. Thus, to a first approximation land-use emissions have the same impact as fossil emissions. But, there are important differences. Land-use emissions do not have the same impact on atmospheric concentrations as fossil emissions because land-use emissions also imply changes in the future behavior of the carbon cycle. A tonne of carbon emitted due to deforestation, for example, is associated with a decrease in forest that would otherwise act as a carbon sink in the future. This effect, however, is not currently captured in GCAM).

Carbon in terrestrial systems can be priced using either a flow approach or a stock approach. The flow approach is analogous to the pricing generally discussed for emissions in the energy sector: landowners would receive either a tax or a subsidy based on the net flow of carbon in or out of their land. If they cut down a forest to grow bioenergy crops, then they would pay a tax on the CO_2 emissions from the deforestation. In contrast, the stock approach applies a tax or a subsidy to landowners based on the carbon content of their land. If the carbon content of the land changes, for example, by cutting forests to grow bioenergy crops, then the tax or subsidy that the landowner receives is adjusted to represent the new carbon stock in the land. The stock approach can be viewed as applying a "carbon rental rate" on the carbon in land. Both approaches have strengths and weaknesses. Real-world approaches may not be explicitly one or the other. By default, GCAM uses the stock approach.

Using Higher Resolution IAMs to Analyze the Impact of Policies to Mitigate Greenhouse Gas Emissions

A Brief Overview of IAMs in Mitigation Policy Analysis

Higher resolution IAMs have been used extensively to estimate the effects of measures to reduce greenhouse gas emissions. Until recently, the great bulk of the literature focused on the analysis of idealized policy instruments, particularly carbon taxes and cap-and-trade policies. For example, an important vein of early analysis focused on the question of emissions trading. In general, this literature showed that emissions mitigation undertaken with tradable permits resulted in lower costs to all parties without any reduction in overall emissions mitigation (see, for example, [39, 78]. The basic architecture of the Kyoto Protocol [40] reflected this line of thought. The application of these idealized pollution pricing mechanisms was inherently straightforward in higher resolution IAMs because these IAMs' representations of the energy and terrestrial systems are all built on economic principles. Furthermore, these mechanisms were of interest because they were theoretically attractive for the efficiency with which they reduced emissions.

In all the stabilization scenarios, the carbon price rises, by design, over time until stabilization is achieved (or the end-year 2100 is reached), and the prices are higher the more stringent is the stabilization level. There are substantial differences in carbon prices between MERGE and MiniCAM stabilization scenarios, on the one hand, and the IGSM stabilization scenarios on the other. Differences between the models reflect differences in the emissions reductions necessary for stabilization and differences in the technologies that might facilitate carbon emissions reductions, particularly in the second half of the century.

Whether for CO_2 or for multiple gases, a major focus of analysis has been to compute minimum-cost emissions trajectories for meeting long-term stabilization goals. The minimum cost is generally calculated on the assumption that all regions of the world undertake emissions mitigation in a coordinated, intertemporal program that reduces emissions in an economically efficient manner. One key characteristic of this pathway is that the marginal cost of emissions mitigation is equal in all sectors and in all regions at any point in time. It also means that the price of CO_2 rises at the rate of interest plus the rate of removal of CO_2 from the atmosphere until stabilization is reached [41]. After stabilization is reached, the CO_2 price no longer rises at this roughly constant rate, but instead is determined so as to ensure that at any point in time emissions match uptake so concentrations remain constant. Examples of classic stabilization CO_2 price pathways are shown in Fig. 8.5.

While mitigation cost may be one of the core questions addressed by the higher resolution IAMs, it is not the only question. A second and complementary set of questions focuses on implications for energy and agricultural systems, the next level of detail upon which higher resolution IAMs focus. How fast must the energy system change? Which technologies need to be deployed and when (see, e.g., [42, 43])? Stabilization of the concentration of CO_2 at any level requires that net anthropogenic carbon emissions must peak and decline indefinitely toward zero [1], but an almost infinite set of combinations of technology could in principle deliver that outcome. For example, fossil fuel use could be replaced with renewable energy forms in combination with energy efficiency improvements. Alternatively, fossil fuels could continue to be deployed in the global energy system in combination with CO_2 capture and storage (CCS), nuclear power, renewable energy, and energy efficiency. The combinations that emerge from different models depend on assumptions about technology performance and availability, scale of the economic system, and climate policy. A wide range of studies has made evolution of the energy system to meet long-term goals a focus of analysis (see, e.g., Fig. 8.6 from [14]).

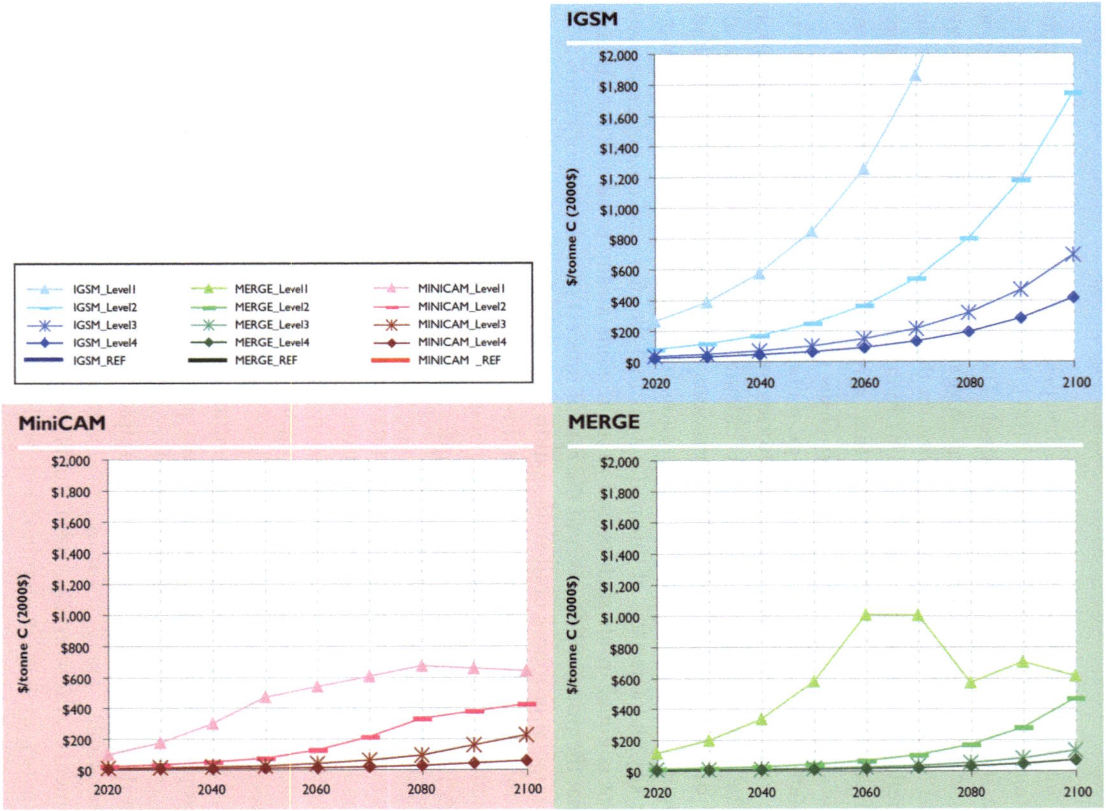

Fig. 8.5 Carbon prices across stabilization scenarios ($/ton C, 2000$) from three higher resolution IAMs leading to stabilization at approximately 750 ppmv CO_2 (Level 4), 650 ppmv CO_2 (Level 3), 550 ppmv CO_2 (Level 2), and 450 ppmv CO_2 (Level 1) (Source: Clarke et al. [14])

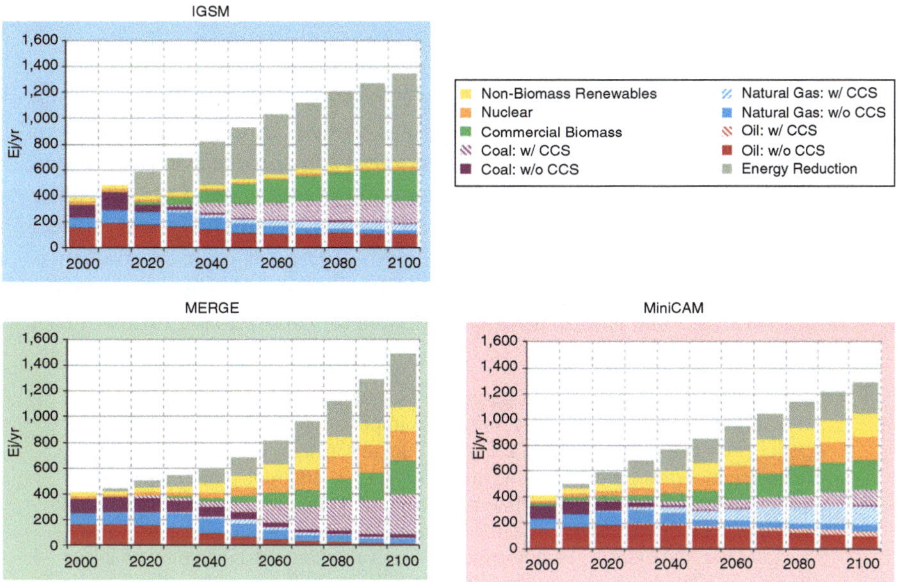

Fig. 8.6 Global primary energy production across scenarios from three higher resolution IAMs leading to approximately 450 ppmv CO_2 (Source: Clarke et al. [14])

Stabilization in IAMs with Multiple Greenhouse Gases

The United Nations Framework Convention on Climate Change (UNFCCC) has as its goal the stabilization of the concentration of greenhouse gases in the atmosphere. As discussed above, examination of the cost of stabilization of CO_2 and other gases has been the focus of a great number of papers utilizing higher resolution IAMs. Early studies focused exclusively on stabilization. However, more recent efforts have explored stabilization considering multiple greenhouse gases [14, 44].

When multiple greenhouse gases are considered simultaneously the problem emerges as to how to compare the greenhouse effects across the various constituents. In terms of climate change, the natural aggregate measure is radiative forcing (see Box 8.2). It is relatively straightforward to compute the radiative forcing for a group of gases, aerosols, and short-lived species and then to estimate what concentration of CO_2 would yield that radiative forcing level if all other species were set at their preindustrial levels. The answer to that question is the CO_2-equivalent concentration for that bundle of gases.

Two approaches have been used to determine the optimal mix of abatement across gases in stabilization. One approach is to minimize the total costs of meeting a long-term radiative forcing target, based on the combined mitigation costs for all greenhouse gases using intertemporal optimization. This is the approach employed by intertemporal optimization models such as MERGE. In this structure, all of the prices of the different greenhouse gases rise at relatively constant rates until

Box 8.2. Radiative Forcing

Most of the Sun's energy that reaches the Earth is absorbed by the oceans and land masses and radiated back into the atmosphere in the form of heat or infrared radiation. Some of this infrared energy is absorbed and reradiated back to the Earth by atmospheric gases, including water vapor, CO_2, and other substances. As concentrations of GHGs increase, there are direct and indirect effects on the Earth's energy balance. The direct effect is often referred to as a radiative forcing, a subset of a more general set of phenomena referred to as climate forcings. The National Research Council [45] offers the following set of definitions:

> Factors that affect climate change are usefully separated into forcings and feedbacks. ... A climate forcing is an energy imbalance imposed on the climate system either externally or by human activities. Examples include changes in solar energy output, volcanic emissions, deliberate land modification, or anthropogenic emissions of greenhouse gases, aerosols, and their precursors. A climate feedback is an internal climate process that amplifies or dampens the climate response to an initial forcing. An example is the increase in atmospheric water vapor that is triggered by an initial warming due to rising carbon dioxide (CO_2) concentrations, which then acts to amplify the warming through the greenhouse properties of water vapor....
>
> Climate forcing: An energy imbalance imposed on the climate system either externally or by human activities.

- *Direct radiative forcing*: A climate forcing that directly affects the radiative budget of the Earth's climate system; for example, added carbon dioxide (CO_2) absorbs and emits infrared radiation. Direct radiative forcing may be due to a change in concentration of radiatively active gases, a change in solar radiation reaching the Earth, or changes in surface albedo. Radiative forcing is reported in the climate change scientific literature as a change in energy flux at the tropopause, calculated in units of watts per square meter (W/m^2); model calculations typically report values in which the stratosphere was allowed to adjust thermally to the forcing under an assumption of fixed stratospheric dynamics.
- *Indirect radiative forcing*: A climate forcing that creates a radiative imbalance by first altering climate system components (e.g., precipitation efficiency of clouds), which then almost immediately lead to changes in radiative fluxes. Examples include the effect of solar variability on stratospheric ozone and the modification of cloud properties by aerosols.
- *Nonradiative forcing*: A climate forcing that creates an energy imbalance that does not immediately involve radiation. An example is the increasing evapotranspiration flux resulting from agricultural irrigation.

Source: Clarke et al. [14], Box 1.1; NRC [45]

Table 8.4 Direct global warming potential coefficients

Industrial designation or common name (years)	Chemical formula	Lifetime (years)	Radiative efficiency (Wm^{-2} ppb^{-1})	IPCC [48] (100^{-year})	20^{-year}	100^{-year}	500^{-year}
Carbon dioxide	CO_2	See notes[a]	[b]1.4×10^{-5}	1	1	1	1
Methane[c]	CH_4	12[c]	3.7×10^{-4}	21	72	25	7.6
Nitrous oxide	N_2O	114	3.03×10^{-3}	310	289	298	153

[a]The CO_2 response function used in this report is based on the revised version of the Bern carbon cycle model (Bern2.5CC) [49] used in IPCC [47] Chap. 10 Global Climate Projections using a background CO_2 concentration value of 378 ppm. The decay of a pulse of CO_2 with time t is given by $a_0 + \sum_{l=1}^{3} a_i \cdot e^{-t/\tau_l}$ where $a_0 = 0.217$, $a_1 = 0.259$, $a_2 = 0.338$, $a_3 = 0.186$, $\tau_1 = 172.9$ years, $\tau_2 = 18.51$ years, and $\tau_3 = 1.186$ years

[b]The radiative efficiency of CO_2 is calculated using the IPCC [50] simplified expression as revised in the TAR, with an updated background concentration value of 378 ppm and a perturbation of +1 ppm (see IPCC [47], Sect. 2.10.2)

[c]The perturbation lifetime for methane is 12 years as in the IPCC [48] (see also [47], Sect. 7.4). The GWP for methane includes indirect effects from enhancements of ozone and stratospheric water vapor (see [47], Sect. 2.10.3.1)

Source: IPCC [43], Table 2.14, pp. 212–213

stabilization is reached, consistent with the general result for minimum-cost CO_2 pathways discussed in the previous section [41], but the rates vary among gases. This leads to different timing of mitigation across gases. Indeed, one of the outcomes of this sort of approach to multi-gas stabilization is that the rate of increase in greenhouse gas prices is higher for gases with shorter lifetimes, with the implication that mitigation for these gases is delayed relative to CO_2. For example, this approach leads to scenarios in which mitigation of CH_4 is relatively modest in the early term and then increases dramatically as the total radiative forcing target gets close.

An alternative, though less rigorous methodology that is used to compare greenhouse gases in multi-gas emissions mitigation programs is the application of Global Warming Potential (GWP) coefficients. This is the approach generally used by dynamic-recursive models such as GCAM. The GWP was developed as an analogue to the Ozone Depletion Potential (ODP) coefficients employed to compare the various stratospheric ozone depleting substances [46]. GWPs are defined as the effect on radiative forcing of the release of an additional kilogram of a gas, relative to the simultaneous release of a kilogram of CO_2, integrated over one of three time horizons: 20 years, 100 years, and 500 years. Values for the GWPs calculated by IPCC Working Group I in the Fourth Assessment Report [47] are given in Table 8.4. GWPs are something of a mixture between the relative contribution of a gas to radiative forcing, which would be better calculated directly if possible, and an incomplete estimate of climate damage associated with the release of an additional kilogram of a greenhouse gas.

The primary virtue in the GWP is its application as an estimate of the relative importance of various greenhouse gases by national, local, and regional parties. Multi-gas policy instruments often employ GWPs as a means of comparing emissions of different greenhouse gases. The ratio of any pair of GWPs serves as the inverse of the relative price of any pair of greenhouse gases.

In application to stabilization studies in IAMs, GWPs yield constant estimates of the relative contributions of various greenhouse gases to climate change. In other words, since the GWPs are assumed to be constant over time, the relative prices of CO_2 and other gases are also constant over time. Hence, in studies that use GWPs to achieve multi-gas stabilization, mitigation for gases with shorter lifetimes generally takes place more quickly that would be the case in models that employ an intertemporal optimization approach. In this sense, although GWPs are a reality in policy design, they are an imperfect tool for comparing greenhouse gases over time. Manne and Richels [52] showed that if the total cost is the only criteria by which emissions pathways are judged then GWPs were not constant, but would rather change systematically with time. Peck and Wan [41] showed that if minimizing the total cost of limiting radiative forcing were the sole criterion by which greenhouse gas concentrations were controlled then the shadow price of each greenhouse gas rises at the interest rate plus the rate of removal from the atmosphere. Hence the corresponding GWP ratio of any two gases changes over time at a rate equal to the removal rate difference between the two gases. This notion is profoundly different than the concept of the GWP as a constant.

Manne and Richels [52] did show that the inclusion of secondary criteria, in addition to limiting radiative forcing, such as limiting the rate of change of radiative forcing, could produce very different GWPs and rates of change in GWPs over time. Some combinations of objective criteria could generate relatively stable GWPs.

The Economic Costs of Implementing the Framework Convention on Climate Change

As mentioned above, estimating the costs of meeting long-term targets is a primary function of IAMs. Typical estimates for global costs of limiting CO_2 equivalent concentrations to alternative levels from the IPCC [43] are shown below for two representative years, 2030 (Table 8.5) and 2050 (Table 8.6).

While the question of the measurement of the economic cost of emissions mitigation has not generated as much debate as questions about discounting, there are important differences in methodology that different modeling teams employ. Perhaps the most commonly used metric comparable across models is the price of carbon. This metric is useful for comparing across models when simple policy instruments to mitigate emissions are employed – specifically either an economy-wide carbon tax or the carbon price emerging from an economy-wide cap-and-

Table 8.5 Estimated global macroeconomic costs in 2030[a] for least-cost trajectories toward different long-term stabilization levels[b, c]

Stabilization levels (ppm CO_2-eq)	Median GDP reduction[d] (%)	Range of GDP reduction[d, e] (%)	Reduction of average annual GDP growth rates[d, f] (percentage points)
590–710	0.2	−0.6 to 1.2	<0.06
535–590	0.6	0.2 to 2.5	<0.1
445–535[g]	Not available	<3	<0.12

[a]For a given stabilization level, GDP reduction would increase over time in most models after 2030. Long-term costs also become more uncertain
[b]Results based on studies using various baselines
[c]Studies vary in terms of the point in time stabilization is achieved; generally this is in 2100 or later
[d]These are global GDP-based market exchange rates
[e]The median and the 10th and 90th percentile range of the analyzed data are given
[f]The calculation of the reduction of the annual growth rate is based on the average reduction during the period till 2030 that would result in the indicated GDP decrease in 2030
[g]The number of studies that report GDP results is relatively small and they generally use low baselines
Source: IPCC [43], SPM, p. 12

Table 8.6 Estimated global macroeconomic costs in 2050 for least-cost trajectories toward different long-term stabilization levels[a]

Stabilization levels (ppm CO_2-eq)	Median GDP reduction[b] (%)	Range of GDP reduction[b, c] (%)	Reduction of average annual GDP growth rates[b, d] (percentage points)
590–710	0.5	−1 to 2	<0.05
535–590	1.3	Slightly negative-4	<0.1
445–535[e]	Not available	<5.5	<0.12

[a]This corresponds to the full literature across all baselines and mitigation scenarios that provide GDP numbers
[b]These are global GDP-based market exchange rates
[c]The median and the 10th and 90th percentile range of the analyzed data are given
[d]The calculation of the reduction of the annual growth rate is based on the average reduction during the period until 2050 that would result in the indicated GDP decrease in 2050
[e]The number of studies is relatively small and they generally use low baselines. High emissions baselines generally lead to higher costs
Source: IPCC [43], SPM, p. 18

trade. As policy assumptions become more complex the usefulness of this metric fades. In fact, in mixed emissions mitigation systems, where only part of the economy is controlled by a tax or cap-and-trade program, the carbon price and real economic cost can move in opposite directions. That is, as more of the high-cost sectors of the economy are controlled with less-efficient nonmarket-based policies, the price of carbon may fall while the total economic cost rises.

A variety of approaches have been applied to obtain the total economic cost. These include integration under the marginal abatement cost schedule, measurement of foregone consumption, and compensated/equivalent variation. Each of these approaches traces its method back to welfare economics. While measures that directly link to welfare functions are in principle best, welfare cannot be directly observed and unless highly unlikely circumstances prevail, Arrow [53] has shown that a welfare function with the properties needed to get a measure on real economic cost cannot exist – a distinct disadvantage for numerical simulations.

While the choice of methodological approach to measuring real economic cost will doubtless affect valuation, two larger sources of variation in cost estimates are the policy instruments applied and the assumed rate of technological improvement. It is well known that different policy instruments can attain the same mitigation level with different costs [54]. Differences in technology assumptions can also produce substantial differences in cost (see, e.g., [42, 55]). Exploring the implication of different policy instruments and technology availability are two important directions of future work by the higher resolution IAM research community.

The principal research question which the higher resolution IAMs addressed has been different from that of the highly aggregated IAMs. Whereas the highly aggregated IAMs focused on the problem of determining the optimal balance between emissions mitigation and adaptation to climate change, the higher resolution IAMs focused more on the cost of implementing a policy to limit emissions, concentrations, or combined radiative forcing of greenhouse gases. The higher resolution IAM community has generally taken an agnostic position on the question of whether the policy instrument or the policy goal in question was desirable or not and simply went about the task of calculating the cost of achieving the given goal of implementing the prescribed policy.

As time has passed, the political conversation has moved away from the question of the use of cap-and-trade to control emissions to consider hybrid policy architectures in which emissions mitigation is pursued through a combination of policy measures some of which differ substantially from the conventional market mechanisms, such as carbon taxes or cap-and-trade. For example, many current emissions mitigation proposals contain renewable portfolio standards (RPS). These policy instruments require a minimum fraction of total power generation to be provided by renewable energy forms such as wind and solar.

There are many reasons for the shift. The prospects for a comprehensive international agreement based on the principles of cap-and-trade have diminished. Many parties in the international negotiations were less concerned with economic efficiency and cost-minimization than they were with a sense of moral obligation to achieve domestic emissions mitigation targets without resorting to emission trading. Within the United States similar forces are at work. Efforts to develop a comprehensive countrywide emissions cap-and-trade system show little prospect for entering into effect. Also, the European Union and Japan have either chosen alternatives to cap-and-trade or employ cap-and-trade within limited sectors of the economy. Such policies have pushed IAMs to develop more sophisticated representations of policies in order to estimate the policy effects [56]. In the same context, the IAMs have begun to explore the implications of international regimes in which nations begin emissions mitigation at different times [57, 58].

Future Directions: Integrating Human Earth Systems with Natural Earth Systems

Integrated Assessment modeling research is a continuously evolving field. As the models have matured and diversified, researchers have pushed the development frontiers in multiple directions simultaneously in order to answer a wide range of research questions. For example, researchers have broadened the scope of the models to include more sectors of the human Earth system such as land use and agriculture. They have expanded coverage of various types of the greenhouse gases by including an increasingly diverse set of their sources and activities. They have also lengthened the time horizon of analysis, pressing past the year 2100 and multiple centuries beyond. At the same time, the researchers have elaborated the key model components by slicing each of them in smaller pieces, for example, by adding finer spatial and temporal resolution and disaggregated representation of technologies.

An increasingly prominent research frontier has been the formal integration with other fields of climate change research, namely climate modeling (CM) and impacts, adaptation, and vulnerability (IAV) research. Although many research questions do not require the use of IPCC-class models of human and natural Earth systems, others cannot be addressed adequately without the development of integrated Earth systems models. The development of integrated Earth systems models opens the door to formally modeling the simultaneous interactions between human activities, climate change, and climate change impacts on human systems.

The Representative Concentration Pathways: An Example of Interactions with Climate Models

The assessment of climate change has traditionally been a linear research process. IA researchers produce emissions scenarios which in turn are transferred to the climate modeling community for use as inputs. The climate modeling community employs these scenarios to force future climate calculations. These climate calculations are then used by IAV researchers to produce estimates of the consequences of climate change. In the past, there has been little communication or feedback between research communities. Each community conducted its research independently and left it for others to figure out how or whether to use it. Beginning in 1990, the integrated assessment modeling community began to interact with the climate modeling community, though interactions with the carbon cycle and other natural Earth system researchers go back even further (see, e.g., [59], and more generally, [60]). Moss et al. [61] provide a succinct history of scenario development, which is summarized in Fig. 8.7.

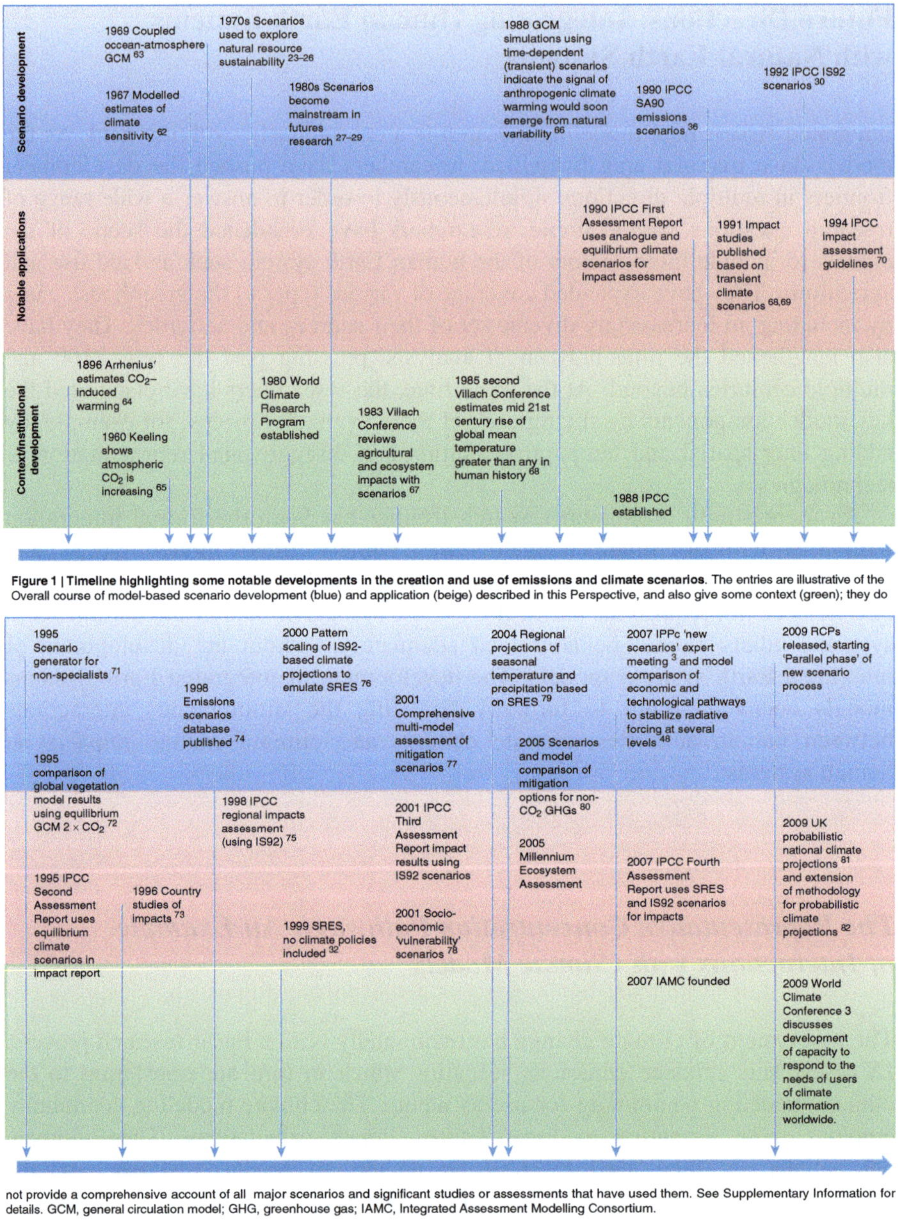

Fig. 8.7 Timeline highlighting some notable developments in the creation and use of emissions and climate scenarios (Source: Moss et al. [61], pp 748–749)

There have been numerous long-term scenarios of global greenhouse gas emissions. Three important benchmarks were the publication of scenarios referred to as SA90 [51], IS92 [62], and SRES [63]. These scenarios are notable in that the climate modeling community used them to simulate potential effect of future emissions paths on the climate system. The earliest scenarios considered only fossil fuel CO_2 emissions. Over time scenarios became richer, including land-use change emissions, non-CO_2 greenhouse gases, and short-lived species. While these scenarios span a wide range of potential future emissions, none considered limitations on emissions, that is, until Moss et al. [61] and the publication of the Representative Concentration Pathways (RCPs).

The RCPs are the most recent set of scenarios developed for use in the climate models. They were chosen to initiate an assessment cycle by providing the climate modeling community with a set of scenarios that were sufficiently differentiated by the end of the century to be scientifically relevant and to provide detailed information on the sources of emissions of greenhouse gases and short-lived species from all anthropogenic sources. RCPs differ from earlier scenario development activities in that they were selected from existing scenarios that were available in the peer-reviewed literature rather than being developed de novo. Selected scenarios from the open literature were named corresponding to their century's end radiative forcing levels: 8.5, 6.0, 4.5, and 2. 6 Wm^{-2} (see Table 8.7).

Subsequent to selection, the four scenarios were updated and harmonized to include the most recent observational data and downscaled to produce harmonized gridded outputs for emissions, land use, and land cover. The resulting time-paths for radiative forcing are given in Fig. 8.8 (The detailed scenario data are available at www.iiasa.ac.at/web-apps/tnt/RcpDb/).

The RCPs differ from previous scenarios employed by the climate modeling community in that they

1. Include scenarios with explicit emissions mitigation
2. Provide geospatially resolved emissions at ½ degree by ½ degree
3. Provide geospatially resolved land use and land cover at ½ degree by ½ degree

The most recent set of scenarios, while highly useful to the climate modeling community, are less useful from the perspective of the impacts, adaptation, and vulnerability community. While the scenarios contain detailed information that would be of interest to climate modelers, they do not carry associated socioeconomic information, or energy or commodity prices.

Furthermore, even if the socioeconomic data were included for these scenarios, each of the scenarios was crafted by a different modeling team, using different assumptions about key socioeconomic and other variables. For instance, it would be difficult, if not impossible to determine if the difference in estimated impacts of climate change associated with RCP4.5 and RCP2.6 was the result of differences in the magnitude of climate change or that of differences in the underlying human Earth systems that characterize the GCAM and IMAGE scenarios, respectively.

In order to establish a framework, in which the human system impact of climate change could be coupled with emissions scenario and climate model, a new

Table 8.7 The four representative concentration pathways

Name	Radiative forcing	Concentration	Pathway	Model providing RCP	References
RCP8.5	>8.5 W/m^2 in 2100	>1,370 CO_2-eq in 2100	Rising	MESSAGE	Riahi et al. [64]
RCP6.0	~6 W/m^2 at stabilization after 2100	~850 CO_2-eq (at stabilization after 2100)	Stabilization without overshoot	AIM	Fujino et al. [65], Hijioka et al. [66]
RCP4.5	~4.5 W/m^2 at stabilization after 2100	~650 CO_2-eq (at stabilization after 2100)	Stabilization without overshoot	MiniCAM (GCAM)	Clarke et al. [14], Smith and Wigley [67]
RCP2.6	Peak at ~3 W/m^2 before 2100 and then decline	Peak at ~490 CO_2-eq before 2100 and then decline	Peak and decline	IMAGE	Van Vuuren et al. [68, 69]

Source: Moss et al. [61], p. 753

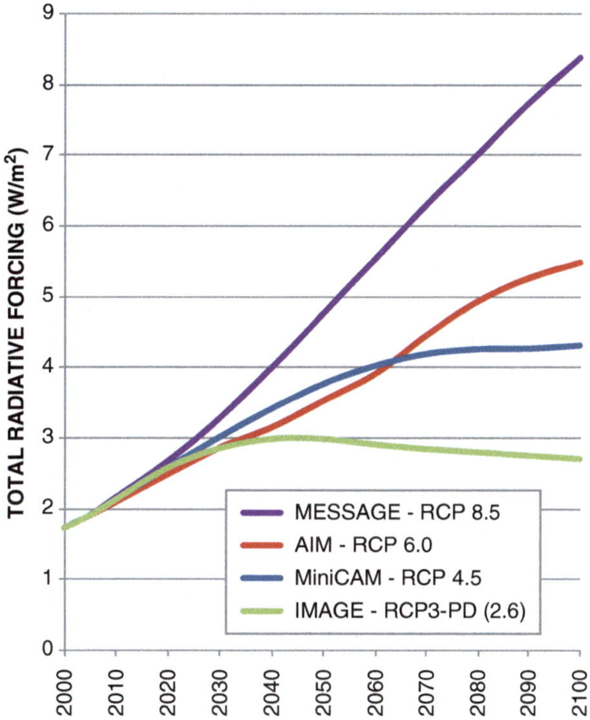

Fig. 8.8 The radiative forcing trajectories of the four RCP scenarios (Source: Moss et al. [61], P. 748–749)

scenario matrix architecture is under development. This architecture would create a suite of scenarios that are defined in terms of two bundles of descriptors: shared socio-ecosystem pathways (SSPs) and shared climate policy assumptions (SPAs).

SSPs have three components: a set of quantitative assumptions that are used by IAMs, such as population and economic growth; a set of quantified assumptions about variables that are not part of IAMs, for example, governance index; and a narrative which describes the general state of the world and its evolution over the course of the twenty-first century.

SPAs define the state of climate policy and its evolution around the world. They are defined with quantitative descriptors, where appropriate, and a qualitative narrative. The quantitative descriptors could be, for example, a limit on radiative forcing, such as was used to define the RCPs. In addition, information regarding the nature of policies that are to be employed to affect the prescribed outcome could be included.

The virtue of harmonizing SPAs with RCPs is that the new scenarios could be coupled smoothly with climate model output from ensemble calculations. This in turn would facilitate analysis that could potentially be fully integrated across three broad research communities: climate modeling, integrated assessment modeling and impacts, and adaptation and vulnerability. Two examples of such scenario matrix architectures can be found in [70, 71].

Climate Impacts in Higher Resolution IAMs

Higher resolution IAMs are increasingly focusing on explicitly modeling the physical impacts of climate change [72]. This work builds on a long tradition of modeling climate impacts in the higher resolution IAM community (see, e.g., [26, 73–75]). However, to date higher resolution IAMs have examined climate impacts using a sequential methodology, that is, they start with emissions, which are assumed to be given by climate models, and then analyzed the consequences of the ensuing climate change.

New model development is increasingly focused on methods and tools that will allow higher resolution IAMs to examine impacts simultaneously with mitigation and therefore to allow the two to interact. For example, there are on-going research efforts that utilize the higher resolution IAMs to study scenarios in which interactions between policies to mitigate emissions through changes in land-use and land cover – e.g., afforestation policies – and adaptive responses to climate change in agricultural sectors are simultaneously examined. Two complementary model development directions are also worthy of note. First, the higher resolution models are beginning to couple with state-of-the-art natural Earth system models (discussed later in this section) and second, they are beginning to move to finer spatial and temporal resolutions.

The increasing attention to climate impacts implies that the higher resolution IAMs will produce new results that will also contribute to the impacts, adaptation, and vulnerability (IAV) research. For nonmarket impacts of climate change, higher resolution IAMs will compute physical consequences, but not necessarily economic damage estimates, as it has generally been the case with climate impacts that the higher resolution IAMs have examined to date. For climate impacts associated with marketable goods and services, economic costs can also be estimated. But, the nonlinear nature of the human and natural Earth system means that separating out the impact of emissions mitigation from the impact of climate change will be nontrivial.

A good example of new work on the interactions between mitigation and impacts within higher resolution IAMs is land use and land cover. Land use will be affected both by a changing climate and by emissions mitigation effort. Mitigation effects will take the form of forest expansion to reduce land-use change emissions along with the use of bioenergy crops for energy production. A changing climate will bring about many changes in the nature of terrestrial systems, including changes in crop yields. All of these dynamics will interact.

To illustrate these interactions, the effects of climate change on crops were modeled as a response function derived from data reported in IPCC [76]. Figure 8.9 shows the distribution of estimates of crop yields for maize and wheat for low and other latitudes.

Both a reference scenario and a policy scenario in which CO_2 concentrations were limited to stay below 500 ppm were presented. Land-use change emissions of CO_2 were recorded for the two scenarios, with and without consideration of climate feedbacks through agricultural crops. These results are displayed in Fig. 8.10.

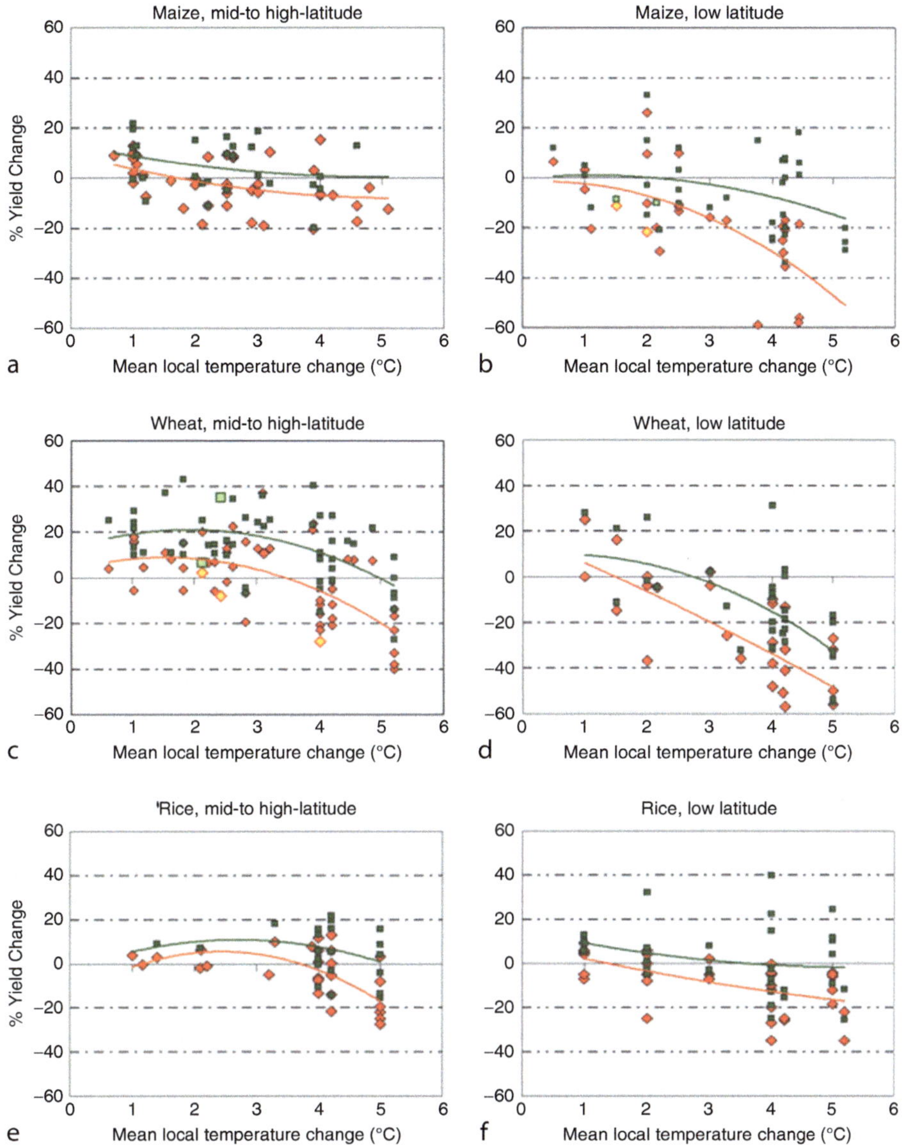

Fig. 8.9 The modeled effects of climate change on crops. Sensitivity of cereal yield to climate change for maize, wheat and rice, as derived from the results of 69 published studies at multiple simulation sites, against mean local temperature change used as a proxy to indicate magnitude of climate change in each study. Responses include cases without adaptation (red dots) and with adaptation (dark green dots). Adaptations represented in these studies include changes in planting, changes in cultivar, and shifts from rain-fed to irrigated conditions. Lines are best-fit polynomials and are used here as a way to summarise results across studies rather than as a predictive tool. The studies span a range of precipitation changes and CO_2 concentrations, and vary in how they represent future changes in climate variability. For instance, lighter-coloured dots in (**b**) and (**c**) represent responses of rain-fed crops under climate scenarios with decreased precipitation. (Source: Parry et al. [76], P. 286)

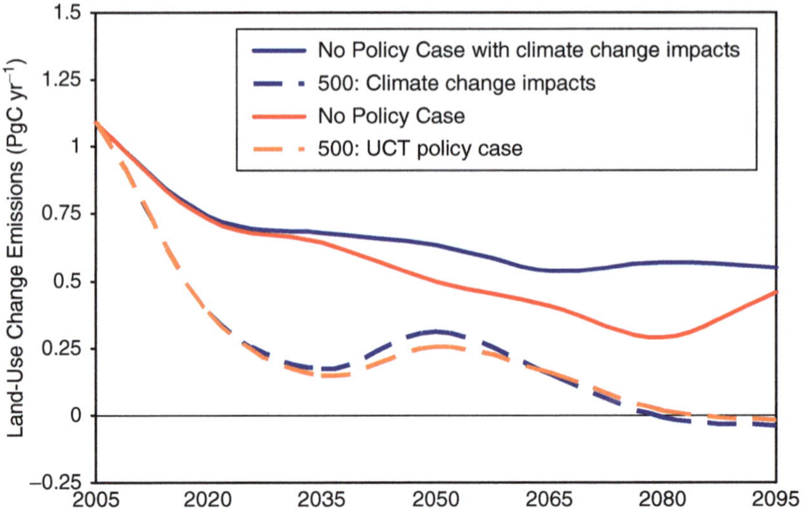

Fig. 8.10 Land-use change emissions of CO_2 under the scenarios with and without consideration of climate feedbacks through agricultural crops

Note that cumulative land-use change emissions vary significantly when climate change effects are considered in the reference scenario, with land-use change emissions significantly higher as a consequence of crop yield reductions in the face of climate change.

Results for the scenario in which CO_2 concentrations were not allowed to exceed 500 ppm exhibit lower emissions than either of the reference scenarios. This is because the mitigation scenario valued terrestrial carbon emissions equally with fossil fuel emissions (results would have been very different had terrestrial carbon not been valued; see also [33, 77]). Equally as interesting, land-use change emissions with and without consideration of climate change effects on crop yields are not significantly different between the two scenarios. This result follows directly from the fact that limiting CO_2 concentrations to 500 ppm would also limit the magnitude of climate change, which in turn moderates the effects on crop yields. The purpose of this example is not so much to showcase results, but rather to motivate the joint consideration of impacts, adaptation, and vulnerability with integrated assessment of emissions mitigation.

Linking Higher Resolution IAMs into integrated Earth System Models (iESMs)

Several research teams have undertaken joint work with the climate modeling community. The IGSM team has developed a relationship with climate researchers at the US National Center for Atmospheric Research (NCAR). The IMAGE team has developed several collaborative relationships including those with the Oak Ridge National Laboratory (ORNL), the Centre National de Recherches Météorologiques Coupled global climate Model (CNRM-CM3) team of France, and other European climate modeling teams to develop coupled scenarios. The MESSAGE integrated assessment modeling team has developed a collaboration with the NASA Goddard Institute for Space Studies climate modeling team. The GCAM team has developed a collaboration with ORNL and the Lawrence Berkeley National Laboratory (LBNL) in the development of a modeling system that joins the Community Earth System Model (CESM) representation of natural Earth systems with the GCAM representation of human Earth systems. To date, the collaborations have produced one-way coupling models, where emission scenarios from IAMs affect the climate, while the resulting climate change does not feedback to emissions. However, current effort is focused around developing a two-way coupled system.

The goal of the joint collaborations is to create a first-generation integrated Earth System Model (iESM) by fully integrating the human dimension from an IAM and a natural dimension from a climate model, that is, to create the capability of simultaneously estimating human system impacts on climate change and climate change impacts on human systems. After creating the capacity to examine the coupled natural and human Earth systems, the project could apply the model to the examination of feedbacks between human systems, the climate systems, and land-use systems. For instance, the policy response of land-use change presented in [33] could be revisited to estimate the magnitude of feedbacks in the system.

Significant effort is required before such research becomes routine. Nonetheless, as the research potential this collaboration opens up is virtually limitless, the importance of integrating human Earth systems with natural Earth systems is sufficiently compelling to drive future collaborations between ESMs and IAMs.

Bibliography

1. Wigley TML, Richels R, Edmonds JA (1996) Economic and environmental choices in the stabilization of atmospheric CO_2 concentrations. Nature 379:240–243
2. Nordhaus WD, Yohe GW (1983) Future carbon dioxide emissions from fossil fuels. Changing climate: report of the carbon dioxide assessment committee. National Academy Press, Washington DC, pp 87–153
3. Nordhaus WD (1993) Optimal greenhouse-gas reductions and tax policy in the "DICE" model. Am Econ Rev 83:313–317
4. Nordhaus WD, Yang Z (1996) A regional dynamic general-equilibrium model of alternative climate-change strategies. Am Econ Rev 86:741–765
5. Dowlatabadi H, Morgan MG (1993) A model framework for integrated studies of the climate problem. Energy Policy 21:209–221
6. Hope C, Anderson J, Wenman P (1993) Policy analysis of the greenhouse effect: an application of the PAGE model. Energy Policy 21:327–338

7. Tol RSJ (1997) On the optimal control of carbon dioxide emissions: an application of FUND. Environ Model Assess 2:151–163
8. Weyant J, Davidson O, Dowlatabadi H, Edmonds J, Grubb M, Parson EA, Richels R, Rotmans J, Shukla PR, Tol RSJ (1996)Integrated assessment of climate change: an overview and comparison of approaches and results. In: Bruce JP, H-sŏng Yi, Haites EF (eds) Climate change 1995: economic and social dimensions of climate change. The contribution of working group III to the second assessment report of the intergovernmental panel on climate change. Cambridge University Press, UK/New York, pp 367–396
9. Parson EA, Fisher-Vanden K (1997) Integrated assessment models of global climate change. Annu Rev Energy Env 22:589–628
10. Nordhaus W (2007) Critical assumptions in the Stern Review on climate change. Science 317:201–202
11. Stern NH (2007) The economics of climate change: the Stern Review. Cambridge University Press, Cambridge, UK/New York
12. Portney PR, Weyant JP (eds) (1999) Discounting and intergenerational equity. Resources for the Future, Washington, DC
13. Dasgupta P, Mäler KG, Barrett S (1999) Intergenerational equity, social discount rates and global warming. In: Portney PR, Weyant JP (eds) Discounting and intergenerational equity. Resources for the Future, Washington, DC
14. Clarke L, Edmonds J, Jacoby H, Pitcher H, Reilly J, Richels R (2007) Scenarios of greenhouse gas emissions and atmospheric concentrations. Synthesis and assessment product 2.1a, report by the U.S. Climate Change Science Program and the Subcommittee on Global Change Research. U.S. Government Printing Office, Washington, DC
15. Edmonds J, Reilly J (1983) A long-term global energy- economic model of carbon dioxide release from fossil fuel use. Energy Econ 5:74–88
16. Edmonds J, Reilly J (1983) Global energy and CO_2 to the year 2050. Energy J 4:21–48
17. Edmonds J, Reilly J (1983) Global energy production and use to the year 2050. Energy 8:419–432
18. Edmonds J, Reilly JM (1985) Global energy: assessing the future. Oxford University Press, New York
19. Brenkert A, Smith S, Kim S, Pitcher H (2003) Model documentation for the MiniCAM. Pacific Northwest National Laboratory. Technical report PNNL-14337
20. Kim SH, Edmonds JA, Smith SJ, Wise M, Lurz J (2006) The object-oriented energy climate technology systems (ObjECTS) framework and hybrid modeling of transportation in the MiniCAM long-term, global integrated assessment model. Energy J 27:63–91
21. Clarke L, Wise M, Kim S, Smith S, Lurz J, Edmonds J, Pitcher H (2007) Model documentation for the minicam climate change science program stabilization scenarios: CCSP product 2.1 a. Pacific Northwest National Laboratory, PNNL-16735
22. Wise MA, Calvin KV, Thomson AM, Clarke LE, Bond-Lamberty B, Sands RD, Smith SJ, Janetos AC, Edmonds JA (2009) The implications of limiting CO_2 concentrations for agriculture, land use, land-use change emissions and bioenergy. Pacific Northwest National Laboratory
23. Sokolov AP, Schlosser CA, Dutkiewicz S, Paltsev S, Kicklighter DW, Jacoby HD, Prinn RG, Forest CE, Reilly JM, Wang C, et al (2005) MIT integrated global system model (IGSM) version 2: model description and baseline evaluation, MIT Joint Program for the Science and Policy of Global Change. Report 124, Cambridge, MA
24. Manne A, Mendelsohn R, Richels R (1995) MERGE: a model for evaluating regional and global effects of GHG reduction policies. Energy Policy 23:17–34
25. Blanford GJ, Richels RG, Rutherford TF (2009) Feasible climate targets: the roles of economic growth, coalition development and expectations. Energy Econ 31:S82–S93
26. Kainuma M, Matsuoka Y, Morita T (eds) (2003) Climate policy assessment: Asia-Pacific integrated modeling. Springer-Verlag, Tokyo

27. Wigley TML, Raper SCB (1992) Implications for climate and sea level of revised IPCC emissions scenarios. Nature 357:293–300
28. Wigley TML, Raper SCB (2002) Reasons for larger warming projections in the IPCC third assessment report. J Climate 15:2945–2952
29. Raper SCB, Wigley TML, Warrick RA (1996) Global sea-level rise: past and future. In: Milliman JD Haq BU (eds) Sea-level rise and coastal subsidence: causes, consequences, and strategies. Kluwer Academic Publishers, Dordrecht, Netherlands
30. Clarke JF, Edmonds JA (1993) Modeling energy technologies in a competitive market. Energy Econ 15:123–129
31. McFadden D (1974) Conditional logit analysis of qualitative choice behavior. In: Zarambka P (ed) Frontiers of econometrics. Academic, New York, pp 105–142
32. McFadden D (1981) Econometric models for probabilistic choice among products. In: Manski C, McFadden D (eds) Structural analysis of discrete data with econometric applications. MIT Press, Cambridge, MA, pp 198–272
33. Wise M, Calvin K, Thomson A, Clarke L, Bond-Lamberty B, Sands R, Smith SJ, Janetos A, Edmonds J (2009) Implications of limiting CO_2 concentrations for land use and energy. Science 324:1183–1186
34. Bouwman AF, Kram T, Goldewijk KK (2006) Integrated modeling of global environmental change: an overview of Image 2.4. Netherlands Environmental Assessment Agency
35. Searchinger T, Heimlich R, Houghton RA, Dong F, Elobeid A, Fabiosa J, Tokgoz S, Hayes D, Yu T-H (2008) Use of U.S. Croplands for biofuels increases greenhouse gases through emissions from land-use change. Science 319:1238–1240
36. Lal R (2004) Soil carbon sequestration impacts on global climate change and food security. Science 304:1623–1627
37. Fargione J, Hill J, Tilman D, Polasky S, Hawthorne P (2008) Land clearing and the biofuel carbon debt. Science 319:1235–1238
38. Luckow P, Wise MA, Dooley JJ, Kim SH (2010) Large-scale utilization of biomass energy and carbon dioxide capture and storage in the transport and electricity sectors under stringent CO_2 concentration limit scenarios. Int J Greenh Gas Control 4:865–877
39. Edmonds JA, Scott MJ, Roop JM, MacCracken CN (1999) International emission trading and the cost of greenhouse gas emissions mitigation. The Pew Center on Global Climate Change, Arlington
40. Kyoto Protocol: the kyoto protocol to the united nations framework convention on climate change. UNEP/WMO, Kyoto
41. Peck SC, Wan YS (1996) Analytic solutions of simple optimal greenhouse gas emission models. In: van Ierland E, Górka K (eds) Economics of atmospheric pollution. Springer, Berlin, pp 113–121
42. Edmonds JA, Wise MA, Dooley JJ, Kim SH, Smith SJ, Runci PJ, Clarke LE, Malone EL, Stokes GM (2007) Global energy technology strategy: addressing climate change phase 2 findings from an international public-private sponsored research program. Pacific Northwest National Laboratory (PNNL), Richland
43. Metz B, Davidson O, Bosch P, Dave R, Meyer L (eds) (2007)Climate change 2007: mitigation of climate change; contribution of working group III to the 4th assessment report of the intergovenmental panel on climate change. Cambridge University Press, Cambridge, UK/New York
44. Weyant JP, Francisco C, Blanford GJ (2006) Overview of EMF-21: multigas mitigation and climate policy. Energy J 27:1–32
45. National Research Council (U.S.) (2005) Climate research committee: radiative forcing of climate change: expanding the concept and addressing uncertainties. Academic, Washington, DC
46. Wuebbles DJ, Edmonds J (1991) Primer on greenhouse gases. Lewis Publishers, Chelsea, Michigan

47. Solomon S, Qin D, Manning M, Marquis M, Averyt K, Tignor M, LeRoy Miller H, Chen Z (eds) (2007) Climate change 2007. The physical science basis: contribution of working group I to the fourth assessment report of the intergovernmental panel on climate change. Cambridge University Press, Cambridge, UK/New York
48. Houghton JT, Meiro Filho LG, Callander BA, Harris N, Kattenburg A, Maskell K (eds) (1996) Climate change 1995: the science of climate change. Cambridge University Press, Cambridge, UK
49. Joos F, Prentice IC, Sitch S, Meyer R, Hooss G, Plattner G-K, Gerber S, Hasselmann K (2001) Global warming feedbacks on terrestrial carbon uptake under the Intergovernmental Panel on Climate Change (IPCC) emission scenarios. Global Biogeochem Cy 15:891–908
50. Houghton JT, Jenkins GJ, Ephraums JJ (eds) (1990) Climate change: the IPCC scientific assessment. Cambridge University Press, Cambridge, UK
51. Houghton JT, Jenkins GJ, Ephraums JJ (eds) (1990) Climate change: the IPCC response strategies. Cambridge University Press, Cambridge, UK
52. Manne AS, Richels RG (2001) An alternative approach to establishing trade-offs among greenhouse gases. Nature 410:675–677
53. Arrow KJ (1950) A difficulty in the concept of social welfare. J Polit Econ 58:328–346
54. Milliman SR, Prince R (1989) Firm incentives to promote technological change in pollution control. J Environ Econ Manag 17:247–265
55. McJeon HC, Clarke L, Kyle P, Wise M, Hackbarth A, Bryant BP, Lempert RJ (2011) Technology interactions among low-carbon energy technologies: what can we learn from a large number of scenarios? Energy Econ 33:619–631
56. Böhringer C, Rutherford TF, Tol RSJ (2009) THE EU 20/20/2020 targets: an overview of the EMF22 assessment. Energy Econ 31:S268–S273
57. Edmonds J, Clarke L, Lurz J, Wise M (2008) Stabilizing CO_2 concentrations with incomplete international cooperation. Clim Policy 8:355–376
58. Clarke L, Edmonds J, Krey V, Richels R, Rose S, Tavoni M (2009) International climate policy architectures: overview of the EMF 22 international scenarios. Energy Econ 31:S64–S81
59. Edmonds JA, Reilly J, Trabalka JR, Reichle DE (1984) An analysis of possible future atmospheric retention of fossil fuel CO_2, TR013, US Department of Energy Carbon Dioxide Research Division, Washington DC
60. Trabalka JR, Reichle DE (eds) (1986) The changing carbon cycle: a global analysis. Springer, New York
61. Moss RH, Edmonds JA, Hibbard KA, Manning MR, Rose SK, van Vuuren DP, Carter TR, Emori S, Kainuma M, Kram T, Meehl GA, Mitchell JFB, Nakicenovic N, Riahi K, Smith SJ, Stouffer RJ, Thomson AM, Weyant JP, Wilbanks TJ (2010) The next generation of scenarios for climate change research and assessment. Nature 463:747–756
62. Leggett J, Pepper WJ, Swart RJ, Edmonds J, Meira Filho LG, Mintzer I, Wang MX, Wasson J (1992) Emissions scenarios for the IPCC: an update. Climate change 1992: The supplementary report to the IPCC scientific assessment. Cambridge University Press, Cambridge, UK/New York
63. Nakicenovic N, Alcamo J, Davis G, de Vries B, Fenhann J, Gaffin S, Gregory K, Grubler A, Jung TY, Kram T (2000) Special report on emissions scenarios: a special report of working group III of the Intergovernmental Panel on Climate Change. Cambridge University Press, Cambridge, UK/New York
64. Riahi K, Grübler A, Nakicenovic N (2007) Scenarios of long-term socio-economic and environmental development under climate stabilization. Technol Forecast Soc 74:887–935
65. Fujino J, Nair R, Kainuma M, Masui T, Matsuoka Y (2006) Multi-gas mitigation analysis on stabilization scenarios using AIM global model. Energy J SI3:343–354
66. Hijioka Y, Matsuoka Y, Nishimoto H, Masui M, Kainuma M (2008) Global GHG emissions scenarios under GHG concentration stabilization targets. J Global Environ Eng 13:97–108
67. Smith SJ, Wigley TML (2006) Multi-gas forcing stabilization with MiniCAM. Energy J SI3:373–392

68. van Vuuren DP, Eickhout B, Lucas PL, den Elzen MGJ (2006) Long-term multi-gas scenarios to stabilize radiative forcing-exploring costs and benefits within an integrated assessment framework. Energy J SI3:201–234
69. van Vuuren DP, Elzen MGJ, Lucas PL, Eickhout B, Strengers BJ, Ruijven B, Wonink S, Houdt R (2007) Stabilizing greenhouse gas concentrations at low levels: an assessment of reduction strategies and costs. Clim Change 81:119–159
70. van Vuuren DP, Riahi K, Moss R, Edmonds J, Thomson A, Nakicenovic N, Kram T, Berkhout F, Swart R, Janetos A, Rose SK, Arnell N (2011) A proposal for a new scenario framework to support research and assessment in different climate research communities. Global Environ Chang (In Press)
71. Kriegler E, O'Neill BC, Hallegatte S, Kram T, Lempert R, Moss RH, Wilbanks TJ (2010) Socio-economic scenario development for climate change analysis. CIRED working paper. Centre International de Recherche sur l'Environnement et le Développement, Paris, France
72. Janetos AC, Clarke L, Collins W, Ebi K, Edmonds J, Foster I, Jacoby HJ, Judd K, Leung L, Newell R, Ojima D, Pugh G, Sanstad A, Schultz P, Stevens R, Weyant J, Wilbanks T (2008) Science challenges and future directions: climate change integrated assessment research. U.S. Department of Energy, Office of Science. http://science.energy.gov/~/media/ber/pdf/Ia_workshop_low_res_06_25_09.pdf. Accessed 5 Dec 2011
73. Leemans R, Eickhout B (2004) Another reason for concern: regional and global impacts on ecosystems for different levels of climate change. Global Environ Change Part A 14:219–228
74. Reilly J, Paltsev S, Felzer B, Wang X, Kicklighter D, Melillo J, Prinn R, Sarofim M, Sokolov A, Wang C (2007) Global economic effects of changes in crops, pasture, and forests due to changing climate, carbon dioxide, and ozone. Energy Policy 35:5370–5383
75. Edmonds JA, Rosenberg NJ (2005) Climate change impacts for the conterminous USA: an integrated assessment summary. Clim Change 69:151–162
76. Parry M, Canzaiani O, Palutikof J, Van der Linden P, Hanson C (eds) (2007) Climate change 2007: impacts, adaptation and vulnerability; Contribution of working group II to the fourth assessment report of the intergovernmental panel on climate change. Cambridge University Press, Cambridge, UK/New York
77. Melillo JM, Reilly JM, Kicklighter DW, Gurgel AC, Cronin TW, Paltsev S, Felzer BS, Wang X, Sokolov AP, Schlosser CA (2009) Indirect emissions from biofuels: how important? Science 326:1397–1399
78. Weyant JP, Hill J (1999) The costs of the Kyoto protocol: a multi-model evaluation; introduction and overview. Energy J 20(Special Issue):vii–xliv
79. Kyle P, Clarke L, Rong F, Smith SJ (2010) Climate policy and the long-term evolution of the US buildings sector. Energy J 31:145–172

Chapter 9
Regional Climate Models

L. Ruby Leung

Glossary

Downscaling	Development of climate information at local or regional scale from coarse resolution data or model outputs; both statistical and dynamical methods can be used.
GCM	Global climate model, a climate model based on the general circulation of the atmosphere, often coupled with models of ocean circulation and sea ice.
Mesoscale	In the atmosphere, mesoscale generally refers to horizontal scales that lie between the scale height of the atmosphere (about 10 km) and the Rossby radius of deformation (tens to hundreds of kilometers).
Nudging	Method to reduce the differences between the simulated and observed or imposed states by applying corrections, usually in the form of tendencies to the prognostic equations, based on the differences.
RCM	Regional climate model (also called nested regional climate model), a climate model applied over a limited area with boundary conditions provided by global models or analyses.

This chapter was originally published as part of the Encyclopedia of Sustainability Science and Technology edited by Robert A. Meyers. DOI:10.1007/978-1-4419-0851-3

L.R. Leung (✉)
Atmospheric Sciences and Global Change Division, Pacific Northwest National Laboratory, Richland, WA 99352, USA
e-mail: Ruby.Leung@pnl.gov

P.J. Rasch (ed.), *Climate Change Modeling Methodology: Selected Entries from the Encyclopedia of Sustainability Science and Technology*,
DOI 10.1007/978-1-4614-5767-1_9, © Springer Science+Business Media New York 2012

Definition of the Subject and Its Importance

Regional climate models are numerical models that simulate the climate of geographic regions typically covering a few thousand square kilometers to a continent. Most regional climate models include models that describe the atmosphere and the underlying land surface, but a few also include models of ocean and sea ice and atmospheric aerosols and chemistry. Given the atmospheric state at the lateral boundaries, regional climate models simulate regional climate in the context of the evolving global climate. Because regional domains cover only a fraction of the globe, it is computationally more feasible to apply regional climate models at higher grid resolution compared to global climate models to better resolve atmospheric and terrestrial processes and how they respond to regional forcings such as topography and land cover/land use. While global climate models are generally applied at grid resolution of a few hundred kilometers, regional climate models have been more commonly applied at grid resolution of a few tens of kilometers. Therefore, a common application of regional climate models is the dynamical downscaling of global climate simulations to provide regional climate information related to climate change projections or climate predictions. As such, regional climate models have served an important function of providing regional climate scenarios needed to assess a wide range of societal relevant climate impacts such as climate change effects on water resources and ecosystems. Regional climate models are also used to study regional climate processes, particularly those that are related to the water cycle that is inherently multi-scale; so explicitly representing finer scale processes is important to simulate its variations at multiple time and space scales.

Introduction

Regional climate models were first developed in the late 1980s to provide a means to simulate climate features that were not well captured by global climate models (GCMs) because of their coarse spatial resolution. Figure 9.1 shows the representation of surface elevation and land cover/land use in climate models of different horizontal resolutions. At 400 km resolution, which was typical for GCMs in the early 1990s, climate models can only resolve very crude topographic variations and land surface heterogeneities to simulate their effects on large-scale and mesoscale circulation. At 50 km resolution, which is a common resolution used in regional climate models even today, models can begin to realistically capture topographic and land cover features important for regional climate.

The first regional climate model (RCM) was developed and applied to the western USA where regional climate is significantly influenced by the complex terrain not well resolved by GCMs [13,17]. The RCM was adapted from a mesoscale or limited-area atmospheric model that was designed for weather forecasting or short-term simulation of a few days. The model was enhanced for

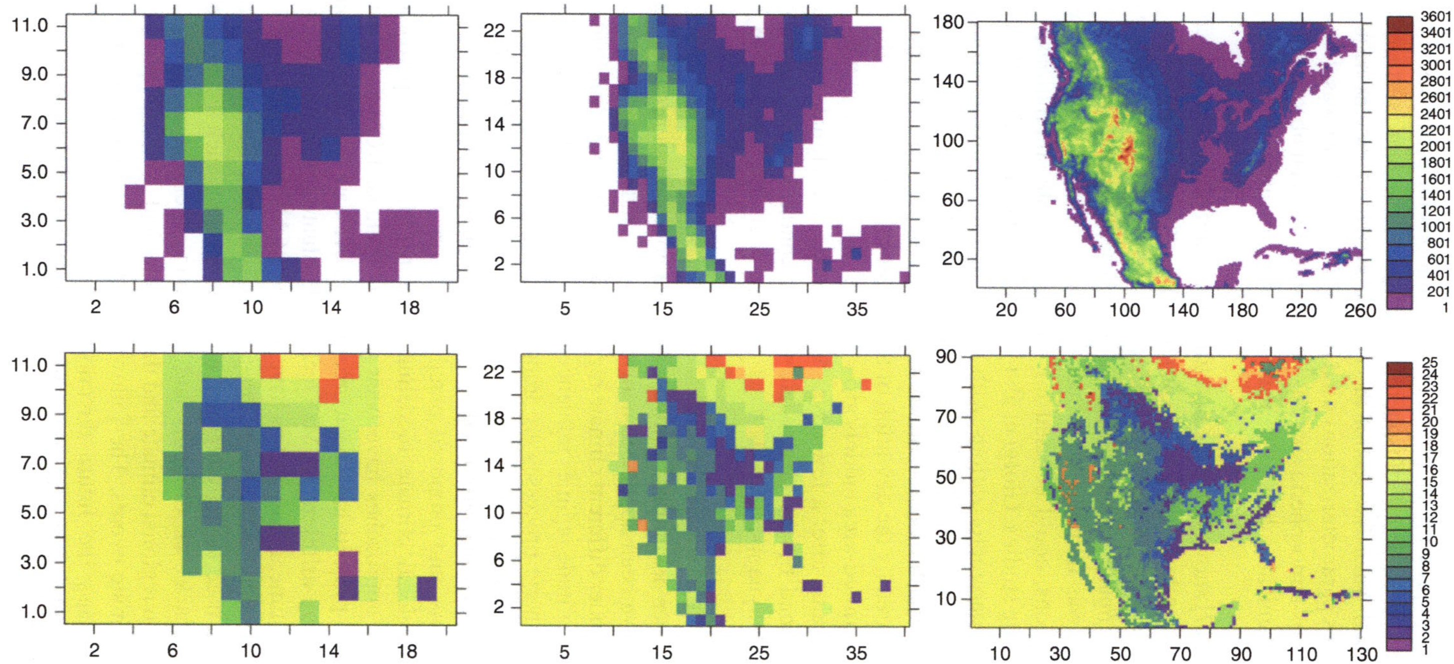

Fig. 9.1 Surface elevation (in meters) (*top row*) and land cover/land use (type) (*bottom row*) represented at 400 km (*left*), 200 km (*middle*), and 50 km (*right*) horizontal resolution in climate models. The land cover/land use types are: *1* urban, *2* dryland crop, *3* irrigated crop, *4* mixed crop, *5* crop/grass, *6* crop/woodland, *7* grass, *8* shrub, *9* mixed shrub/grass, *10* savanna, *11* deciduous broadleaf, *12* deciduous needleleaf, *13* evergreen broadleaf, *14* evergreen needleleaf, *15* mixed forest, *16* water bodies, *17* herbaceous wetland, *18* wooded wetland, *19* barren/sparsely vegetated, *20* herbaceous tundra, *21* wooded tundra, *22* mixed tundra, *23* bare ground tundra, *24* snow/ice, *25* playa. The x- and y-axes show the number of grid points in the domain at the three spatial resolutions

climate simulation by improving the physics representations for processes such as radiative transfer and biosphere-atmosphere exchange at the land surface that governs the energy and water budgets of the climate system. This was achieved by adopting the physics parameterizations used in a GCM. The RCM was driven at the lateral boundaries by atmospheric analysis [17] that provides an observationally constrained and dynamically balanced atmospheric state and global climate simulations [13].

Giorgi and Bates [17] showed, for the first time, that limited-area models could be used to produce long-term (more than a month) continuous simulations, as opposed to prior applications that use limited-area models to simulate weather for just a few days. By comparing the regional simulations with observations and the GCM simulations, it was demonstrated that a mesoscale weather model, with appropriate modifications, could be used for regional climate simulations. Following these pioneering studies, Giorgi et al. [19] further enhanced their RCM by updating the physics parameterizations with newer options available from the GCM, and explored model sensitivity to physics parameterizations and methods of assimilating the lateral boundary conditions. Giorgi and Mearns [16] showed that errors (e.g., measured by the deviation of the model solution from the driving large-scale fields) in limited-area models grow initially during model spin up, but reach an asymptotic value after a few days. At this stage, the climate simulated by the models is defined by the large-scale driving conditions and the model internal physics and dynamics, as well as the regional forcings within the model domain.

Subsequent to the early studies by Giorgi and his colleagues, more regional climate models have been developed following a similar approach and development path. These models have been applied to many regions around the world to assess their simulation skill under different climate regimes such as the monsoon, arid and semiarid deserts, mid-latitude regimes influenced by synoptic systems, and the high latitudes where cryospheric processes are important. As regional climate models became more widely used, questions have been raised about the validity and usefulness of the approach that prompted a series of studies to vigorously assess the various assumptions, and proposed practical or more mathematically well-posed solutions to regional climate modeling (section "Modeling Approach"). Different datasets and approaches have been used to evaluate RCMs, and large model intercomparison projects have been organized to evaluate and intercompare simulations produced by different RCMs (section "Evaluating Regional Climate Models"). At the same time, many studies have applied RCMs to simulate regional climate change that provided insights on climate change impacts. Regional climate models have also been used to study regional climate processes such as the role of land-atmosphere feedbacks on droughts and monsoon precipitation, effects of aerosols and land use on regional climate and the hydrological cycle, and processes leading to extreme climate events. The following sections provide a synopsis of these topics, and discuss the future directions in regional climate modeling. Examples of RCM applications are given in section "Application of Regional Climate Models."

Modeling Approach

How Do Regional Climate Models Work

Regional climate models are numerical models that simulate the climate of a specific region. Although some regional climate models, or regional earth system models, are beginning to include models of ocean, sea ice, and atmospheric aerosol and chemistry coupled to the atmosphere and land components, this review focuses mainly on regional climate models that traditionally include only atmosphere and land components with prescribed sea surface temperature and sea ice.

Similar to global atmospheric models, regional climate models numerically and simultaneously solve the equations of the conservation of energy, momentum, and water vapor that govern the atmospheric state. These equations are based on the Navier-Stokes equations for fluid flow (conservation of momentum) with approximations that apply to the atmosphere, the thermodynamic energy equation (conservation of energy), the continuity equation (conservation of mass), and the equation of state (ideal gas law). These partial differential equations are cast in various forms for different conservative properties and integrated forward in time using dynamical solvers. The solvers are applied to three-dimensional computational domains that are divided horizontally with grid spacing of a few to tens of kilometers and vertically into tens of vertical layers with a model top near 10–50 hPa. In regional climate models, solving these equations on limited-area domains require lateral boundary conditions, which can be derived from global climate simulations or global analyses to describe the large-scale atmospheric states. This method of simulating regional climate using limited-area models with prescribed lateral boundary conditions is called nesting (Fig. 9.2), so regional climate models are also called nested regional climate models to distinguish them from other dynamical frameworks such as global variable resolution or global stretched-grid models that simulate

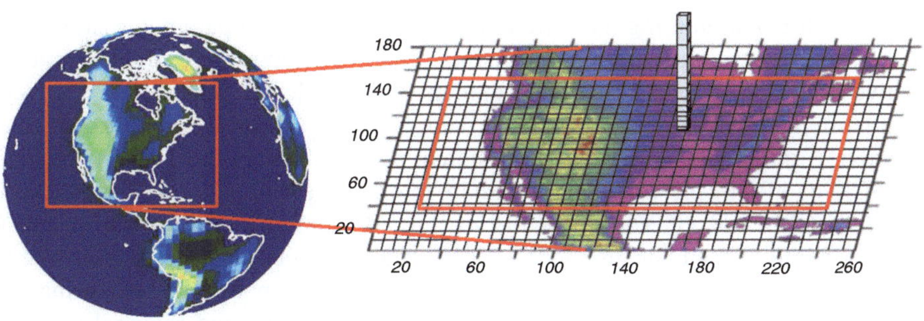

Fig. 9.2 A schematic showing the nesting of a regional climate model within a global climate model. The *right hand* figure shows the regional domain over North America with the horizontal grid (*black lines*), boundary of the buffer zone (*red box*), and a vertical column indicating the atmospheric layers represented by the model

regional climate for specific regions through regional refinement within the global domain.

The most commonly used lateral boundary treatment in nested regional climate models involves the relaxation of the interior flow in the vicinity of the boundary, called the lateral boundary buffer zone, to the prescribed flow [8]. In most models, the same treatment is also applied to the thermodynamics variables. When applied to RCMs, increasing the width of the lateral boundary buffer zone allows stronger control of the lateral boundary conditions to keep the simulated large scales closer to the global simulations or analyses that provide the lateral boundary conditions. Some RCMs have the capability to use nesting to further zoom into smaller regions with increasing grid resolutions. As computational resources increased over time, more RCMs are now formulated using non-hydrostatic dynamics, as the mean vertical motion of the air column within a model grid cell can no longer be assumed negligible at higher grid resolution. In contrast, most GCMs use hydrostatic solvers because the hydrostatic assumption is valid in coarser grids.

Besides numerically solving the momentum, thermodynamics, and continuity equations, climate models, global or regional, include parameterizations of physical processes such as radiative transfer, convection, cloud microphysics, land surface and biosphere-atmosphere exchange, and boundary layer turbulence. These parameterizations calculate the diabatic heating, moistening, and momentum changes due to the various processes. The resulting tendencies or rates of change are included as sources and sinks in the equations of energy, momentum, and water vapor to drive the atmospheric circulation.

Traditionally, GCMs use more sophisticated parameterizations of slow physical processes such as radiation and land surface for more accurate simulations of the global energy budgets, while limited-area models that are developed mainly for weather forecasting and short-term simulations emphasize detailed parameterizations of fast physical processes such as cloud microphysics and turbulence transfer. To simulate regional climate, both fast and slow physical processes are important because of the short spatial scale and long time scale of interest. Therefore many RCMs have adapted parameterizations of slow processes from GCMs, while maintaining the suite of the relatively detailed parameterizations of fast processes used in weather forecasting. Sharing of physics parameterizations between the global and regional models is considered desirable to reduce inconsistency between the simulated and driving large-scale conditions (see section "Modeling Issues" for a discussion of potential issues caused by mismatch of GCM and RCM solutions) and facilitate interpretation of differences simulated by the RCMs and GCMs. Since the first RCM (section "Introduction"), most RCMs developed and in use today still include subsets of physics parameterizations that are adapted from their host GCMs. Driven by high performance computing and the need to improve accuracy, both global and regional climate models are including more and more sophisticated parameterizations for all physical processes, which together with increasing model resolution, demand significant advances in high performance computing to support climate modeling research.

Modeling Issues

The climate of a region is determined by the large-scale atmospheric circulation as well as regional forcings such as topography within the region, and how they interact through various physical and dynamical processes. For example, the regional climate of the US Great Plains is strongly influenced by atmospheric circulation that brings moisture from the Gulf of Mexico during summer. How much precipitation is produced over land depends on moisture convergence, which is influenced not only by large-scale circulation patterns, but mesoscale features such as the Great Plain Low Level Jet, propagating disturbance from the Rocky Mountain, and local moisture sources from the land surface also play an important role. Therefore in the nested regional climate modeling approach, regional climate simulations depend on both the lateral boundary conditions that control the large-scale circulation, regional topography and land cover/land use features being resolved by the model, as well as physics parameterizations that ultimately determine the local changes in the energy, moisture, and momentum as influenced by the large-scale circulation and regional forcings.

Because of the dependence on large-scale circulation, large biases in global climate simulations used to provide lateral boundary conditions could have detrimental effects on the regional climate simulations under the nesting approach. Even if the global climate simulations were perfect, the lateral boundary conditions do not uniquely define the regional climate because the associated boundary value problem (i.e., solving the hyperbolic equations) is ill posed. Relaxation methods such as proposed by Davies [8] convert the hyperbolic equations to the well-posed parabolic form. However, mismatches between the large-scale circulation simulated by the regional models and the imposed atmospheric states at the lateral boundaries that may result from differences in grid resolution, physics, and dynamical formulations between the global and regional models can induce errors that propagate to the interior of the domains and contaminate the regional simulations [56]. This issue also leads to the sensitivity of the simulated regional climate to the domain size and locations of the lateral boundaries – an undesirable feature as it introduces uncertainties to the simulation results.

To address the validity of the nested regional climate modeling approach, a series of idealized numerical experiments have been designed and performed to assess the various assumptions used in regional climate modeling. The idealized experimental framework, known as "Big Brother Experiments (BBE)" [10], addresses modeling issues specifically related to the nested regional climate modeling approach. The Big Brother Experiment protocol consists of performing a high-resolution global climate simulation, referred as the Big Brother, BB, that serves as reference against which a regional climate simulation, referred as the Little Brother, LB, would be compared (Fig. 9.3). The BB, with proper spatial filtering to remove the fine scales to emulate coarse resolution global climate simulations, provides lateral boundary conditions for driving the LB. The differences between the climate simulated by the LB and the reference BB could

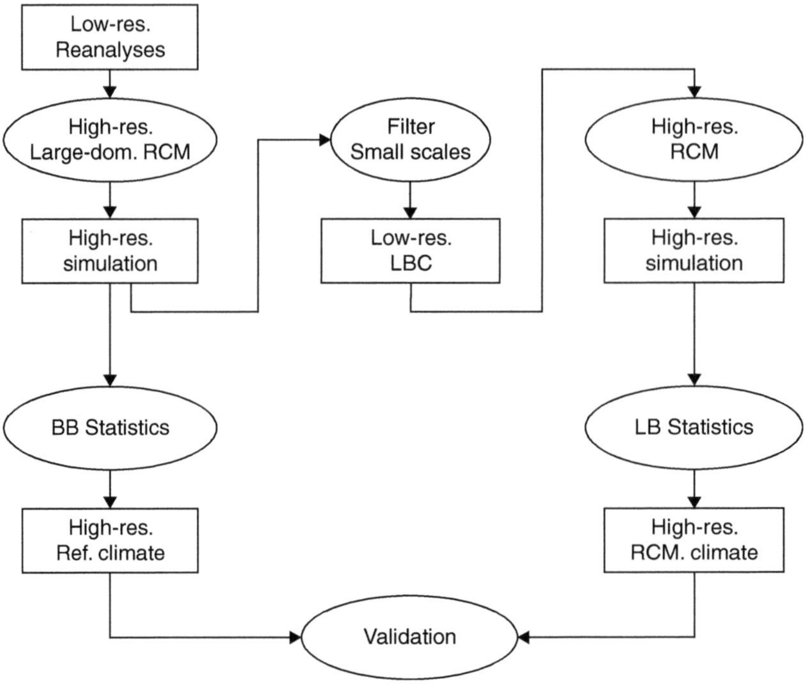

Fig. 9.3 Flow chart of the Big Brother Experiment (BBE). The high-resolution large domain RCM simulation is used as a virtual reality to evaluate the high-resolution simulation generated by the same RCM for a smaller domain achieved through nesting (Source Laprise et al. [31] © 2008 *Meteor. Atmos. Phys.*)

be attributed to the nesting approach of the limited-area model. Unlike the evaluation of real-world simulations that depends on the fidelity of model physics and availability of observational data, the idealized BBE framework allows different nesting-specific issues (e.g., the relaxation treatment and width of the buffer zone, frequency of LBC update) to be evaluated regardless of limitations of model physics and data because deficiencies of the nesting approach can be identified and quantified based on the comparison of the LB and BB alone.

A series of studies using the BBE protocol has been performed, focusing on different modeling issues specific to the nested regional modeling approach. As summarized by Laprise et al. [31], the BBE shows that the LB is capable of generating small-scale features absent from the lateral boundary conditions, and the small-scale features are consistent with the BB. These results demonstrate that the nested regional climate modeling approach does work as designed. That is, given large-scale conditions provided by the GCMs at the lateral boundaries, the RCMs can downscale to produce finer scale features absent from the GCMs. Moreover, the fine scales produced by the RCMs are consistent with what the GCMs would generate if they were applied at similar spatial resolution as the regional models. However, the small scales are not uniquely defined by the lateral boundary conditions and the domain-specific regional forcings, as the interactions between the two can be sensitive to small perturbations in the initial conditions that alter the time evolution of the small scales. The variations produced in regional simulations by perturbations in the initial conditions have been called "internal variability," as they relate to internal processes rather than external or LBC forcings. This issue has also been investigated by others (e.g., [3, 9, 21, 27]) who found that model internal variability depends on factors such as seasons, atmospheric flow regimes, and domain size. This puts a caveat on using single member short (seasonal and sub-seasonal) RCM simulations for model evaluation or hypothesis testing, as internal variability may overwhelm the signals (e.g., model errors or model response to external perturbations) being sought.

To address the issue of internal variability, ensemble modeling with perturbed initial conditions can be performed to quantify the internal variability and its impacts on model errors or model response. Alternatively, different techniques have been developed to constrain the large scales simulated by the regional models by the global climate simulations or global analyses throughout the regional domains. Spectral nudging [2, 28, 44, 54] is one example of such techniques. With spectral nudging, both the regional climate simulation and the global analyses or global climate simulations that provide lateral boundary conditions are decomposed into different spectral components in space. The simulated large-scale spectral components are nudged toward that of the global data using relaxation to provide stronger large-scale constraints on the regional climate simulations than that imposed by the lateral boundary conditions alone. These methods reduce the mismatch between the simulated large scales and the imposed lateral boundary conditions that contaminate the regional simulations. They also reduce internal variability, so simulation with a single initial atmospheric condition may be sufficient to assess model errors or estimate model response to external forcings. On the other hand, the degree of nudging to be applied to constrain the large scales can be rather arbitrary. Also, one may argue that by nudging the large scales of the regional climate simulations toward the global climate simulations, these methods increase the dependence of the regional simulations on the skill of the global models and eliminate the potential for the regional models to improve the large scales through upscaling of mesoscale features that are better resolved by the regional models [43].

Besides some form of large-scale nudging applied throughout the regional model domain, some studies have proposed a different mode of simulating regional climate by applying regional climate models with frequent initialization of the atmosphere to simulate short time segments that are then concatenated to compose the long-term regional climate simulations [45, 47]. This method takes advantage of the time period of limited error growth shortly after model initialization so the mismatch between the simulated and imposed large scale is small even without additional constraints on the large scale in the interior of the model domain. Two-way nesting of global and regional models has also been proposed as an approach to reduce large-scale inconsistency that may develop in one-way nested regional climate models because the upscaled influences of the regional models are included in the global models through feedbacks [40]. Such an approach has only been evaluated in a few studies [26, 41], and the results have been encouraging.

In summary, although the limited-area or nested modeling approach upon which regional climate models are based is an ill-posed boundary value problem, practical solutions such as the relaxation boundary treatment and spectral nudging of the large scale throughout the regional climate model domain have been developed and found to work well for a large number of cases. Furthermore, idealized experiments have confirmed most of the assumptions used in regional climate modeling [31]. However, uncertainty in regional climate simulations remain, owing in part to issues such as physics parameterizations, model resolutions, and initial conditions that are common to both global and regional climate models, and issues such as dependence on the lateral boundary condition, boundary treatment, regional domain size and location, and use of interior nudging that are specific to the nested regional modeling. Reviews and discussions of these issues can be found in Giorgi and Mearns [16], Laprise et al. [31], Leung et al. [35], and Wang et al. [55]. More research is needed to better understand the sensitivity of regional climate simulations to different factors and develop ways to reduce the uncertainty introduced by the nested modeling framework.

Evaluating Regional Climate Models

Model evaluation is important for assessing and documenting model skill and how it may evolve over time as changes and improvements are added to the models. It also provides information needed to understand model behaviors and diagnose model biases, and to assess uncertainties associated with the regional climate simulations. Model evaluation is achieved primarily by comparing model simulations with observations. The most common observation data used in evaluating regional climate simulations are atmospheric data such as 500 hPa height and upper level winds from global analyses, and surface temperature and precipitation from surface meteorological stations (e.g., Climatic Research Unit (CRU) and University of Delaware (UD) datasets), satellite-derived data (e.g., Tropical Rainfall Measurement Mission (TRMM)), and integrated station/ satellite products (e.g., Global Precipitation Climatology Project (GPCP) [24] and Climate Prediction Center (CPC) Merged Analysis of Precipitation CMAP) [59]. These data are typically spatially interpolated to uniform latitude/longitude grids.

Both surface temperature and precipitation have high spatial variability due to surface topography and other factors. The effect of topography is relatively easy to account for in surface temperature as it varies with altitude more or less according to the standard temperature lapse rate, but its influence on precipitation is more spatially variable depending on a number of factors such as wind direction and surface slope and aspect. Statistical methods such as Parameter-elevation Regressions on Independent Slopes Model (PRISM) [7] have been developed to account for surface topographical effects in gridded precipitation data. There is a continuing need to develop high temporal and spatial resolution datasets for evaluating regional climate models. Recent efforts in Europe [22] and Asia [60] have made great strides in providing high resolution (0.1° and 25/50 km resolution for Europe and 0.25° and 0.5° for Asia) gridded daily precipitation data for model evaluation and analysis, although differences among datasets can still be substantial in mountain areas due to measurement methods, retrieval algorithms, grid resolutions, and whether topographic effects are explicitly accounted for.

By comparing observed and simulated surface temperature, precipitation, and atmospheric fields, model biases can be identified. However, determining the sources of model errors and thereby providing guidance on reducing model biases requires more information. Observations that can be used to diagnose model errors are more limited. For example, to understand model biases in surface temperature, it is useful to know which components (e.g., net shortwave and longwave radiation and sensible and latent heat fluxes) of the surface energy budgets may be in error. Ground-based measurements of the surface energy fluxes are limited both spatially and temporally. However, some flux data are available from a global network (FLUXNET) of about 400 micrometeorological tower sites that provide continuous measurements, some dating back to 1996. There is a challenge in relating point measurements of surface fluxes with model simulations that represent grid box averages. Satellite retrievals of radiation fluxes are available globally for recent decades, but large differences exist among different datasets such as Clouds and the Earth's Radiant Energy System (CERES) and International Satellite Cloud Climatology Project (ISCCP).

Diagnosing errors in precipitation is even more challenging because precipitation is the end product of many interactive processes. Although precipitation is more directly related to clouds, measurements of cloud macrophysical and microphysical properties are limited. Cloud climatologies are available from ISCCP and CERES, but the grid resolution is relatively coarse (280 km for ISCCP and 1° for CERES) compared to regional models. Furthermore, errors in simulating clouds may be reflecting other problems because myriads of processes can influence the formation and evolution of clouds. Higher temporal and spatial resolution precipitation can provide a means to evaluate temporal variability from diurnal to seasonal, and probability distribution of precipitation rates, which can provide important clues to processes that may not be well represented in models. Some surface hydrological variables such as river runoff and snowpack may also be used to infer model biases in precipitation or combinations of precipitation and temperature biases.

Besides advances in the development of datasets for model evaluation, the methods used to evaluate models have also become more sophisticated. In the 1990s, comparisons of observations and model simulations were mostly limited to seasonal/annual and regional averages, but more studies now also compare observed and simulated temporal and spatial variability such as interannual variability and spatial distributions. With more studies producing longer regional climate simulations, more aspects of the simulations such as diurnal variability, extreme statistics, regime-specific features, frequency distributions, co-variability of different variables (e.g., between temperature and precipitation), and parameters that reflect the strengths of feedback processes have been evaluated (e.g., comparing land-atmosphere coupling strengths between models).

Although model evaluation studies are broadly aimed at understanding and quantifying model biases so model improvements can be made, some efforts also evaluate specific aspects such as precipitation and runoff [32], wind resources [57] of the regional simulations to provide practical guidance on their usefulness to provide climate information for impact assessments and resource management or planning. To support more detailed analyses, the requirements on model outputs have significantly increased as higher temporal frequency model outputs (e.g., hourly and daily) and more simulated state variables and tendencies are becoming more commonly archived.

Besides comparing model simulations with observations, model intercomparison can add significant information to understand and characterize model differences and uncertainties. The Atmospheric Model Intercomparison (AMIP) project [15] was initiated in the early 1990s to determine the systematic errors of global atmospheric models used to simulate long-term climate. Since the first AMIP project, many intercomparison projects have been developed to evaluate climate models used in different simulation modes. Similar coordinated projects have also been initiated to intercompare regional climate simulations since the mid-1990s. The first of such projects is the Project to Intercompare Regional Climate Simulations (PIRCS) [53]. PIRCS includes two phases, with the first phase focusing on simulations of two anomalous years, the 1988 drought and 1993 flood in the US Great Plains, and the second phase comparing multiyear simulations over North America. All simulations were driven by global reanalysis data and observed sea surface temperature. Besides regional climate models, one global stretched-grid model also participated in PIRCS for comparison between two dynamical frameworks for regional climate modeling. Following PIRCS, several intercomparison projects were developed to compare regional climate simulations over the Arctic (ARCMIP) (http://curry.eas.gatech.edu/ARCMIP/) and East Asia (RMIP) [14]. More discussions of intercomparison projects that focus on climate change simulations are provided in section "Dynamical Downscaling."

Applications of Regional Climate Models

Climate Process Studies

An important application of regional climate models is to advance the understanding of regional climate processes. In this context, regional climate models are often used to test hypotheses of how different regional forcings or feedback mechanisms play a role in regional climate variability and change.

For example, Leung et al. [34] used long-term simulations of the western USA to investigate the role of topography on precipitation spatial distribution during El-Nino and La-Nina events. Comparing precipitation during El-Nino years with the 20-year simulated climatology, they found a positive-negative-positive anomaly pattern in the Olympic Mountains and the west side and east side of the Cascades Mountains in the US Pacific Northwest. The pattern was found to be a result of the interactions between the large-scale atmospheric circulation that are influenced by the ENSO conditions and the orientation of the mountains. With atmospheric flow assuming a more southwesterly rather than a westerly direction during El-Nino years, the rain shadow created by the north-south oriented Cascades Mountains is reduced, resulting in more precipitation reaching the lee side of the mountains. Such regional anomaly patterns are generally not found in global reanalyses or global climate simulations because of their coarser resolution, but are consistent with observed precipitation and streamflow anomalies in the region.

Hughes and Hall [25] performed regional climate simulations for the western USA to investigate large-scale and local controls on Santa Ana winds in Southern California. Using a simulation at 6 km resolution, their analysis showed that both large-scale anomaly corresponding to a high pressure over the Great Basins, and local thermodynamic forcing due to surface temperature gradient between the cold desert (Mojave Desert) and warm ocean create pressure gradients that drive offshore winds. The latter was found to be particularly important in determining the timing of Santa Ana winds, which occur more frequently during December when the temperature gradient between the desert and Pacific coast is the largest.

The role of soot on mountain snowpack and hydrology was investigated by Qian et al. [48] using regional climate simulations with and without soot deposition in western USA. Their study shows that soot-induced snow-albedo perturbations increase the surface net solar radiation flux during late winter to early spring. This increases the surface air temperature and reduces snow accumulation and spring snowmelt, causing a trend toward earlier snowmelt. Snow-albedo feedback was found to play an important role in amplifying the soot effects in the mountains.

Riddle and Cook [50] used regional climate simulations to study the mechanism of abrupt rainfall transition in the Greater Horn of Africa. The yearly monsoon jump of about 20° latitude during April and May was found to coincide with abrupt circulation changes associated with the Somali jet that develops during that time. The cross-equatorial branch of the Somali jet brings moisture to the southern slopes of the Ethiopian plateau, which then produces the abrupt rainfall transition in the region.

To investigate why temperature in the central USA has cooled by 0.2–0.8°C in the late twentieth century, instead of warmed as in most continental regions, Pan et al. [46] used a regional climate model and found that under a global warming scenario, increased moisture from the Gulf of Mexico due to warming and increasing occurrence of the Great Plain Low Level Jet (LLJ) in the south and decreasing occurrence in the north enhances atmospheric moisture convergence and cloudiness and precipitation in the central USA. These changes replenish soil moisture during summer, which increases late-summer evapotranspiration and suppresses daytime maximum temperature, and hence the "warming hole." Because of coarse resolution, most GCMs cannot simulate the observed "warming hole" in the late twentieth century.

Regional climate models also offer great potentials to understand the mechanisms of extreme events and their projected changes in the future. For example, Seneviratne et al. (2006) performed two regional climate simulations with and without land-atmosphere interactions to investigate the role of land-atmosphere feedbacks on heat waves in Europe. They showed that soil moisture – temperature and soil moisture – precipitation feedbacks increase summer temperature variability in central and eastern Europe. Under climate change, the region of stronger land-atmosphere coupling shifts northward in response to greenhouse warming to central and eastern Europe, and enhances summer temperature variability and increases the potential for more frequent heat waves in that region.

In the above examples, high resolution is important for the model to reproduce the observed climatology of temperature, precipitation, wind, or snowpack, which in the western USA, Europe, and the Greater Horn of Africa depend strongly on regional orography. High resolution is also important for simulating soot deposition caused by anthropogenic emissions in cities being carried to the mountains downwind, or LLJ and its effects on cloudiness and precipitation. Successful simulations of the base states and model ability to simulate regional forcings and feedback mechanisms (e.g., snow-albedo, soil moisture – temperature feedbacks, LLJ – precipitation coupling) are critical for assessing their role in the observed regional climate phenomena.

Dynamical Downscaling

Dynamical downscaling is an important application of regional climate models, which aims to provide more spatially resolved climate predictions or projections provided by GCMs. Most of the downscaling applications to date are related to climate change projections. Early efforts described the use of an individual RCM to dynamically downscale climate change projections by a specific GCM. Typically only a single emission scenario such as the business-as-usual scenario (1% increase of CO_2 per decade) was used. Although GCMs generally produce simulations that cover preindustrial to 2100, RCM simulations are usually performed only for two time segments of 10–30 years corresponding to a current and a future time period.

Giorgi et al. [18] reported the first set of studies on using a regional climate model to dynamical downscale climate change scenario for Europe and the western Mediterranean basin. The GCM and RCM they used had a spatial resolution of R15 (about 400 km) and 70 km, respectively. The current and future climate corresponds to the equilibrium conditions simulated by the GCM using $1 \times CO_2$ (preindustrial level) and $2 \times CO_2$ (doubling of preindustrial level), respectively. Although the GCM generally reproduced the basic seasonal migration patterns of storm tracks, significant biases were also found in large-scale features such as the location and strength of the North Atlantic jet, cold tropospheric temperature and low tropospheric relative humidity, and underprediction of summer precipitation. Overall, the RCM was found to inherit most of the large-scale biases from the GCM, but the spatial distribution of temperature and precipitation was better simulated due to topographic effects. In addition, the RCM produced more spatially refined temperature and precipitation change scenarios. The RCM also simulated significant sub-GCM-scale changes in surface hydrological variables such as snow depth and runoff.

Following a similar approach, Leung and Ghan [36] used a regional climate model driven by GCM $1 \times CO_2$ and $2 \times CO_2$ simulations to produce climate change scenarios for the western USA. However, much more spatially resolved simulations of temperature and precipitation were produced by using a subgrid parameterization of orographic precipitation and vegetation [38, 39]. This method divides a model grid cell into subgrid surface elevation and vegetation classes based on high resolution (1 km) DEM and vegetation data. The influence of topography and vegetation on atmospheric and land surface processes is represented through a parameterization that accounts for orographic effects on clouds, which then affect precipitation and surface hydrology. During postprocessing, surface temperature and precipitation, among other variables, simulated for each subgrid class are mapped geographically to 1 km resolution based on the DEM and vegetation data. This approach greatly improves the simulation of surface temperature, precipitation, and snowpack compared to the GCMs. Their results show that snowpack will potentially be reduced by up to 50% under a $2 \times CO_2$ scenario. They also found a strong elevation dependence of climate change signals in temperature, precipitation (amount and phase), snow cover, and runoff (see also [1, 20] for a discussion of elevation dependence of climate change signals in mountainous regions).

In the 2000s, as more GCM transient simulations became available and the regional modeling community has grown, more studies have been published that investigated the potential effects of climate change in different climatic regimes or geographical locations. Figure 9.4 shows an example of cold season heavy precipitation (95th percentile) simulated by a GCM and an RCM driven by the GCM boundary conditions, using the same models described by Leung et al. [32], except for a change in the regional domain to cover the conterminous USA. Comparison of the simulated and observed heavy precipitation shows that the RCM reproduced the observed spatial distribution of heavy precipitation better than the GCM primarily because of the increased spatial resolution. As regional climate information is useful for assessing climate impacts and addressing climate adaptation, many studies that involve the use of regional climate models for producing regional climate change scenarios included scientists and stakeholders of the specific regions being studied to focus on subjects both scientifically interesting and societally relevant. The regional human resources and knowledge base that have been tapped have proven beneficial and contributed to more diverse analyses and applications of the climate change results. More examples of these efforts have been summarized in Christensen et al. [5].

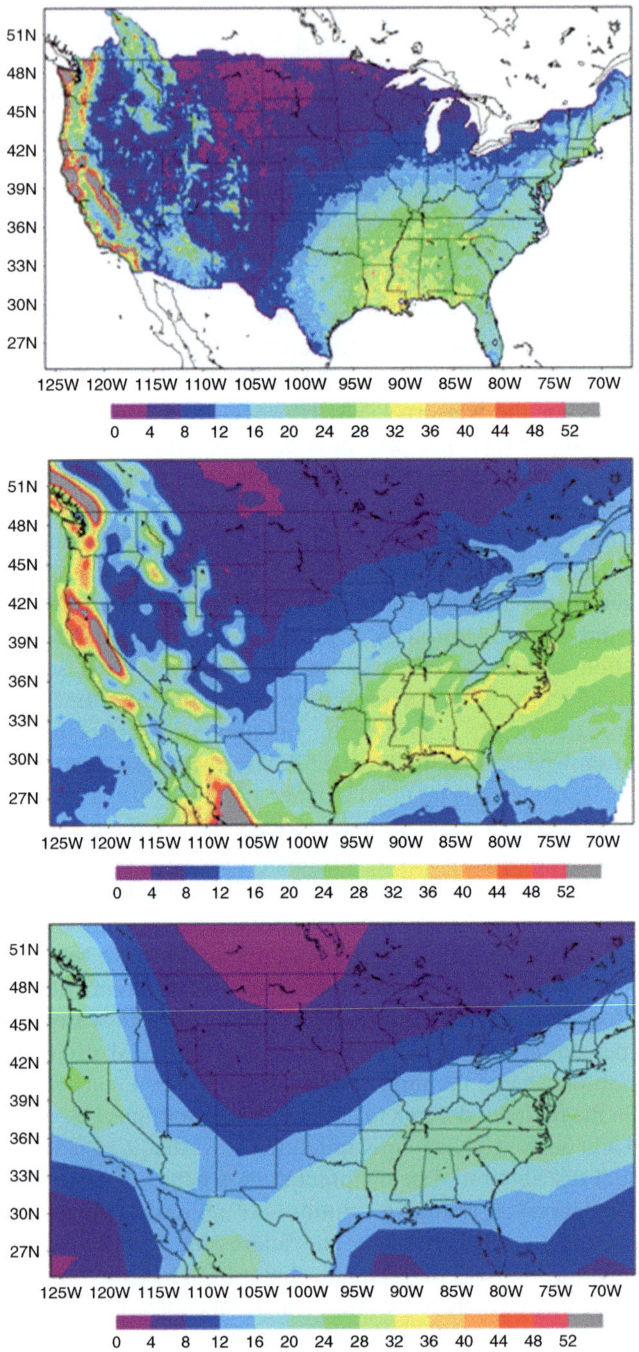

Fig. 9.4 An example of cold season heavy (95th percentile) precipitation simulated by a GCM (*top*) and an RCM (*middle*) and comparison with observation (*bottom*) over the USA. The prominent effects of topography are well captured by the RCM at 36 km grid resolution compared to the GCM, which was applied at roughly 250 km resolution

Besides individual efforts of using a particular RCM to downscale climate predictions or projections from a particular GCM, larger efforts have also been coordinated to develop ensembles of dynamically downscaled simulations. Common to these coordinated efforts is the objective to fill the gap in providing regional climate change scenarios for different geographic regions and to improve the characterization of uncertainty of the scenarios. To this end, an ensemble modeling approach is used in which multiple RCMs are nested within multiple GCMs to generate a matrix of regional climate change scenarios to facilitate the interpretation and characterization of uncertainty of regional climate change. These efforts also enable large, multi-model datasets to be archived following common protocols similar to the AMIP and CMIP efforts adopted by the GCM community over the last two decades.

In Europe, two large coordinated projects, PRUDENCE [5] and ENSEMBLES [23], intercompared regional climate models driven by global reanalysis as well as global climate simulations for the current and future climate. PRUDENCE designed, executed, analyzed, and synthesized regional climate scenario development for Europe. In brief, four GCMs and ten RCMs were involved to produce regional climate scenarios at 50 km resolution, but a few scenarios at 20 km resolution were also produced. Two time slices were simulated by each RCM, corresponding to 30 years of control and future (2071–2100) conditions. Two emission scenarios, A2 and B2, were considered, and some GCMs and RCMs provided multiple ensemble members (using different initial conditions) for assessing internal variability. Although only 28 combinations out of the full matrix of GCM, RCM, and scenario combinations were performed, PRUDENCE provided sufficient model outputs to evaluate the variance due to the four sources of uncertainty: GCM, RCM, scenario, and sampling. Figure 9.5 summarizes the surface temperature and precipitation changes simulated by the regional models for Europe.

One of the main conclusions of PRUDENCE is that the largest source of uncertainty resides in the GCM boundary conditions applied to the RCM [11]. The choice of RCM becomes more important, however, for certain subregions or seasons (summer in particular). Furthermore, many local features and aspects of extremes can vary substantially between RCMs [30] to alter the climate change signal from that simulated by the driving GCM. For example, RCM simulations performed at higher resolution (12 km vs 25 km) reduced the magnitude of future summer drying over southern Europe [4]. This effect could be attributed to the diminished control of the LBCs on RCM simulations during summer, and the general tendency of RCM to produce more precipitation at higher resolutions (e.g., [33, 49]).

Building on the foundation of PRUDENCE, ENSEMBLES is the largest and most comprehensive RCM comparison project conducted to date. Focusing again on Europe, ENSEMBLES utilized 15 GCMs and 11 RCMs to create a large GCM-RCM matrix for a single emission scenario (A1B). Simulations were also performed with reanalysis boundary conditions at 25 km and 50 km horizontal resolution. Interestingly, higher resolution (25 km vs 50 km) did not improve the simulation of large-scale weather types [51] or seasonal precipitation [49] by

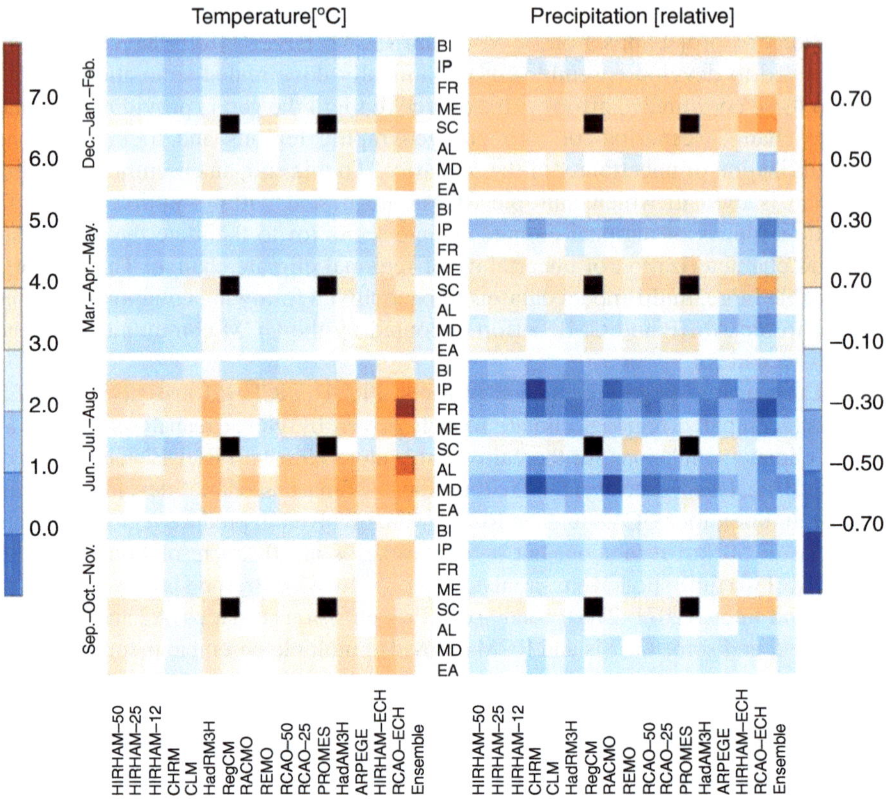

Fig. 9.5 An overview of seasonal changes in surface temperature (degree C) (*left*) and precipitation (relative change) (*right*) simulated by the PRUDENCE regional climate models for different analysis areas (row) and models (column). The analysis areas are: *BI* British Isles, *IP* Iberian Peninsula, *FR* France, *ME* Mid-Europe, *SC* Scandinavia, *AL* Alps, *MD* Mediterranean, *EA* Eastern Europe. Results from 17 regional simulations (some are produced by the same model at different resolutions) are shown, but some simulations did not cover certain geographical areas (shown by the *black squares*) (Source Christensen and Christensen [4] © 2007 *Climatic Change*)

many RCMs, suggesting that physics and/or downscaling approach (i.e., dynamical framework) may be more important than resolution. The most novel aspect of the ENSEMBLES project is the construction and use of a set of metrics to weight models according to their performance to construct an ensemble mean [6]. However, application of the weights to the GCM-forced RCM simulations for the twentieth century did not substantially improve the performance of the multi-model temperature or precipitation mean over the unweighted multi-model mean when averaged over Europe. This suggests that more research is needed to further explore the productive use of ensembles of climate change scenarios to reduce uncertainty.

The North American Regional Climate Change Assessment Program (NARCCAP) [42] is another coordinated project similar to PRUDENCE and

ENSEMBLES, but with a geographic focus on North America. The NARCCAP GCM-RCM matrix includes mapping 4 GCMs with 6 RCMs statistically for a more balanced design for uncertainty analysis. In addition, two high-resolution time-slice global simulations are included for comparison with the RCM simulations over North America. More recently, CORDEX (a COordinated Regional Downscaling EXperiment) has been developed to coordinate regional climate change scenario development for all continents around the world, and to foster international collaborations and promote interactions and communications between the various communities involved in scenario development and applications. The CORDEX design is similar to the multi-GCM/RCM matrix used in PRUDENCE, ENSEMBLES, and NARCCAP, but an additional level of uncertainty being assessed is model dependence on climate regimes and/or geographic locations. Therefore, an important CORDEX effort is to develop and compare climate simulations across different continents. Additionally, CORDEX encourages the development of Regional Analysis and Evaluation Teams to develop a set of regionally specific metrics for model evaluation, collect observational data, design experiments to investigate the added value of RCMs, and evaluate the ensemble of simulations from CORDEX.

Besides climate change, dynamical downscaling has also been applied to the area of seasonal climate predictions, but to a much lesser extent compared to downscaling of climate change simulations. The Multi-Regional Ensemble Downscaling (MRED) is a coordinated project in which multiple RCMs were used to downscale global seasonal climate forecasts for the USA (http://ecpc.ucsd.edu/projects/MRED/). Dynamical downscaling has also been used to develop regional analysis for studying climate variability and trends. Unlike regional analysis such as the North American Regional Reanalysis that assimilates observation data in regional models driven by global analysis, the dynamical downscaling approach assimilates only global analysis, but no additional observational data within the regional model domains to generate regional climate information. As examples, Sotillo et al. [52] used a regional model to downscale global reanalysis to generate a high-resolution 44-year atmospheric analysis for the Mediterranean Basin. Kanamitsu and Kanamaru [29] used a regional climate model at 10 km resolution driven by a global reanalysis in the California Reanalysis Downscaling at 10 km (CaRD10) project to produce 57 years of regional analysis for California.

Although numerous studies that evaluated different aspects of regional climate simulations using observations have demonstrated some skill in simulating regional temperature and precipitation, the skill depends very much on the large-scale data used to drive the model, the model physics, and how the models were configured. More recently, besides asking what ability RCMs have in reproducing observed climate features, the question of whether dynamical downscaling adds values to global climate simulations has become an important topic. Essentially, this begs the question of whether the additional step of running regional climate models as a means to dynamically downscale global climate simulations indeed provides additional (useful) information not available from the global climate simulations.

One way to address this question is to define and apply various metrics to quantitatively measure the added skill or added information provided by the regional models. For example, spatial filters can be used to partition the model-simulated variability to a larger scale that is resolved by the global simulation and a smaller scale that is beyond the limit resolved by the global simulation. The amount of finer scale variability generated by the regional models is considered value added since it provides climate information beyond what the global simulations provide (e.g., [12]). Similarly, spectral decomposition can be applied to simulated quantities such as different components of the surface water budgets and forecast skill to determine the added value of regional modeling.

Another aspect of evaluating the value added by RCMs is to compare dynamical downscaling with statistical downscaling, which is computationally a much cheaper method to produce regional climate information. To date, comparison of dynamical and statistical downscaling methods is limited to a few studies. Wood et al. [58] represent an early effort to apply a simple statistical downscaling method called Bias Correction Spatial Disaggregation (BCSD) to not only global simulations, but also regional climate simulations. The latter is a hybrid approach that combines dynamical and statistical downscaling. Comparing statistically downscaled simulations driven by global and regional simulations with the global and regional simulations, this study showed that hydrologic response to climate change can be enhanced using the hybrid approach compared to applying statistical downscaling directly to the GCM outputs because the RCM simulated larger warming in mountainous areas as a result of snow-albedo feedbacks, which are not captured by GCM or statistical downscaling.

Future Directions

In summary, both idealized experiments and real case applications have demonstrated that nested regional climate modeling is a viable approach for regional climate simulations. However, applications of regional climate models must be exercised with care because many factors can introduce uncertainty in the simulated results. These factors, which include domain size and location, physics parameterization, model resolution, lateral boundary condition and treatment, and use of interior nudging, must be carefully assessed before proceeding to long-term climate simulations. More research is also needed to better understand the sensitivity of regional climate simulations to those factors and develop ways to characterize and reduce uncertainty introduced by the nested modeling framework. As computing resources allow global models to be applied at higher and higher spatial resolution, and alternative approaches such as global variable resolution models become feasible, more research is needed to evaluate and compare different approaches to modeling regional climate to establish their validity and usefulness in addressing different aspects of climate research and applications.

Acknowledgments I would like to thank my colleagues at the Pacific Northwest National Laboratory and my collaborators over the years, whose research has inspired me. I also thank them for sharing their ideas, knowledge, and results with me.

Bibliography

1. Beniston M, Diaz HF, Bradley RS (1997) Climatic change at high elevation sites: an overview. Clim Chang 36:233–251
2. Castro CL, Pielke RA Sr, Leoncini G (2005) Dynamical downscaling: an assessment of value added using a regional climate model. J Geophys Res 110. doi:10.1029/2004JD004721, D05108
3. Caya D, Biner S (2004) Internal variability of RCM simulations over an annual cycle. Clim Dyn 22(1):33–46
4. Christensen JH, Christensen OB (2007) A summary of the PRUDENCE model projections of changes in European climate by the end of this century. Clim Chang 81:7–30. doi:10.1007/s10584-006-9210-7
5. Christensen JH, et al (2007) Regional climate projections. In: Solomon S, Qin D, Manning M, Chen Z, Marquis M, Avery KB, Tignor M, Miller HL (eds) Climate change 2007: the physical science basis, contribution of Working Group I to the Fourth Assessment Report of the intergovernmental panel on climate change. Cambridge University Press, Cambridge, UK/New York
6. Christensen JH, Rummukainen M, Lenderink G (2009) Formulation of very high-resolution regional climate model ensembles for Europe, chapter 5. In: van der Linden P, Mitchell JFB (eds) ENSEMBLES: climate change and its impacts: summary of research and results from the ENSEMBLES project. Met Office Hadley Centre, Exeter, 160pp
7. Daly C, Neilson RP, Phillips DL (1994) A statistical-topographic model for mapping climatological precipitation over mountanious terrain. J Appl Meteor 33:140–158
8. Davies HC (1976) A lateral boundary formulation for multi-level prediction models. Quart J Roy Meteor Soc 102:405–418
9. de Elía R, Laprise R, Denis B (2002) Forecasting skill limits of nested, limited-area models: a perfect-model approach. Mon Weather Rev 130:2006–2023
10. Denis B, Laprise R, Caya D, Côté J (2002) Downscaling ability of one-way-nested regional climate models: the Big-brother experiment. Clim Dyn 18:627–646
11. Déqué M, Rowell DP, Lüthi D, Giorgi F, Christensen JH, Rockel B, Jacob D, Kjellström E, Castro M, van den Hurk B (2007) An intercomparison of regional climate simulations for Europe: assessing uncertainties in model projections. Clim Chang 81:53–70
12. Di Luca A, de Elía R, Laprise R (2011) Assessment of the potential added value in multi-RCM simulated precipitation. Clim Dyn. doi:10.1007/s00382-011-1068-3
13. Dickinson RE, Errico RM, Giorgi F, Bates GT (1989) A regional climate model for the western United States. Clim Chang 15:383–422
14. Fu C, Wang S, Xiong Z, Gutowski WJ, Lee D-K, McGregor JL, Sato Y, Kato Hi, Kim J-W, Suh M-S (2005) Regional climate model intercomparison project for Asia. Bull Amer Meteorol Soc 86. doi:10.1175/BAMS-86-2-257
15. Gates WL (1992) AMIP: the atmospheric model intercomparison project. Bull Amer Meteorol Soc 73:1962–1970
16. Giorgi F, Mearns LO (1999) Introduction to special section – regional climate modeling revisited. J Geophys Res 104(D6):6335–6352
17. Giorgi F, Bates GT (1989) On the climatological skill of a regional model over complex terrain. Mon Weather Rev 117:2325–2347
18. Giorgi F, Marinucci MR, Visconti G (1990) Use of a limited area model nested in a general circulation model for region climate simulation over Europe. J Geohys Res 95:18,413–18,431

19. Giorgi F, Marinucci MR, DeCanio G, Bates GT (1993) Development of a second generation regional climate model (REGCM2): cumulus cloud and assimilation of lateral boundary conditions. Mon Weather Rev 121:2814–2832
20. Giorgi F, Hurrell JW, Marinucci MR, Beniston M (1997) Elevation signal in surface climate change: a model study. J Clim 10:288–296
21. Giorgi F, Bi X (2000) A study of internal variability of a regional climate model. J Geophys Res 105:29503–29521
22. Haylock MR, Hofstra N, Klein Tank AMG, Klok EJ, Jones PD, New M (2008) A European daily high-resolution gridded dataset of surface temperature and precipitation. J Geophys Res (Atmos) 113(D20119). doi:10.1029/2008JD10201
23. Hewitt CD, Griggs DJ (2004) Ensembles-based predictions of climate changes and their impacts. EOS 85:566
24. Huffman GJ, Adler RF, Morrissey M, Bolvin DT, Curtis S, Joyce R, McGavock B, Susskind J (2001) Global precipitation at one-degree daily resolution from multisatellite observations. J Hydrometeor 2:36–50
25. Hughes M, Hall A (2010) Local and synoptic mechanisms causing Southern California's Santa Ana winds. Clim Dyn 34. doi:10.1007/s00382-009-0650-4
26. Inatsu M, Kimoto M (2009) A scale interaction study on East Asian cyclogenesis using a general circulation model coupled with an interactively nested regional model. Mon Weather Rev 137. doi:10.1175/2009MWR2825.1
27. Ji YM, Vernekar AD (1997) Simulation of the Asian summer monsoons of 1987 and 1988 with a regional model nested in a global GCM. J Climate 10:1965–1979
28. Kanamaru H, Kanamitsu M (2007) Scale-selective bias correction in a downscaling of global analysis using a regional model. Mon Weather Rev 135:334–350
29. Kanamitsu M, Kanamaru H (2007) 57-Year California reanalysis downscaling at 10 km (CaRD10) part 1. System detail and validation with observations. J Climate 20:5527–5552
30. Kjellström E, Bärring L, Jacob D, Jones R, Lenderink G, Schär C (2007) Modelling daily temperature extremes: recent climate and future changes over Europe. Clim Chang 81:249–265
31. Laprise R, de Elía R, Caya D, Biner S, Lucas-Picher Ph, Diaconescu EP, Leduc M, Alexandru A and Separovic L (2008) Challenging some tenets of regional climate modelling. Meteor. Atmos Phys 100, Special Issue on Regional Climate Studies, 3–22. doi:10.1007/s00703-008-0292-9
32. Leung LR, Qian Y, Bian X, Washington WM, Han J, Roads JO (2004) Mid-century ensemble regional climate change scenarios for the western United States. Clim Chang 62(1–3):75–113
33. Leung LR, Qian Y (2003) The sensitivity of precipitation and snowpack simulations to model resolution via nesting in regions of complex terrain. J Hydrometeorol 4(6):1025–1043
34. Leung LR, Qian Y, Bian X, Hunt A (2003) Hydroclimate of the western United States based on observations and regional climate simulation of 1981–2000. Part II: mesoscale ENSO anomalies. J Clim 16(12):1912–1928
35. Leung LR, Mearns LO, Giorgi F, Wilby R (2003) Workshop on regional climate research: needs and opportunities. Bull Amer Meteorol Soc 84(1):89–95
36. Leung LR, Ghan SJ (1999) Pacific northwest climate sensitivity simulated by a regional climate model driven by a GCM. Part I: control simulations. J Clim 12(7):2010–2030
37. Leung LR, Ghan SJ (1999b) Pacific Northwest climate sensitivity simulated by a regional climate model driven by a GCM. Part II: 2xCO2 simulations. J Clim 12(7):2031–2053
38. Leung LR, Ghan SJ (1998) Parameterizing subgrid orographic precipitation and surface cover in climate models. Mon Weather Rev 126(12):3271–3291
39. Leung LR, Ghan SJ (1995) A subgrid parameterization of orographic precipitation. Theor Appl Climatol 52:95–118
40. Leung LR, Kuo Y-H, Tribbia J (2006) Research needs and directions of regional climate modeling using WRF and CCSM. Bull Am Meteorol Soc 87(12):1747–1751

41. Lorenz P, Jacob D (2005) Influence of regional scale information on the global circulation: a two-way nesting climate simulation. Geophys Res Lett 32:L18706. doi:10.1029/2005GL023351
42. Mearns LO, Gutowski W, Jones R, Leung R, McGinnis S, Nunes A, Qian Y (2009) A regional climate change program for North America. Eos Trans AGU 90:311–312
43. Mesinger F, Brill K, Chuang H, DiMego G, Rogers E (2002) Limited area predictability: can upscaling also take place? Research activities in atmospheric and oceanic modelling. Report No. 32, WMO/TD – No. 1105, 5.30–5.31
44. Miguez-Macho G, Stenchikov GL, Robock A (2004) Spectral nudging to eliminate the effects of domain position and geometry in regional climate model simulations. J Geophys Res 109(D13):D13104. doi:10.1029/2003JD004495
45. Pan Z, Takle E, Gutowski W, Turner R (1999) Long simulation of regional climate as a sequence of short segments. Mon Weather Rev 127:308–327
46. Pan Z, Arritt RW, Takle ES, Gutowski WJ Jr, Anderson CJ, Segal M (2004) Altered hydrologic feedback in a warming climate introduces a "warming hole". Geophys Res Lett 31:L17109. doi:10.1029/2004GL020528
47. Qian J-H, Seth A, Zebiak S (2003) Reinitialized versus continuous simulations for regional climate downscaling. Mon Weather Rev 131:2857–2874
48. Qian Y, Gustafson WI Jr, Leung LR, Ghan SJ (2009) Effects of soot-induced snow albedo change on snowpack and hydrological cycle in Western U.S. based on WRF chemistry and regional climate simulations. J Geophys Res 114:D03108. doi:10.1029/2008JD011039
49. Rauscher SA, Coppola E, Piani C, Giorgi F (2009) Resolution effects on regional climate model simulations of seasonal precipitation over Europe. Clim Dyn. doi:10.1007/s00382-009-0607-7. 28
50. Riddle EE, Cook KH (2008) Abrupt rainfall transitions over the Greater Horn of Africa: Observations and regional model simulations. J Geophys Res 113:D15109
51. Sanchez-Gomez E, Somot S, Déqué M (2008) Ability of an ensemble of regional climate models to reproduce weather regimes over Europe-Atlantic during the period 1961–2000. Clim Dyn 33(5):723–736. doi:10.1007/s00382-008-0502-7
52. Sotillo M, Ratsimandresy A, Carretero J, Bentamy A, Valero F, Gonzalez-Rouco F (2005) A high-resolution 44-year atmospheric hind-cast for the Mediterranean basin: contribution to the regional improvement of global reanalysis. Clim Dyn 25:219–236
53. Takle ES, Gutowski WJ Jr, Arritt RW, Pan Z, Anderson CJ, Silva R, Caya D, Chen S-C, Christensen JH, Hong S-Y, Juang H-MH, Katzfey JJ, Lapenta WM, Laprise R, Lopez P, McGregor J, Roads JO (1999) Project to intercompare regional climate simulations (PIRCS): description and initial results. J Geophys Res 104:19,443–19,462
54. von Storch H, Langenberg H, Feser F (2000) A spectral nudging technique for dynamical downscaling purposes. Mon Weather Rev 128:3664–3673
55. Wang Y, Leung LR, McGregor JL, Lee D-K, Wang W-C, Ding Y, Kimura F (2004) Regional climate modeling: progress, challenges, and prospects. J Meteor Soc Jpn 82(6):1599–1628
56. Warner TT, Peterson RA, Treadon RE (1997) A tutorial on lateral conditions as a basic and potentially serious limitation to regional numerical weather prediction. Bull Amer Meteor Soc 78(11):2599–2617
57. Winterfeldt J, Weisse R (2009) Assessment of value added for surface marine wind speed obtained from two regional climate models. Mon Weather Rev 137:2955–2965
58. Wood AW, Leung LR, Sridhar V, Lettenmaier DP (2004) Hydrologic implications of dynamical and statistical approaches to downscaling climate model outputs. Clim Chang 62(1–3):189–216
59. Xie P, Yatagai A, Chen M, Hayasaka T, Fukushima Y, Liu C Yang S (2007) A gauge-based analysis of daily precipitation over East Asia. J Hydrometeor 8:607–626
60. Yatagai A, Arakawa O, Kamiguchi K, Kawamoto H, Nodzu MI, Hamada A (2009) A 44-year daily gridded precipitation dataset for Asia based on a dense network of rain gauges. SOLA 5:137–140. doi:10.2151/sola.2009-035

Chapter 10
Climate Change Projections: Characterizing Uncertainty Using Climate Models

Ben Sanderson and Reto Knutti

Glossary

Bayes' Theorem	A law in probability theory relating the probability of a hypothesis given observed evidence to the often easier to characterize probability of that evidence given the hypothesis. The theorem states that the conditional "posterior" probability of an event A given an event B is equal to the "prior" probability of A multiplied by the likelihood of B given A is true, normalized by the prior probability of B.
Climate sensitivity	The equilibrium global mean near surface air temperature response in Kelvin to a sustained doubling of the atmospheric carbon dioxide concentration.
CMIP-3	The Coupled Model Intercomparison Project Phase 3, a set of coordinated model experiments using General Circulation Models from the world's major modeling centers.
Detection and attribution	A process whereby spatial "fingerprints" associated with individual climate forcing factors (such as aerosol or greenhouse gas concentrations) are identified and used to quantify whether an observed change exceeds the

This chapter was originally published as part of the Encyclopedia of Sustainability Science and Technology edited by Robert A. Meyers. DOI:10.1007/978-1-4419-0851-3

B. Sanderson (✉)
National Center for Atmospheric Research, NCAR, 1850 Table Mesa Dr,
80305 Boulder, CO, USA
e-mail: bsander@ucar.edu

R. Knutti
Institute for Atmospheric and Climate Science, ETH, Universitätstrasse 16,
Zürich, Switzerland
e-mail: reto.knutti@env.ethz.ch

P.J. Rasch (ed.), *Climate Change Modeling Methodology: Selected Entries from the Encyclopedia of Sustainability Science and Technology*,
DOI 10.1007/978-1-4614-5767-1_10, © Springer Science+Business Media New York 2012

	range of natural internal climate variability (detection) and to attribute it to different causes, that is, different forcings (attribution).
Empirical model	A model based on fitting empirical data, and thus makes no attempt to justify its representations of the system with any physical basis.
General circulation model (GCM)	A three-dimensional mathematical model for the atmosphere and possibly the ocean, land, and sea ice.
Initial condition ensemble	A number of simulations using a single climate model, each with a small, unique perturbation to the initial state.
Last glacial maximum (LGM)	A period in the most recent ice age lasting several 1,000 years, peaking approximately 20,000 years ago at the maximum extent of the ice sheets.
Lead time	The period in between the time at which the forecast is made and the time to be forecasted.
Multi-model ensemble (MME)	A collection of structurally different models from a range of institutions used to perform a coordinated set of experiments.
Parameter space	The multidimensional domain created by considering the possible values of a number of parameters within a model.
Perturbed physics ensemble	A set of climate simulations generated by taking a single physical model and altering uncertain parameters within a range of plausibility.
Prior probability (marginal probability)	The probability of an event before any additional data is considered in a Bayesian sense.
Posterior probability	The probability of an event after considering additional relevant evidence in a Bayesian sense.
Systematic error	The difference between a model simulation and observations or a poorly represented process which is not reducible by parameter tuning.

Definition of the Subject and Its Importance

The atmosphere, ocean, land surfaces, and ice sheets of the Earth are highly complex and coupled systems, with physical laws which describe behavior from the microscopic to the planetary scale. General Circulation Models are computational analogs for these physical systems, which can be used to study how these systems might behave when boundary conditions are changed (e.g., by increasing the concentration of atmospheric greenhouse gases).

Inherent in the design of such models are a myriad of choices when deciding which components of the system are to be modeled and how to represent processes which cannot be currently modeled explicitly. In order to have any confidence in the ability of our models to have value for simulating aspects of future climate change, it is necessary for those models to reproduce observable properties of the physical system. However, model errors in the simulation of the past or present are likely to be smaller than errors in future projections because model developers can use observations and historical records in the development of their code. Additionally, some processes may not be observable or testable yet, because they might only take place in a warmer (or otherwise changed) world.

One way to characterize at least some of the uncertainty in future projections is to produce an ensemble of climate simulations, each making different but reasonable assumptions about their representation of physical processes.

In recent years, a number of groups in the international climate science community have produced General Circulation Models of the earth system, each making different choices about model complexity, resolution, and parameterization of processes which occur at scales finer than those resolved. By conducting coordinated experiments with each of these models, it has become possible to examine some of the effect that such choices have on uncertainty in future climate simulations. However, the sheer volume of data and range of models available from such an ensemble presents a new challenge for the science to address: How can a spread of non-independent "best guesses" be combined to produce meaningful statements of uncertainty which are relevant to climate-related policy decisions?

Introduction

In 1979, Jule Charney chaired a committee on anthropogenic global warming, producing a report [1] providing a brief overview of the state of the science of climate change. At the time, two General Circulation Models were available for consideration, one led by Syukuro Manabe and the other by James Hansen. The report produced an estimate for the climate sensitivity (the equilibrium global mean temperature change to a doubling of the atmospheric carbon dioxide concentrations) based on the mean result of these two models. In comparing the predicted future climate of these two models, the report stated:

> We conclude that the predictions of CO_2-induced climate changes made with the various models examined are basically consistent and mutually supporting. The differences in model results are relatively small and may be accounted for by differences in model characteristics and simplifying assumptions.

This, in many ways, represents the first effort to combine multiple results from an ensemble of climate model simulations, and the statements made using those models are still relevant to ensemble modeling. A better understanding in the uncertainties in the simulations and increased confidence can be claimed if an ensemble of somewhat independent models produces common features in its simulations, and if the origins of the differences between simulations may be traced back to physical characteristics.

When presented with a range of simulations of future climate, one must make judgments on many levels on how that ensemble should be interpreted: How should model agreement, or lack of it, translate into a degree of confidence in the simulations? Should all models be treated equally, and if not then how should one distinguish between them? If some processes are absent from some or all of the simulations, how can the projections be updated to account for these "known and unknown unknowns"? Should each ensemble member be interpreted to be an estimate of the "truth" with some unknown error, or should the "true" earth system be considered as a potential member of the ensemble? Although some of these questions verge on the philosophical, the judgments made in answering them can have large effects on the results themselves that are obtained and the degree of uncertainty in those results.

In recent years, there have been systematic efforts to explore and characterize uncertainty using large ensembles of increasingly complex models of the earth system. These model simulations have been coordinated and analyzed to help in characterizing climate change in a series of assessment reports for the Intergovernmental Panel on Climate Change (IPCC). In 1990, 1995, 2001, and 2007, a selection of GCMs were assembled from various major modeling groups around the world to compare simulations of past, future, and other idealized scenarios of climate change. Through the successive decades, model complexity and scope have increased; the early GCMs of Manabe [2] and Hansen [3] modeled atmospheric dynamics and radiative transfer, with a simplistic representation of the hydrological cycle. By the time of the First Assessment Report of the IPCC [4] (FAR) in 1991, models included clouds, a land surface model, and prescribed ice cover. For the Second Assessment Report [5] (SAR), some models also included a representation of the ocean and interactive sea ice. In the Third Assessment Report [6] (TAR), some models considered the effects of volcanic eruptions and aerosol emissions, with a fully dynamical representation of the oceans. By the time of most recent Fourth Assessment Report [7] (AR-4), some models were beginning to include an explicit representation of the carbon cycle in the earth system. Today's models continue to model additional components of the earth system, such as interactive vegetation, dynamically resolved ice sheets, a coupled carbon-nitrogen cycle and full atmospheric chemistry. In parallel with all of these improvements, the last few decades have also seen a continued increase in model resolution. Where the models used in the FAR split the earth into cells as large as 500 km on a side, models for the AR4 can resolve at a scale of a few tens of kilometers.

This entry focuses on the advantages and additional complexities which one must consider when studying a range of different model simulations. Rather than giving a comprehensive description of the results of the models assessed in the successive generations of the IPCC, this entry will discuss the added technical and conceptual challenges encountered when considering the results of a range of non-independent models and how a range of simulations may be combined into best estimates and uncertainties for future climate evolution.

Projection Uncertainty and the Need for Ensembles

Empirical and Physical Models

In 150 AD, Ptolemy devised a model of the motion of planets in the solar system by describing a system of concentric, geocentric circles (or "*deferents*") on which were mounted smaller circles ("*epicycles*") on which the planets themselves were mounted. This system thus had a large number of degrees of freedom (the diameters and speeds of rotation of each of the deferents and epicycles), which could be finely tuned to reproduce the motions of the bodies in the night sky. Such was the predictive power of this approach, that variations of this simple model were accepted until Copernicus' heliocentric model was published in 1543. Although Copernicus' model fits the established view of the universe more closely, both of these models were *empirical* in that they were not based on any physical principles at that time. However, even without a physical basis, Galileo was able to validate the Copernican model by studying the phases of the planet Venus – which was only consistent with the heliocentric formulation. It was not until Newton's law of universal gravitation that the model could be given a physical underpinning.

Any model of a physical system is an approximate representation of the truth. It should be able to reproduce some behavior of that system, and it might do this empirically like Ptolemy's model or by explicitly simulating physical processes within the true system like an orbital system based upon Newtonian gravitation. A model, whether empirical or physical, cannot ever be validated in the strict sense of showing it to be a wholly correct representation of the true system; it can only be evaluated by reproducing some output not used in the tuning of the model itself. This was true of Galileo's observation of the phases of Venus – information not used in the tuning of the Ptolemic model. However, any empirical model becomes very sensitive to changing boundary conditions. For example, if the mass of the Sun were to instantly double, the Copernican model of the solar system would be a very poor approximation of planetary motion, whereas a model based upon Newtonian mechanics would capture enough of the necessary physics to remain useful.

These fundamental principles are relevant to methodologies for simulating the climate today. If the simplest, zero dimensional empirical model of the climate is taken to be:

$$C\frac{dT'}{dt} = F' - \lambda T'$$

where T' is the global mean temperature difference from an equilibrium state, F' is the additional radiative forcing to the planet (i.e., the change in the top of atmosphere radiative balance caused by a forcing, e.g., increased CO_2), C is the effective heat capacity of the system, and λ is the global sensitivity parameter. This equation has two free parameters, C and λ which may be tuned such that the model can fit an observed past time series of F' and T', that of the twentieth century, for example.

The model can then be evaluated by predicting a previously unseen time period, such as the last glacial maximum. This evaluation, if successful, would give more confidence in the model but would not necessarily make it trustworthy for a prediction of the future – where the boundary conditions are outside those seen in both the training and validation period.

The added advantage of using a GCM to simulate future climate is that model simulations are in theory more trustworthy because they are based upon physical principles, which it is believed can reproduce observed climate by coupling underlying physical laws that are known to be true. However, this view is often overoptimistic; although some components of the modeled climate, such as the equations of motion in the atmosphere or the instantaneous radiative forcing due to a change in atmospheric carbon dioxide concentrations are well understood and consistently implemented in different GCMs, there are other processes such as convection which cannot explicitly be resolved with current computing resources. These processes and their effects on the large scale climate must be approximated with uncertain parameters that must be estimated by tuning the model to reproduce some observed features of the climate. What this means, in practice, is that a GCM is neither only an empirical nor an explicitly physical model; it is a hybrid of the two where model developers face many arbitrary choices in parameterizing processes which cannot be explicitly resolved. The necessity for the tuning process reintroduces some of the problems encountered with an empirical model, with the possibility of false confidence in model performance by over-tuning the model to reproduce past climate. The ambiguity in these parameterizations justifies the existence of multiple models for the same purpose [8]; each of these models is seen as a plausible approximation of the climate system given the imperfect understanding, the uncertainties in observations and the computational constraints.

Types of Uncertainty and the Need for Ensembles

Although weather and climate simulations share some properties (sometimes they are conducted with the same model), the limiting uncertainties are very different. Climate represents the distribution of all possible states in which one expects to find a system, whereas weather is the specific evolution of the system from a given initial state. A model-based weather forecast is a *prediction*, in that the initial state of the simulation is as close as possible to observations and the absolute errors grow rapidly afterwards. In weather prediction, these errors occur as specific weather systems evolve from the initial state. Because the atmosphere is a chaotic system, very small errors in the estimate of the initial state can result in very large differences in the distributions of weather systems a few days later. Initial condition uncertainty is evaluated by repeating simulations with a range of slightly different initial conditions to form "Initial Condition Ensembles." The spread of these ensembles initially grows rapidly but eventually saturates when the "memory" of the initial state is lost (this timescale is longer in the oceans, perhaps up to 10 years for North Atlantic ocean temperatures [9]).

On decadal to century timescales, the mean and spread of an initial condition ensemble represents a *projection* of the future climate state, although this spread is only a small fraction of the total error (sometimes known as "uncertainty of the first kind"). The second kind of uncertainty relates to the boundary conditions of the problem, some of which are naturally occurring such as the level of incoming solar radiation or volcanic activity, while others are dependent on anthropogenic factors such as the future emissions of greenhouse gases or aerosols. To address this uncertainty, one must perform a range of simulations using different plausible scenarios for changes in boundary conditions. The results of any simulation are therefore conditional on assumptions made about future human behavior. There is currently little real skill in forecasting future volcanic activity and changing solar activity so simplistic but plausible scenarios for these quantities are often used (such as repeating past values). However, in most future scenarios these represent a relatively small fraction of the total anthropogenic climate forcing.

Figure 10.1 shows the relative sources of error in a climate model projection as a function of the lead time [11]. For lead times of less than a decade, the uncertainty in the initial state combined with chaotic error growth and natural patterns of variability are the dominant sources of error but on the scale of several decades or more, it is the future emissions scenario which dominates the uncertainty. Predictions on all timescales, however, are subject to model uncertainties. These arise when a climate model contains parameterizations for unresolved or missing processes. Parameterizations take large-scale quantities resolved by the model, such as temperature, wind speed, and humidity, and relate them to unresolved processes, such as convective mass flux and cloud profiles. Although these parameterizations are usually constructed from physical underpinnings and evaluated with observed data, they introduce some unavoidable uncertainty when a range of parameter values might be physically plausible. GCMs are often subject to a tuning of parameter values to reproduce features of the observed climate, but with tens or hundreds of uncertain parameters this process is time consuming and can yield multiple solutions because of the computational cost, a systematic tuning of all parameters is unfeasible.

One method of quantifying the parameter uncertainty problem is to construct a "Perturbed Physics Ensemble" (PPE) using a single GCM. This process has been attempted using several major climate models [11–13] and involves taking a subset of important unknown parameters within the GCM and perturbing them within the bounds of physical plausibility. Such experiments might perturb, for example, a parameter which states the necessary humidity required for the formation of cloud. By varying this parameter, one can change dramatically how the model distributes clouds both in the present day and the future. These changes can affect the strength of global feedbacks which can change, for example, the amount of warming that the model predicts for a given rise in greenhouse gases. An example of a PPE is the "climateprediction.net" experiment [13], which used idle time in volunteer's computers to perform perturbed simulations of future climate. Incorporating this range into an uncertainty estimate for predictions of future

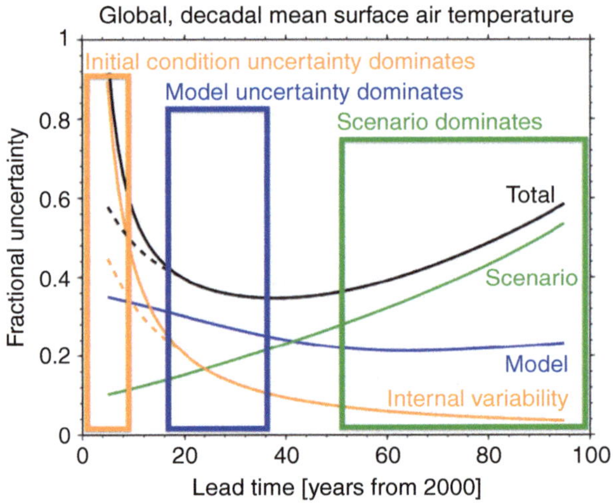

Fig. 10.1 A figure showing the fractional sources of uncertainty in a climate model projection as a function of time. The *orange* "internal variability" *line* shows the errors due to uncertain knowledge of the initial state of the system, while the *dotted line* shows the potential reduction in error if effort is made to assimilate ocean observations into the model at the start of the simulation. The *green line* shows the fractional error due to the unknown future emissions of climate altering gases, while the *blue line* shows the error due to the model imperfections. The *boxes* show where different types of uncertainty are dominant for a projection of future climate (Reproduced from [10])

climate requires a framework for joint consideration of each model's performance in simulating past and present climate as well as its future response.

The remaining model uncertainties are due to so-called "systematic" or structural errors arising from the model design, that is, the choices of which processes to model, the resolution of the model, the numerical schemes, and the specific form of the parameterization scheme. The structural differences between different GCMs provide a lower bound on the extent of the structural difference between any one GCM and the true climate system, but in reality the models in an ensemble such as those used for the IPCC reports share many common properties in terms of resolution, numerical methods, missing components, and parameterization schemes which might make all the models subject to similar errors. Nevertheless, considering a range of GCMs which make different modeling assumptions is an essential step when evaluating the robustness of any prediction of future climate change because it places a lower bound on the uncertainty arising from the choices made by model developers.

Multi-model and Perturbed Physics Ensembles

When making predictions of a future climate state, there is a wealth of evidence to suggest that considering a combined prediction using multiple, somewhat independent models yields more accurate results than any single model [14–16].

Additionally, the spread of simulations provides a measure of robustness in the prediction. The following section describes some reasons for the increased performance of multi-model and perturbed physics ensemble forecasts, together with some of the complexities arising from their analysis.

Range of Ensemble Responses

The spread of results from an ensemble of climate simulations is dependent upon the experimental design, or lack of it. A perturbed physics ensemble (PPE) has the luxury of allowing some control of the distribution of models in the parameter space of the model, though the structure of the underlying model places a fundamental limit on the range of observable behavior in the ensemble. For example, if a PPE is created by perturbing cloud parameters in a GCM which has no parameterization for cirrus clouds formed by gravity waves, then there is no way that such an ensemble can include uncertainty about that process. Designers of such experiments must also be aware that the decisions of how to sample the parameter space of a model will directly influence the distribution of future climate simulations [17]. In contrast, multi-model ensembles (MMEs) such as the Coupled Model Intercomparison Project (CMIP-3), explore "systematic" model differences, which sample models with different representations of the physical system, rather than simply varying parameters in a single model. These are "ensembles of opportunity" where multiple modeling groups run coordinated experiments but the ensemble itself is not sampled in any systematic fashion. Nor is the ensemble randomly sampled because each modeling group will tune their model to minimize model differences from observations, thus creating an ensemble of "best-guesses." This is quite different from the PPE case where the model is intentionally detuned to produce a wide range of behavior. Evidence for this can be seen by examining the spread of climate sensitivity in both a multi-model and a perturbed physics ensemble (Fig. 10.2). When considering a range of observational constraints on climate sensitivity, it is apparent that the multi-model values tend to cluster about the most likely value, whereas the perturbed physics ensemble contains models which span the full range of uncertainty in climate sensitivity. Although impossible to verify, it also is possible that there is also a component of social anchoring [18] which draws multi-model sensitivities toward the mean value as any group which finds their model to be an outlier may have to defend why this is the case, whereas a model with the consensus value of sensitivity is less likely to be questioned.

The Ensemble Mean State

In various fields, it has been shown that the combined performance of multiple models can exceed that of an individual ensemble member. Examples of this can be seen in models of crop yield [20], disease modeling [21], and even in the optimization routines

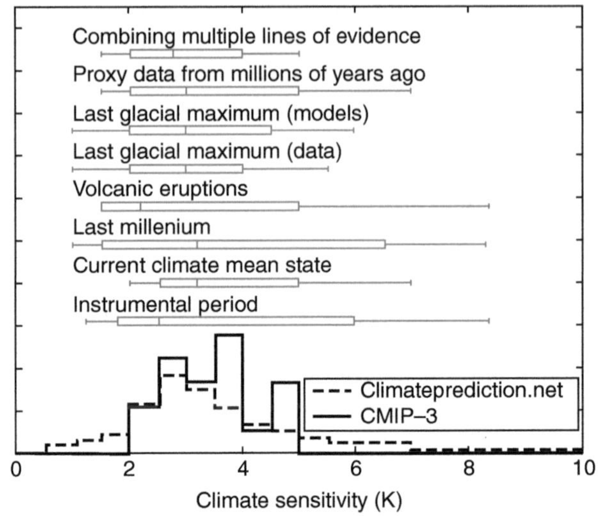

Fig. 10.2 Distribution functions of climate sensitivity (an estimate of the equilibrium response of a model to a doubling of CO_2) for models in the CMIP-3 ensemble (hinting at the range of responses from an MMF analysis), compared with a selection of models from the climateprediction.net project (hinting at the range of uncertainty from a PPE analysis). *Box and whisker plots* show estimates of the most likely values, together with 66th and 90th percentiles of likelihood for climate sensitivity taken from various lines of observational evidence (Adapted from [19]). *Histograms* represent the fraction of models in each 0.5 K bin of climate sensitivity for the atmosphere-only components of 19 models in the CMIP-3 archive and for a 2,000 member subset of the climateprediction.net ensemble [13]

used for movie recommendations based upon past viewing choices [22]. Similarly in seasonal climate predictions, it has been shown that the multi-model ensemble means yield better forecasts, in general, than using only initial condition ensembles from a single model [16]. A multi-model study incorporating a set of initial conditions for each model is often referred to as a "super-ensemble." The accuracy of the model mean often performs best in multivariate applications, that is, a single model may show increased skill in predicting one particular diagnostic, but when many variables are considered in the same metric the ensemble mean prediction tends to show greater skill than any individual model [23].

This effect can also be seen in GCM simulations of recent past climate. Figure 10.3 shows successive generations of the CMIP ensemble evaluated using a multivariate error metric comparing twentieth century observations to model simulations of that period for a variety of model diagnostics. The figure shows that model errors in simulating the current climate have decreased over time but also that for each generation of the ensemble, the multi-model mean results in a model-data discrepancy almost as good, or better than the best performing ensemble member. Various studies have found that both in detection and attribution studies [24] and in simulations of recent climate [25] that a multi-model mean provides a better multivariate simulation than any individual model.

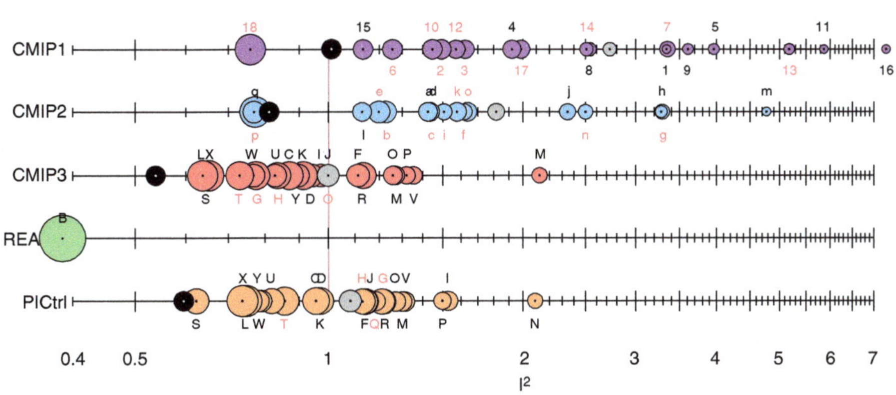

Fig. 10.3 A comparison of model errors reproduced from Reichler and Kim ([26], corrigendum) multivariate errors are evaluated for successive generations of the coupled model intercomparison project (CMIP). *Colored circles* represent individual models in the ensemble, whereas *black circles* show the performance of the multi-model mean. REA represents the NCEP reanalysis and the CMIP-3 PICNTRL is the performance of the preindustrial control simulations when evaluated against present-day observations

This approach is common in the reports of the IPCC, where an unweighted mean of future model simulations is used to show a "best-guess" simulation of future climate, while the degree of model spread is used to estimate some measure of the significance of the result. There are more sophisticated methodologies that one may use for model combination, involving Bayesian methodologies [27] or model weighting [28], but the correct implementation and interpretation of such studies is subject to some debate. It has been shown that the ranking of model performance within a multi-model ensemble such as CMIP-3 is often highly dependent on the choice of metric used to evaluate the model. A metric based on the model's ability to reproduce observed variability will produce a different ranking than a metric which evaluates the model simulation of the mean state [30], and the performance of different models on these two metrics are very weakly correlated (Fig. 10.4). In addition, violation of the model "democracy" (one model, one vote) in the IPCC process is potentially controversial, as choices of how to weight models could be interpreted as a political statement [31].

The question of why multi-model means perform better than individual models is a complex one. Certainly, the mean is not in itself a self-consistent representation of a physical system and is therefore not subject to many of the restrictions that apply when tuning one model to reproduce an observed climate. As an example, a single model may be tuned in different ways to reproduce two different observed values "A" and "B," but it might be impossible to tune the model to reproduce "A" and "B" simultaneously. However, if different models in the ensemble make different choices about the relative importance of "A" and "B," it is likely that the ensemble mean will be close to the observed values in the case of a large ensemble. Clearly, real GCMs have a large number of observable diagnostics to

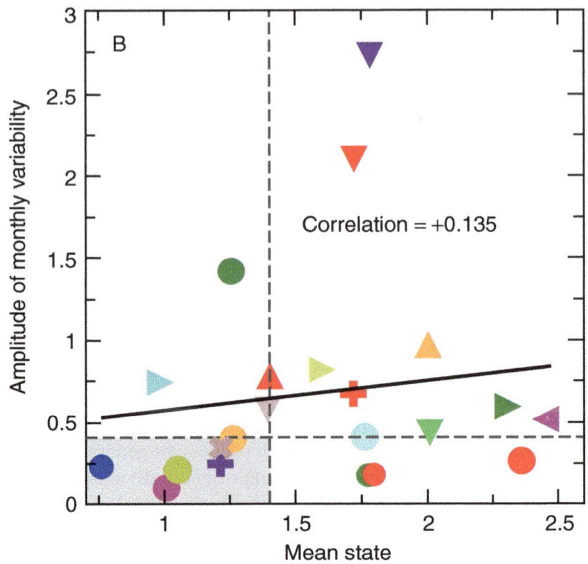

Fig. 10.4 The relationship between model skill in reproducing the mean climate state and skill in reproducing patterns of natural variability for models in the CMIP-3 ensemble. Each point represents a single model in the CMIP-3 archive, and errors are averaged over a large number of diagnostics. The *black line* shows the fitted least-squares regression (Reproduced from [29])

reproduce and a large number of tuning parameters, but it remains true that the multi-model mean is less restricted by model structure than any individual model in the ensemble. Another interpretation is that some of the model biases are random perturbations about the truth (i.e., each model reproduces the observations with some pattern of bias that is characteristic to that model and but different in each model), such that averaging many models reduces the magnitude of the biases. In the limit of completely random independent biases, the average would be perfect for an infinite number of models.

In some cases, the multi-model mean can indicate behavior unrepresentative of any of the models within the ensemble. Figure 10.5 shows the distribution of expected percentage precipitation change per unit global temperature increase in the current dry season for various models within the CMIP3 archive. Each individual model shows a wide distribution of change with some regions showing up to 30% decrease in precipitation for every degree rise in global mean temperature. If the models are averaged together in advance, however, the resulting multi-model mean has no region which displays this extreme decrease in precipitation in the dry season. The multi-model mean is thus not representative of the findings of the individual ensemble members in the respect that it fails to recover the extremes of the distribution of precipitation change. The reason for this discrepancy is, at least partially, a difference of the spatial representation of precipitation patterns in different ensemble members. Different models have different resolution, representations of orography, and parameterizations for precipitation. When combined this gives each models unique spatial modes of variability for precipitation. This allows each model to display extreme future drying in some specific regions, but critically those regions are not necessarily identical in all models in the

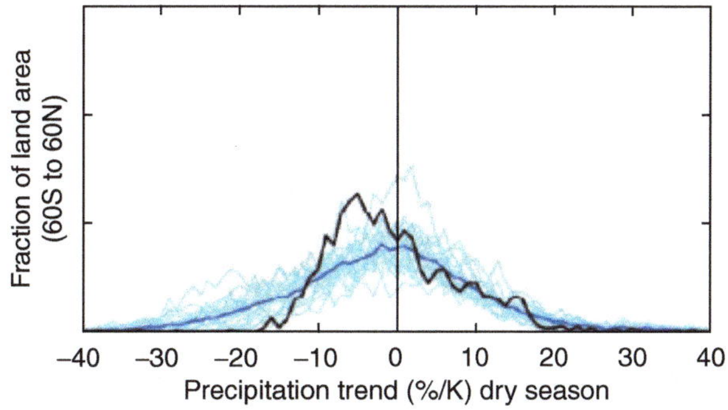

Fig. 10.5 This plot, from Knutti [31] shows the fraction of land area between 60°N and 60°S experiencing a given change in precipitation in the dry season. Precipitation change is measured in percent per unit global temperature rise in Kelvin measured over the period 1900–2100 relative to the 1900–1950 average. Each *light blue line* represents a single CMIP3 ensemble member, the *dark blue line* marks the average of all distributions. The *black line* shows the precipitation change in the multi-model mean. The expected absolute precipitation change in the multi-model mean is about 30% smaller than in any single model

ensemble, effectively smearing out the small scales and the extremes of the distribution. Thus, although the mean result of a large ensemble may provide a reduction in model bias, the averaging process itself may create an unrepresentative forecast.

Model Independence

Given a set of truly independent models distributed about the truth, one would always be able to improve simulation quality by increasing the number of models in the ensemble as truly independent errors would tend to cancel. Any study which treats CMIP ensemble members as independent realizations of a possible future is implicitly making this assumption, but one can make statistical arguments to show that the models are not distributed in a way which would be consistent with this assumption [36]. To illustrate this visually, Fig. 10.6 shows maps of temperature and precipitation from a selection of models in the CMIP-3 archive, all of which could be used with equal weight in producing a multi-model mean. However, one can see instantly that the two GFDL models have very similar biases in surface temperature, even though they are submitted as separate models to the archive. The temperature biases in the other two models shown have very different spatial patterns. The precipitation plots, however, show that there are some common biases in all four of the models. There are many reasons why these common biases might exist; all models in the CMIP-3 ensemble cannot explicitly resolve features smaller than about half a degree, which renders them incapable of simulating behavior such as atmospheric blocking or the response to local orography. Models may also share parameterization schemes and be tuned to reproduce the same observations [32],

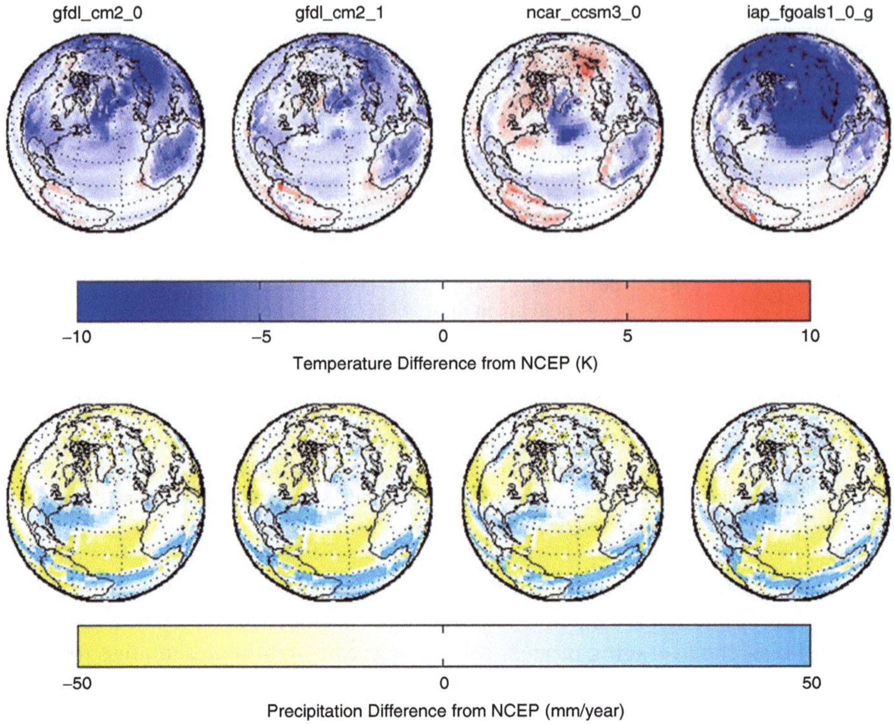

Fig. 10.6 Temperature and precipitation maps of the North Atlantic region from four models submitted to the CMIP-3 archive. Each map shows the 1980–2000 averages for June, July, and August – expressed as a difference between the model simulation and the NCEP reanalysis for the same period. The *top row* shows the anomaly for surface temperatures, while the *bottom row* shows the anomaly for annual total precipitation

and in some cases the same model can be submitted to the ensemble at multiple resolutions meaning that models can share considerable parts of code, making it very likely that model biases will be correlated. In summary, it is both expected and evident that the current generation of climate models does not provide an independent sample of estimates distributed about an underlying truth, and it is unlikely that increasing the number of similar models in the ensemble would drastically increase the accuracy of combined predictions.

Model Validation and Tuning

GCMs are frequently tuned by minimizing differences between simulations of the past century and observations. The observations can be in the form of data from satellites and in situ measurements or may be expressed as reanalyzed products

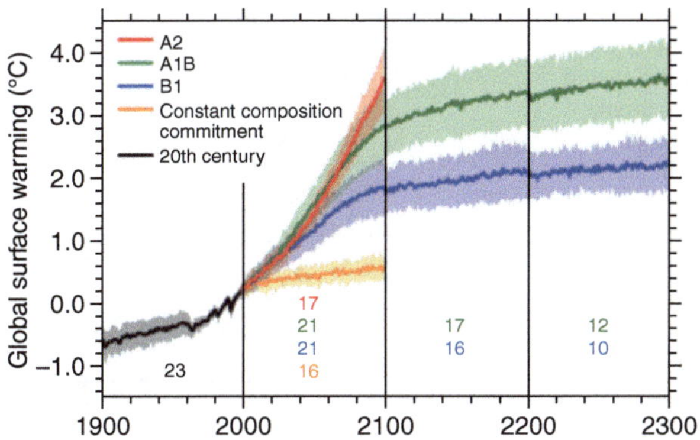

Fig. 10.7 A figure reproduced from the IPCC AR-4 report (Fig. 10.4) showing the mean and inter-model spread of simulations in the CMIP-3 model archive for simulations of the twentieth century, together with the simulations of three different scenarios for periods after the year 2000. Global mean temperatures are shown relative to the 1990–2000 mean. In each case, the *line* represents the multi-model mean and the shading shows the 1 standard deviation ensemble spread

which attempt to incorporate information from both of these. Simulations of earlier periods may also be evaluated against proxy data (estimates of temperature or rainfall etc., produced from tree rings, ice cores, etc.), although the long simulations and necessary model reconfiguration for these periods often mean they do not form part of the active model development process. Because models are tuned to agree with data over the twentieth century, they tend to agree with each other for this time period. There is little spread in the model simulations over the twentieth century. Figure 10.7 taken from the IPCC AR-4 report shows that the models behave similarly throughout the twentiethcentury when compared to any one of the scenarios for the twenty-first century. The reader should not attach any significance to the absolute values of the global mean temperature time series, which are expressed as anomalies with respect to the 1980–2000 mean for all models.

The remarkable consistency of the global mean temperature evolution in the twentiethcentury in the current generation of GCMs is made possible through the various degrees of freedom the models have in fitting this well-observed period. The response of any model is governed by a combination of transient ocean heat uptake, climate sensitivity, and the radiative forcing to the system, which effectively makes the problem poorly constrained with multiple ways to fit the twentieth century global mean temperature time series [33]. A study by Kiehl (2007) concluded that models produced this agreement by compensating between differences in climate sensitivity with differences in aerosol forcing. Figure 10.8 shows both the climate sensitivities and the later twentieth century anthropogenic forcing of climate in a selection of GCMs [34]. It is apparent that those models with a larger anthropogenic climate forcing in the twentieth century have a smaller climate sensitivity, allowing the models to successfully reproduce the twentieth century temperature record (a weak correlation between aerosol forcing and climate sensitivity is also seen in the CMIP-3 archive used for the AR4 report [35]).

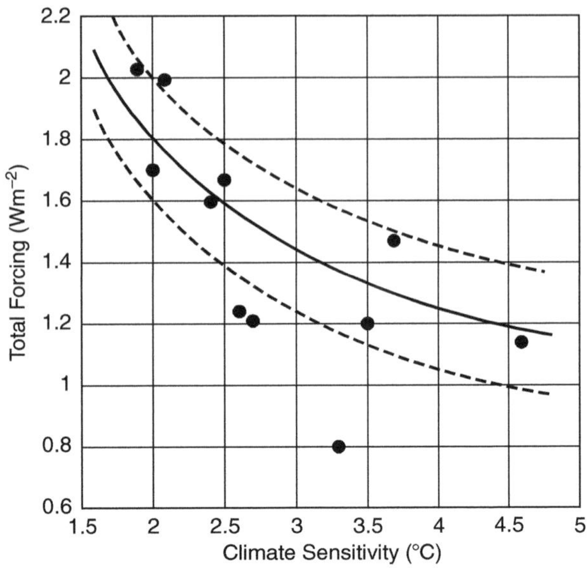

Fig. 10.8 A figure reproduced from Kiehl [35] which shows the relationship between climate sensitivity and total anthropogenic forcing of climate in the late twentieth century in a set of GCMs, represented by the *black dots*. The *solid line* represents a theoretical relationship between the two quantities necessary to produce the warming observed over the twentieth century, the *dashed lines* show the uncertainty in this relationship due to uncertainty in transient ocean heat uptake

In each of the twenty-first century scenarios illustrated in Fig. 10.7, the aerosol concentrations are predicted to decrease as increasingly stringent clean air legislation comes into effect. Meanwhile, all the scenarios show a continuing increase in greenhouse gases throughout the twenty-first century, which makes the climate sensitivity of the models the primary factor influencing their future evolution as the total anthropogenic forcing increases. The differing climate sensitivities amongst the CMIP3 models thus cause a larger spread in the twenty-first century simulations than for the twentieth century simulations. However, it should be noted that most AR4 models included the "direct" radiative effect of aerosols, but not their indirect effect on cloud properties. This means that eliminating the correlation between climate sensitivity and aerosol forcing would not necessarily reduce projection uncertainty, and the success of most models in simulating the twentieth century may be partly spurious [36].

An additional problem lies in the lack of independent data with which to tune and verify the models. In many cases, model quality metrics are based upon mean state and variability data from the latter twentieth century data, which is very likely to be used in the development of parameterizations and tuning of the model. For example, most models use satellite data products to tune the top of atmosphere energy fluxes, and these products are often considered to be one of the more robust constraints when evaluating a model quality metric. In addition, models may often be evaluated against reanalyses, rather than the observational data itself. Reanalysis products are model simulations strongly "nudged" to reproduce an incomplete set of observations, effectively filling in the gaps with self-consistent model data output. This process introduces an additional layer of complexity, because the reanalysis climate will contain features both of the constraining observations and the underlying model. For fields where real data is sparse or where data is not assimilated directly (such as precipitation metrics), the reanalysis output might have much more dependence on the underlying model than on any real-world data. As a result, when using reanalysis data as a constraint for multiple models, those models with a similar representation of the hydrological cycle to that used in the reanalysis will appear to perform better.

The model development process involves a considerable amount of value judgment, as a model serves many purposes and some compromise between the many different plausible performance metrics must be made. The relatively small number of degrees of freedom available to model developers makes it impossible to perfectly match a large number of observable quantities simultaneously, which means that there may be multiple possible parameter combinations which are equally valid. Each of these combinations, although they fit historical observations equally well, may have different projections of future climate change if they exhibit different climate sensitivities or aerosol responses.

In the past, model tuning has largely been a time-consuming process of expert judgment and trial and error, which leads to some uncertainty of what errors in a simulation are irreducible through parameter adjustment. Although not yet used operationally, various techniques have been proposed to automate this tuning process. One technique uses an optimal gradient descent approach to minimize some multivariate error metric [34]. This approach can yield multiple solutions, as the response surface in the parameter space of the model may show local minima. Another approach involves using a preexisting perturbed physics ensemble and fitting a nonlinear response surface [37] to interpolate between the sampled points in the parameter space. This effectively produces a "model emulator" which can predict the point in parameter space which minimizes model error, but the result is dependent on the parameter space being sufficiently densely sampled to capture the dominant features. One can also combine the predictions from a range of plausible perturbed models. The ensemble Kalman filter [38] approach has been used [39] to create a set of valid perturbed versions of a single climate model, but is subject to uncertainty that there is an unknown systematic error in the climate model which cannot be corrected by parameter modification.

A final problem lies in the incompleteness of the model representation of the climate. Many current models, for example, cannot simulate the indirect effect of aerosols on cloud amounts. Tuning an incomplete model to reproduce the observed radiative balance at the top of the atmosphere therefore involves overcompensating the cloud amounts by artificially enhancing other processes, which arguably makes the representation of the current and future state less accurate.

Statements of Probability

Multi-model Ensembles

As indicated throughout this entry, the production of a probabilistic statement for future climate from a multimember or perturbed physics ensemble has no clearly established methodology and requires a priori assumptions to be made. Arguably the simplest assumption that can be made is one of model equality, using the democratic "one model, one vote" approach [40]. In such a method, the probability of a future event is estimated by the fraction of models in which the event happens. This hypothesis can be tested by cross-validation within an unused subset of the ensemble. However, this approach is limited by the implicit assumption that the ensemble is a random sample of plausible estimates of the true climate, where the various arguments in section "Projection Uncertainty and the Need for Ensembles" suggest this assumption may not be valid.

The next logical step is therefore to consider some measure of model skill as a weighting for each model, producing an estimate of future climate as a median of model predictions, such that models with a small bias are given a greater weight [41]. Such approaches are always highly dependent on the exact choice of metric used to evaluate the model weighting [29].

Many studies have adopted Bayesian methodologies, where prior beliefs about the range of future climate change are updated with information from models and/or observations. One example [42] takes a prior probability distribution for current and future regionally averaged climate signal (or the corresponding climate change signal) and updates this using information from models and observations. Priors can be chosen to be uninformative (flat over a large range of possible values) so that the final PDF shape is mainly influenced by the information from models and observations. The likelihood of each model simulation of the past and the future is then represented as the realization from PDFs centered around the unknown "true" present and future climate, as if the ensemble were a sample from a large idealized population of possible models. The width of the PDFs is in turn estimated jointly with the climate signals. Its magnitude depends on that model bias compared to the consensus estimate of the present day and future state. Markov Chain Monte Carlo techniques are used to approximate the result of Bayes theorem applied to priors and likelihood, allowing a joint probability distribution for the "true" climate states and the unknown parameters characterizing the model distributions to be estimated. From it, the PDF of the regional climate change signal is also straightforward to derive.

A Bayesian approach has been also applied at the grid-point scale by representing the entire field of future climate anomaly for each model in terms of a truncated set of basis functions combined with some noise estimate [43], such that each model has its own low-dimensional set of coefficients to describe the pattern of climate change. The advantage of this approach is that a similar Bayesian methodology may be applied to derive estimates for the "true" values of the coefficients, which when recombined with the basic functions results in PDFs for climate change at the grid-point level.

An issue with both traditional weighting schemes and the Bayesian approaches is the way in which outliers are treated – the so-called "convergence criterion". In the case of a large PPE, such as the climateprediction.net experiment, the logic in down-weighting outliers assumes that there is some significance in the consensus mean projection, errors are distributed randomly and that models which deviate strongly from the consensus are somewhat less trustworthy. However, in a small ensemble of best-guesses such as CMIP-3, this argument is subject to question. It is possible that a single model in the ensemble is able to simulate processes which are not simulated in other models. This model is arguably more trustworthy than the rest of the ensemble and yet it would be down-weighted through the application of a simple convergence criterion.

Another issue with all of the methods discussed thus far is the assumption of model independence. It can be shown [44] that the width of the final PDF using a Bayesian methodology is inversely proportional to the number of the models considered in the ensemble. Whilst this would be true if all models were independent estimates of a true climate, it has been demonstrated that this not a valid assumption [32]. Although some statistical methodologies have endeavored to artificially reduce the more obvious interdependencies of the CMIP-3 ensemble [45], there is at present no generally accepted methodology for doing so. The Bayesian techniques that have been developed so far tend to produce a PDF narrower than the spread of the original ensemble, as the independence assumption causes uncertainties to decrease with added ensemble members.

A completely different approach to producing model projections is to statistically "calibrate" models, where a relationship is established between model simulations and observations over an observed period. Once this relationship has been determined, it may be applied to future climate projections to produce a "calibrated" estimate of the true future response. This approach assumes, of course, that the relationship between the projections and the true response will remain constant in the future. This approach has been applied to large scale metrics such as past and future sea-ice loss [46], as well as more complex statistical multivariate approaches which find the best fitting relationship between modes of variability in model simulated and observed past climate, again using those relationships to produce a calibrated future projection [47, 48].

Finally, some "detection and attribution" studies [49] determine spatial patterns of climate changes associated with different atmospheric forcings, using observations to determine whether models are over- or under-representing those changes in past simulations. This allows future model projections to be rescaled in light of the observations. One of the major uncertainties in such approaches is in the derivation of the calibration coefficients themselves, and whether the calibration is valid when applied to a future planet in a very different state. These uncertainties tend to result in wider PDFs than Bayesian methodologies [33].

Perturbed Physics Ensembles

While "one model, one vote" may be a questionable assumption in a multimodel ensemble, it is quite ostensibly wrong in a perturbed physics ensemble where some models have vastly inaccurate simulations of the mean climate [50]. PDFs of future climate derived from a perturbed physics ensemble have therefore often been forced to take a different approach.

Most studies thus far arising from PPEs have focused on producing PDFs for climate sensitivity, and have broadly fallen into three categories: weighting of the parameter space, using the ensemble to establish relationships between observable quantities and unknowns such as climate sensitivity, or a traditional Bayesian technique. An example of the first approach [11] takes a PPE and ascribes each model a weighting, based upon model skill in reproducing the observed climate. By interpolating between the sampled points in the parameter space, one can then produce a weighted integral of the unknown quantity (e.g., climate sensitivity). It is argued, however [51], that the PDF obtained from such an approach is fundamentally dependent upon the prior assumptions made in sampling the original parameters.

A second approach of finding relationships between observable and unknown quantities has been demonstrated using both linear [52] and nonlinear [53] transfer functions. In each case, the ensemble is used to derive some predictors which internally estimate the climate sensitivities of ensemble members. These regression coefficients can then be used together with observations of the true climate state to make a prediction of the true climate sensitivity. Clearly, these predictions are subject to uncertainty in the observational state and in the internally derived prediction error, both of which may be estimated relatively easily. The major "unknown unknown" in such an approach is the systematic or irreducible error of the underlying model, that is, how much additional uncertainty arises when the predictor is applied to the real world. A lower bound of this quantity may be obtained by examining the skill of the predictor when applied to a multimodel ensemble such as CMIP-3, but this will not account for common errors arising from lack of resolution or simulated processes.

The final approach to be considered is the use of an ensemble Kalman filter [40]. The ensemble is used together with observations to update prior beliefs about several unknown model parameters. The ensemble Kalman filter then involves an iterative process forming an idealized ensemble of plausible perturbed models. Once again the methodology is sensitive to assumptions about model error, which scale the relative importance of the model-observation discrepancies forming the overall cost function. By assuming model errors are small, the resulting idealized ensemble will be more tightly clustered about the observed state. The distribution of climate sensitivities in this idealized ensemble is then deemed to approximate a PDF for the sensitivity. One advantage of such a technique is that the predictions may be validated by producing a hindcast for the past (the Last Glacial Maximum, in this case). The LGM simulation can then be used to produce an out of sample weighting for the optimized ensemble.

Future Directions

The analysis of climate simulations from multiple models is still a problem in its relative infancy. Various techniques have been proposed in this entry, each making different assumptions about model independence, prior distributions, systematic model errors, and about what statistical framework is appropriate. These choices remain, at present, somewhat subjective and often yield different probability distributions for unknown climate variables. The apparent contradictions between the methodologies can be understood, however, in light of the assumptions made. In contrast to a numerical weather forecast where thousands of verification cases are available to test the forecast skill, the climate projections for a century into the future are making a statement about a situation never observed before and where no model evaluation is possible. Because there is only a single realization of the future, any statement of probability expresses a degree of belief in a Bayesian sense of how different future outcomes are supported by current evidence (models, data, methods), and is therefore inherently subjective.

Clearly, any projection (and the uncertainty associated with it) must be tailored in a fashion useful to decisions on policy and planning for a changing climate. Policymakers tend to push for increases in precision, but this can lead to decreases in real accuracy if predictions are overconfident [54]. There is arguably little point in providing PDFs of future change for planning purposes if the width of those PDFs are massively sensitive to either subjective decisions or unknown errors, and the raw collection of "best-guesses" from the different models is as useful a way as any to present the ensemble of forecasts. One inherent danger with this approach, however, is the tendency to see the multi-model distribution as a discrete probability distribution for future climate. As is seen, the lack of model independence, the fact that all models are neglecting certain sources of uncertainty (e.g., the carbon cycle climate feedback uncertainty) and the fact that every modeling group will tend to submit only a best-guess climate together implies that the true uncertainty may be larger than that indicated by the spread of model simulations.

Future generations of multi-model ensembles are also likely to introduce more complex "Earth System Models," at least for some ensemble members. These models, in addition to atmosphere, ocean, land, and sea ice components are likely to introduce fully coupled carbon–nitrogen cycles, chemistry, urban, and ecosystem models into the simulation. These components of future uncertainty have not been thoroughly explored in previous generations of the CMIP experiments, and are likely to increase the spread of simulation response for the coming century. Although this could be perceived to indicate an increase in uncertainty, it is more accurately converting an "unknown unknown" into a parametric uncertainty. If different models include different components of the earth system in their models, it will also become more difficult to compare them on a like-with-like basis, as is mostly possible today. However, this underlines the importance with each generation of climate models of recognizing the uncertainties associated with what is omitted, as well as those arising from the simulations themselves.

Although there may be some use in overall metrics of model skill [55], it is likely that projections of specific phenomena will benefit from tailored metrics to rank the performance of different models (e.g., El Niño or future sea ice extent). This will also require a proper assessment of which subset of models to use for each particular application based upon both past model skill and physical plausibility [31]. In addition, the community may benefit more from a diverse range of model predictions, where each model may be evaluated on its own performance, in place of a group of models which are artificially clustered toward a mean response leaving no way of simulating the extremes or boundaries of future climate change.

In this entry both multi-model ensembles and perturbed physics ensembles have been discussed, but there is little discussion on how the information from the two may be combined. Indeed, at present there is little to no literature on how one may combine the parametric uncertainty sampled in a PPE with the inter-model systematic differences in a multi-model ensemble. This presents a fundamental problem in that current PDFs from both of these techniques cannot incorporate the best estimates of systematic and parametric uncertainty. Future analyses must combine these various uncertainties in order to make statements about model robustness. Currently, the ability to conduct such an analysis is limited because only a small subset of the models in the CMIP-3 archive have produced a perturbed physics ensemble, and for those ensembles which do exist, the experiments have not been conducted in any coordinated fashion.

Despite all of the challenges associated with combining and interpreting results from multiple climate models, the presence of coordinated ensembles of projections provides an invaluable insight into the magnitude of some of the uncertainties which are inherent in every simulation conducted, and the ensemble provides a unique opportunity to understand why models differ. As time goes on, the length of good quality observations will increase allowing better evaluation of the transient behavior of the models (a better metric for future transient response than those based upon the model simulation of the base climate [56]). In addition, as more components of the climate system are simulated, although model convergence is not expected (at least in the short term), one can be confident that at least "unknown unknowns" in future predictions can be represented in the form of parametric uncertainty.

Finally, possibly the greatest single uncertainty in future climate remains that of human behavior. Certainly in the case of the CMIP-3 ensemble, the spread in twenty-first century simulations due to different emission scenarios generally exceeded that of the inter-model spread to any particular scenario. Simple models of the climate have already been coupled to socioeconomic models [57–59], but little progress to date has been made in coupling socioeconomic models to GCMs. As a result, potential complex feedbacks between climate change and human behavior have not been sampled in any systematic framework. Nevertheless, although an integrated treatment of uncertainty in future climate projections may seem some way off, the use of multi-model ensembles will continue to frame at least some of those uncertainties in a systematic framework, providing a robustness which would be impossible with any single model, however complex that model may become.

Bibliography

Primary Literature

1. Ad Hoc Study Group on Carbon Dioxide and Climate (1979) Carbon dioxide and climate: a scientific assessment. National Academy of Sciences, Washington, DC
2. Manabe S et al (1979) A global ocean-atmosphere climate model with seasonal variation for future studies of climate sensitivity. Dyn Atmos Oceans 3:393–426
3. Hansen JE et al (1983) Efficient three-dimensional global models for climate studies: models I and II. Mon Weather Rev 111:609–662
4. Houghton JT, Jenkins GJ, Ephraums JJ (eds) (1991) Scientific assessment of climate change – report of working group I. Cambridge University Press, Cambridge, p 365
5. Houghton JT, Meira Filho LG, Callender BA, Harris N, Kattenberg A, Maskell K (eds) (1995) Contribution of working group I to the second assessment of the Intergovernmental Panel on Climate Change. Cambridge University Press, Cambridge, p 572
6. Houghton JT, Ding Y, Griggs DJ, Noguer M, van der Linden PJ, Xiaosu D (eds) (2001) Contribution of working group I to the third assessment report of the Intergovernmental Panel on Climate Change (IPCC). Cambridge University Press, Cambridge, p 944
7. Solomon S, Qin D, Manning M, Chen Z, Marquis M, Averyt KB, Tignor M, Miller HL (eds) (2007) Contribution of working group I to the fourth assessment report of the Intergovernmental Panel on Climate Change, 2007. Cambridge University Press, Cambridge/New York
8. Parker WS (2006) Understanding pluralism in climate modeling. Found Sci 11:349–368
9. Collins M, Allen MR (2002) Assessing the relative roles of initial and boundary conditions in interannual to decadal climate predictability. J Climate 15:3104–3109
10. Hawkins E, Sutton R (2009) The potential to narrow uncertainty in regional climate predictions. BAMS 90:1095–1107
11. Murphy JM et al (2004) Quantifying uncertainties in climate change from a large ensemble of general circulation model predictions. Nature 430:768–772
12. Stainforth DA et al (2005) Uncertainty in predictions of the climate response to rising levels of greenhouse gases. Nature 433:403–406
13. Annan J, Hargreaves J (2006) Using multiple observationally-based constraints to estimate climate sensitivity. Geophys Res Lett 33(4):L06704
14. Palmer TN, Doblas-Reyes FJ, Hagedorn R, Weisheimer A (2005) Probabilistic prediction of climate using multi-model ensembles: from basics to applications. Philos Trans R Soc B 360:1991–1998
15. Weigel AP, Liniger MA, Appenzeller C (2008) Can multi-model combination really enhance the prediction skill of probabilistic ensemble forecasts? Q J R Meteorol Soc 134:241–260
16. Hagedorn R, Doblas-Reyes FJ, Palmer TN (2005) The rationale behind the success of multi-model ensembles in seasonal forecasting. Part I: basic concept. Tellus 57A:219–233
17. Frame DJ, Booth BBB, Kettleborough JA, Stainforth DA, Gregory JM, Collins M, Allen MR (2005) Constraining climate forecasts: the role of prior assumptions. Geophys Res Lett 32: L09702. doi:10.1029/2004GL022241
18. van der Sluijs J et al (1998) Anchoring devices in science for policy: the case of consensus around climate sensitivity. Soc Stud Sci 28(2):291–323
19. Knutti R, Hegerl GC (2008) The equilibrium sensitivity of the Earth's temperature to radiation changes. Nat Geosci 1:735–743
20. Cantelaube P, Terres J-M (2005) Seasonal weather forecasts for crop yield modelling in Europe. Tellus Ser A 57:476–487. doi:10.1111/j.1600-0870.2005.00125.x
21. Thomson MC, Doblas-Reyes FJ, Mason SJ, Hagedorn R, Connor SJ, Phindela T, Morse AP, Palmer TN (2006) Malaria early warnings based on seasonal climate forecasts from multi-model ensembles. Nature 439:576–579. doi:10.1038/nature04503

22. Schclar A et al (2009) Ensemble methods for improving the performance of neighborhood-based collaborative filtering. In: Proceedings of the third ACM conference on recommender systems, ACM, New York, 23–25 Oct 2009, pp 261–264
23. Hagedorn R, Doblas-Reyes FJ, Palmer TN (2005) The rationale behind the success of multi-model ensembles in seasonal forecasting – I. Basic concept. Tellus A57:219–233. doi:10.1111/j.1600-0870.2005.00103.x
24. Gillett NP, Zwiers FW, Weaver AJ, Hegerl GC, Allen MR, Stott PA (2002) Detecting anthropogenic influence with a multi-model ensemble. Geophys Res Lett 29:1970. doi:10.1029/2002GL015836
25. Lambert SJ, Boer GJ (2001) CMIP1 evaluation and intercomparison of coupled climate models. Clim Dyn 17:83–106. doi:10.1007/PL00013736
26. Reichler T, Kim J (2008) How well do coupled models simulate today's climate? Bull Am Meteorol Soc 89:303–311
27. Robertson AW, Lall U, Zebiak SE, Goddard L (2004) Improved combination of multiple atmospheric GCM ensembles for seasonal predition. Mon Weather Rev 132:2732–2744. doi:10.1175/MWR2818.1
28. Krishnamurti TN, Kishtawal CM, Zhang Z, Larow T, Bachiochi D, Williford E, Gadgil S, Surendran S (2000) Multimodel ensemble forecasts for weather and seasonal climate. J Climate 13:4196–4216. doi:10.1175/1520-0442(2000) 013!4196:MEFFWAO2.0.CO;2
29. Santer BD, Taylor KE, Gleckler PJ, Bonfils C, Barnett TP, Pierce DW, Wigley TML, Mears C, Wentz FJ, Brüggemann W, Gillett NP, Klein SA, Solomon S, Stott PA, Wehner MF (2009) Incorporating model quality information in climate change detection and attribution studies. PNAS 106:14778–14783
30. Knutti R (2010) The end of model democracy? Clim Change 102(3–4):395–404. doi:10.1007/s10584-010-9800-2
31. Knutti R, Furrer R, Tebaldi C, Cermak J, Meehl GA (2010) Challenges in combining projections from multiple climate models. J Climate 23:2739–2758
32. Tebaldi C, Knutti R (2007) The use of the multi-model ensemble in probabilistic climate projections. Philos Trans R Soc A 365(1857):2053–2075
33. Knutti R, Stocker TF, Joos F, Plattner G-K (2002) Constraints on radiative forcing and future climate change from observations and climate model ensembles. Nature 416:719–723. doi:10.1038/416719a
34. Jackson C et al (2004) An efficient stochastic Bayesian approach to optimal parameter and uncertainty estimation for climate model predictions. J Climate 17(14):2828–2841
35. Kiehl JT (2007) Twentieth century climate model response and climate sensitivity. Geophys Res Lett 34:22710
36. Knutti R (2008) Why are climate models reproducing the observed global surface warming so well? Geophys Res Lett 35(18):5
37. Sanderson BM et al (2008) Constraints on model response to greenhouse gas forcing and the role of subgrid-scale processes. J Climate 21(11):2384–2400
38. Evensen G (2003) The Ensemble Kalman Filter: theoretical formulation and practical implementation. Ocean Dyn 53:343
39. Annan JD et al (2005) Parameter estimation in an intermediate complexity earth system model using an ensemble Kalman filter. Ocean Model 8:135
40. Raisanen J, Palmer TN (2001) A probability and decision-model analysis of a multimodel ensemble of climate change simulations. J Climate 14(15):3212–3226
41. Giorgi F, Mearns LO (2002) Calculation of average, uncertainty range, and reliability of regional climate changes from AOGCM simulations via the 'reliability ensemble averaging' (REA) method. J Climate 15:1141
42. Tebaldi C et al (2004) Regional probabilities of precipitation change: a Bayesian analysis of multimodel simulations. Geophys Res Lett 31:24213
43. Furrer R, Sain S, Nychka D, Meehl G (2007) Multivariate Bayesian analysis of atmosphere–ocean general circulation models. Environ Ecol Stat 14:249–266

44. Lopez A et al (2006) Two approaches to quantifying uncertainty in global temperature changes. J Climate 19:4785
45. Smith R, Tebaldi C, Nychka D, Mearns L (2009) Bayesian modeling of uncertainty in ensembles of climate models. J Am Stat Assoc 104:97–116
46. Boé J et al (2009) September sea-ice cover in the Arctic ocean projected to vanish by 2100. Nat Geosci 2(4):1–3
47. Greene AM et al (2006) Probabilistic multimodel regional temperature change projections. J Climate 19:4326
48. Buser CM et al (2009) Bayesian multi-model projection of climate: bias assumptions and interannual variability. Clim Dyn 33(6):849–868
49. Stott PA, Kettleborough JA (2002) Origins and estimates of uncertainty in predictions of twenty first century temperature rise. Nature 416:723–726
50. Sanderson BM et al (2008) Towards constraining climate sensitivity by linear analysis of feedback patterns in thousands of perturbed-physics GCM simulations. Clim Dyn 30(2–3):175–190
51. Frame DJ et al (2005) Constraining climate forecasts: the role of prior assumptions. Geophys Res Lett 32(9):L09702
52. Piani C et al (2005) Constraints on climate change from a multi-thousand member ensemble of simulations. Geophys Res Lett 32(23):L23825
53. Knutti R et al (2006) Constraining climate sensitivity from the seasonal cycle in surface temperature. J Climate 19(17):4224–4233
54. Dessai S et al (2008) In: Adger N, Lorenzoni I, O'Brien K (eds) Climate prediction: a limit to adaptation. Living with climate change: are there limits to adaptation. Cambridge University Press, Cambridge, pp 49–57
55. Gleckler PJ et al (2008) Performance metrics for climate models. J Geophys Res 113:D06104
56. Allen MR, Frame DJ (2007) ATMOSPHERE: call off the quest. Science 318:582–583
57. Edmonds J, Wise M, Pitcher H, Richels R, Wigley T, MacCracken C (1997) An integrated assessment of climate change and the accelerated introduction of advanced energy technologies. Mitig Adapt Strateg Glob Change 1:311–339
58. Messner S, Strubegger M (1995) User's guide for MESSAGE III, WP-95-69. International Institute for Applied Systems Analysis, Laxenburg
59. Bouwman AF, Kram T (2006) Integrated modelling of global environmental change. An overview of IMAGE 2.4. Netherlands Environmental Assessment Agency (MNP), MNP publication number 500110002/2006, Bilthoven

Books and Reviews

Kharin VV, Zwiers FW (2002) Climate predictions with multimodel ensembles. J Climate 15(7):793–799
Knutti R et al (2008) A review of uncertainties in global temperature projections over the twenty-first century. J Climate 21:2651–2663

Chapter 11
Climate Predictions, Seasonal-to-Decadal

Lisa Goddard

Glossary

Climatology	Reference period used to describe the characteristics of the climate, such as the mean annual cycle, or the expected statistics of weather or of year-to-year climate variability. The World Meteorological Organization recommends the most recent three full decades; e.g., in 2009, the WMO climatology period would cover 1971– 2000.
External forcing	Factors that influence the climate system but are not explicitly driven by the climate system, such as human emissions of greenhouse gases, changes in the sun's radiation, and volcanic emissions.
Forecast	The guidance offered by a forecaster or forecast center on the future climate conditions. A forecast could be based on a single prediction, but typically is a distilled product that involves recalibrated model predictions and often multiple prediction inputs.
Internal variability	The chaotic evolution of a fluid, such as the ocean or atmosphere, due to nonlinear dynamics that are sensitive to small uncertainties or variations in initial conditions. Depending on timescale, internal variability may refer to that generated internally to the atmosphere, to the ocean, or due to

This chapter was originally published as part of the Encyclopedia of Sustainability Science and Technology edited by Robert A. Meyers. DOI:10.1007/978-1-4419-0851-3

L. Goddard (✉)
Forecasting and Prediction Group, International Research Institute for Climate Society (IRI), The Earth Institute of Columbia University, Lamont Campus, 228 Monell Bldg. 61 Route 9W, 10964-8000 Palisades, NY, USA
e-mail: goddard@iri.columbia.edu

P.J. Rasch (ed.), *Climate Change Modeling Methodology: Selected Entries from the Encyclopedia of Sustainability Science and Technology*,
DOI 10.1007/978-1-4614-5767-1_11,

	ocean–atmosphere interaction. It is the part of the seasonal-to-decadal climate that is not deterministically predictable.
Prediction	The future climate conditions indicated by a single prediction model, which could be statistical or dynamical. These differ from climate change projections in that information of the climate state at or near the initial time of the forecast is highly relevant to its future evolution.
Teleconnections	Climate variability in one region that is driven remotely by climate variability in another region. This typically refers to regional patterns of climate anomalies over land and/or oceans that result from specific ocean phenomena, such as during El Niño events.

Definition of the Subject and Its Importance

Seasonal-to-decadal climate prediction seeks to quantify the likely evolution or change of the climate system over a specific time horizon of months to years. Climate predictions based on dynamical models incorporate all relevant processes to the extent possible, including anthropogenic climate change, but most importantly those processes that govern the likely evolution of natural climate variability. The predictions, if well calibrated, describe the probability of a given magnitude of change in the mean climate or changes in the characteristics of the weather over the forecast period. For example, a seasonal forecast for next winter might indicate a greater likelihood for the seasonal mean temperatures to be colder than usual, or might indicate the likelihood for more frequent storms over the 3-month period.

Seasonal prediction is a fairly well-established enterprise with a number of forecast centers around the world issuing real-time seasonal predictions–based dynamical models [1]. Increasingly, national meteorological and hydrological services create seasonal forecast products based on their own statistical or dynamical prediction tools and/or incorporate predictions from the international centers. Decadal prediction is a much newer endeavor and is still considered experimental [2]. Only a few groups have attempted decadal-scale climate predictions intended to capture the evolution of natural decadal variability for the coming decade from a global circulation model [3–5], and although the results indicate there may be added information from these predictions relative to the more familiar climate change projections of the Intergovernmental Panel on Climate Change (IPCC), it is not clear that the added information results from better prediction of the decadal-scale climate variability that would exist even in the absence of increasing greenhouse gases.

Climate forecasts are potentially valuable to society on seasonal-to-interannual timescales to inform resource management, planning decisions, and on decadal timescales to inform longer-term plans and infrastructure investment. Even in the climate change context, decadal prediction could prove important, as the climate experienced regionally for the coming decade(s) will likely be some combination of anthropogenic climate change and natural decadal variability. Decisions and investments related to climate change adaptation typically apply to the next 10–20 years into the future, rather than 80 years into the future. Thus better information on evolution of the climate and changes in risks of climate extremes can lead to more appropriate planning. However, climate predictions are necessarily probabilistic, and in the case of decadal predictions are yet to be established as skillful. Thus it is important that decision systems be designed and optimized to account for the inherent uncertainty in future climate, that can still allow benefits to be realized in times of favorable climate and losses to be mitigated in times of adverse climate.

Introduction

Climate varies on all timescales, from seasonal variations to millennial ice ages. Prediction of the climate at timescales that are relevant to societal decisions, but extending beyond weather forecasts, has been roughly broken into three classes: seasonal-to-interannual prediction that addresses the changes in seasonal climate and its weather characteristics a couple months to a year in the future, decadal prediction, sometimes referred to as near-term climate change prediction that addresses changes in the mean climate and its characteristics for a couple years to a couple decades into the future, and climate change projections that consider changes in the mean climate and its characteristics 50–100 years in the future.

The seasonal-to-interannual timescale dominates the climate that is experienced locally. On a local-to-regional scale, year-to-year variability almost always explains the majority of the variance in the observed climate (e.g., Fig. 11.1). Year-to-year variability is where most impacts are experienced. However, it is the superposition of the three climate timescales that can lead to changes or trends in the frequency of adverse years. Extreme examples are potentially the protracted drought conditions in the western United States from the mid-1990s to the early twenty-first century [6], the 2003 European heat wave [7], or the extremely active hurricane season 2005 [8], which was accompanied by many land-falling hurricanes in the United States such as Katrina.

The primary difference between prediction of climate variability on different timescales is the drivers, or phenomena, associated with those impacts. This leads to differences in the way prediction systems are designed to predict the climate fluctuations and associated impacts on different timescales. Seasonal-to-interannual prediction is an initial value problem; by initializing the climate system close to the observed state at the beginning of the prediction, a dynamical model will aim to capture the likely evolution of the climate system. At the other end of the time spectrum, climate change projection is a boundary value problem, which means that

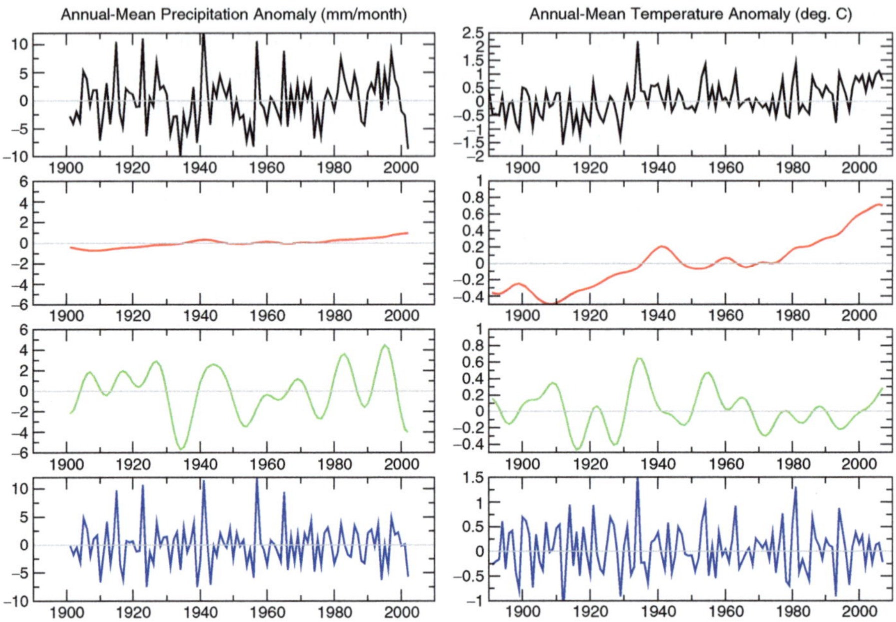

Fig. 11.1 Example of simple decomposition of (**a**) temperature and (**b**) precipitation averaged over the state of Colorado in the United States. The *top* panels (*black*) show the observed annual mean time series. The second panels (*red*) represent "climate change" time series, in which the climate changes are consistent with the globally averaged temperature, obtained by decadally filtering the time series and regressing it against a similar low-frequency filtered time series of globally averaged temperature. The *third* panels (*green*) represent the "natural decadal variability," which is low-frequency time series that is not coincident with globally averaged temperature changes, obtained as the difference between the low-frequency filtered time series and the "climate change" time series. The *bottom* panels (*blue*) represent the year-to-year variability on top of the low-frequency changes, which is the difference between the full time series and the low-frequency filtered time series. Note that there is no attribution to anthropogenic changes or physical phenomena in any of these time series. Details are likely to change with different filtering parameters and with different approaches to estimate global warming

the driver of the climate change is external to the climate system and imposed upon it. Anthropogenic increases in greenhouse gases are due to man's activities and are not part of the natural climate system. Climate change projections depend on correctly projecting the changes in the Earth's atmospheric composition and the subsequent changes in the Earth's energy balance. Decadal prediction lies at the intersection between seasonal-to-interannual prediction and climate change projection; it is an initial value as well as a boundary value problem. Decadal prediction depends both on initializing the climate system close to the observed state, especially the slowly evolving components, and on correctly representing the changes in Earth's energy budget.

This is not to say that predictions on longer timescales do not contain the higher frequency phenomenon. However, there is a predictability limit for natural climate variability, which refers to how far into the future some aspect of climate variability can be predicted before the uncertainty, or range of possibilities, approaches the climatological uncertainty. At that point little to no predictive information remains. The limit of predictability is not necessarily a fixed quantity. It changes with the phenomenon, but also changes with time, meaning that at some times a phenomenon will be more predictable than others and thus the evolution can be predicted farther into the future. It is not possible to determine what the true limit of predictability is or should be [9]. The model(s) that can predict the phenomenon with the greatest fidelity when compared to observations over some long history containing many realizations of the phenomenon determine the *current* limit of predictability.

In order to make a prediction one must first determine what is to be predicted. If the aim is to predict local-to-regional scale climate over land, one must know the driver of that climate variability. Numerous research and prediction studies have demonstrated that it is the large-scale variability in the pattern of surface temperature, and in particular the sea surface temperatures that drive the predictable aspect of changes in the atmospheric circulation and thus regional temperature and precipitation. But what drives that sea surface temperature variability? The sea surface temperatures must be predicted if it is hoped to predict the associated terrestrial climate impacts. Once the ocean phenomena or processes relevant to sea surface temperature variations are identified, the climate models must be capable of simulating those. Furthermore, if the prediction of some phenomenon from a particular model is to provide actionable information, then the phenomenon must be predictable above the other ongoing processes in the climate system; in other words, the signal of the phenomenon must be predictable above the background noise of the climate system. In the next section, an example of this process of identification, model validation, and prediction based on the El Niño-Southern Oscillation (ENSO) phenomenon and seasonal climate prediction is presented.

Brief History of ENSO Prediction: Impacts of the ENSO phenomenon have been experienced for centuries, long before the phenomenon itself was identified. The peoples of Peru used the term El Niño to refer to the expected changes in the local climate and fish stocks associated with a seasonal reversal of the current system off the coast of western South America, because these changes occur near the end of the year at a time near Christmas (El Niño is Spanish for the Christ child). However, they also noted that warm seasonal waters associated with the change of currents, would occasionally be very warm and would also bring abundant rainfall. It is these extreme years, which recur about every 3–7 years that are now called El Niño events. Farmers in drought-prone regions of the Andes even developed a method to predict the coming of the increased rainfall during these events by monitoring the visibility of a star in the Pleiades constellation [10]. What they were observing was the shift of convection from the western tropical Pacific into the central Pacific in concert with the development of an El Niño event (Fig. 11.2).

Sir Gilbert Walker could be said to be the pioneer of seasonal forecasting as he sought to quantify the atmospheric component of ENSO, the Southern Oscillation,

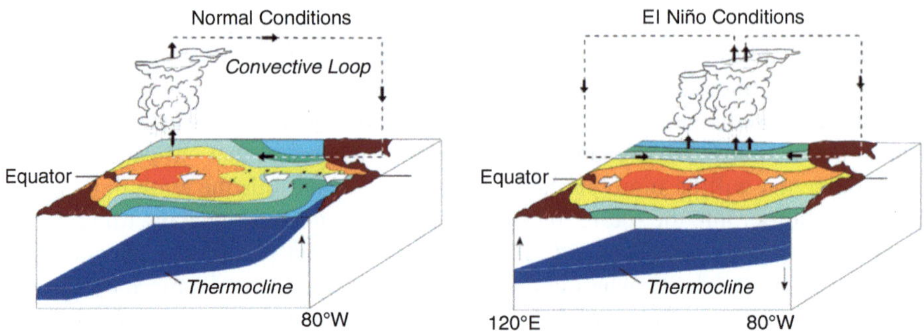

Fig. 11.2 Schematic drawing of the tropical Pacific ocean–atmosphere state during (**a**) average or neutral conditions in which trade winds blow east to west, pushing warm surface waters to the western Pacific, which pushes down the thermocline (separation between warm upper ocean and cold deep ocean) and concentrates the deep convection in the western Pacific; (**b**) El Niño conditions in which the thermocline becomes deeper in the eastern Pacific and warm water moves westward, which weakens the east–west Trade Winds and allows the convection to move into the central Pacific (Source: http://www.tao.noaa.gov/proj_overview/tao_tour_ndbc.shtml)

and its relationship to regional climate variability, such as the devastating droughts in India [11]. To accomplish this he examined correlations between 32 stations across the world for fields of sea level pressure, temperature, rainfall, and riverflow. He discovered that negative excursion of the Southern Oscillation Index was associated with increased likelihood for drought over India; his empirical model has not been much improved upon over the last century for that region. Researchers have continued to improve upon the foundation that Walker laid for ENSO teleconnections (Fig. 11.3). Maps that show significant correlation between regional temperature and precipitation changes to ENSO events for specific 3-month average seasons are widely used to illustrate ENSO's global reach [12, 13]. However, these teleconnection patterns represent expectations based on statistics and are not guaranteed to occur in any specific event; the probabilistic likelihood of a regional impact [14] is a further refinement of the climate anomalies due to ENSO, and something that climate prediction models should be expected to replicate in their ensemble distributions over time.

It was not until the second half of the twentieth century that researchers discovered that the Southern Oscillation was associated with changes in the large-scale sea surface temperature pattern over the tropical Pacific; it was the coupled interaction between the east–west sea surface temperature gradient and the low-level winds between the high and low pressure centers of the Southern Oscillation that led to the growth of Niño events [15]. It was soon after recognized that the change in the winds due to the changes in sea surface temperatures, associated with the Southern Oscillation, modified the distribution of the upper-ocean mass field below the surface [16], and that the adjustment of these perturbations to the mass field could lead to the eventual decay of the El Niño event and possible initiation of the opposite phase, La Niña.

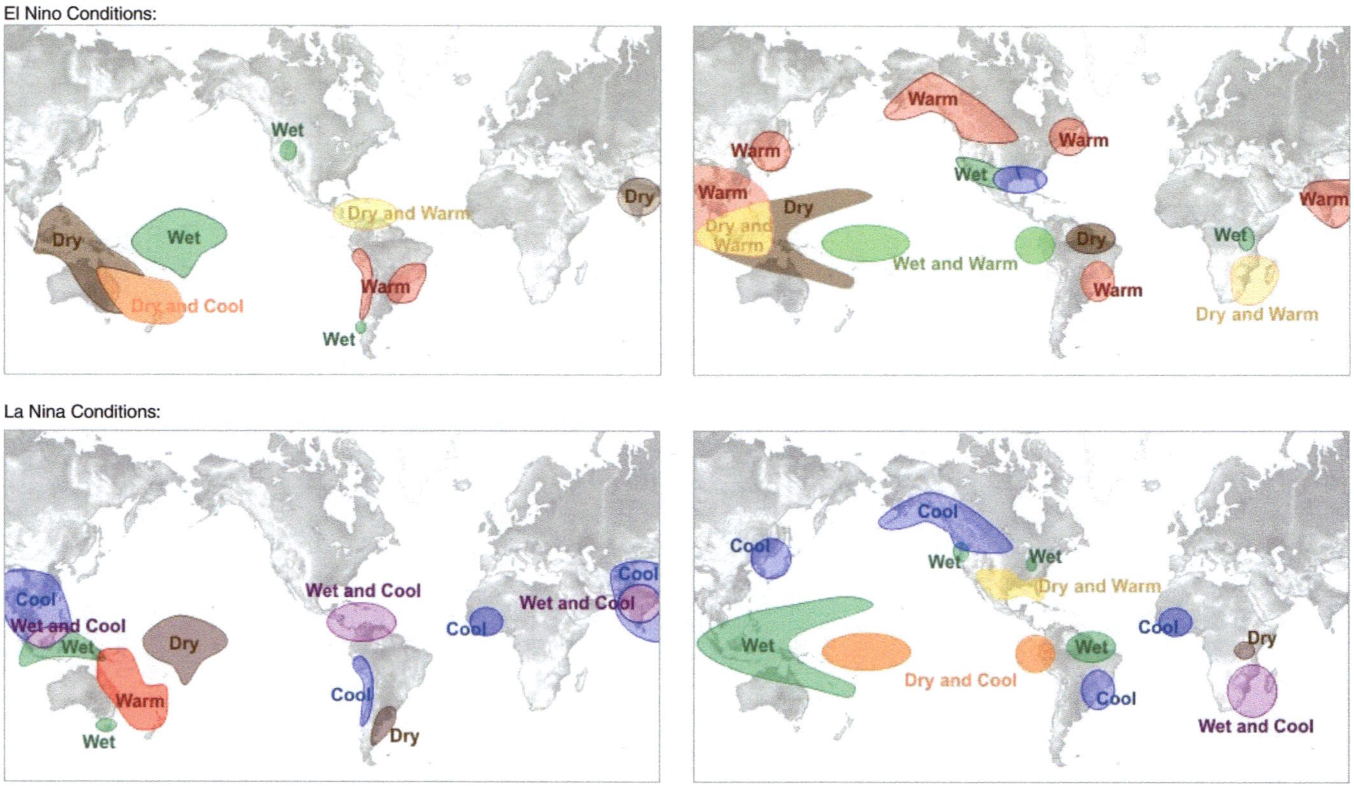

Fig. 11.3 Expected climate anomalies for (**a**) El Niño events during northern hemisphere summer when events are developing, and (**b**) El Niño events during northern hemisphere winter when events are mature; and for (**c**) La Niña events during northern hemisphere summer, and (**d**) La Niña events during northern hemisphere winter (Source: http://www.srh.noaa.gov/jetstream//tropics/enso_impacts.htm)

In the 1980s simple dynamical models [17–19] were developed that simulated the coupled air-sea processes central to ENSO and reinforced the theory that had been informed primarily by observations. The first experimental El Niño forecast was published in 1986 [20], using one of these simplified dynamical models. Since then more complex models have been built that capture not only the dominant processes behind ENSO but also provide a more complete representation of the climate system to better capture uncertainties in ENSO. These models also simulate the atmospheric teleconnections that lead to changes in sea surface temperatures of other ocean basins and to changes in the terrestrial climate. These are the impacts that served as the initial motivation for the study of ENSO. Finally, through the process of identification of a primary driver of seasonal climate variability and the dominant physical processes behind it, and the development of models that could simulate and predict this driver and its teleconnections, seasonal prediction was born.

The prediction of decadal-scale climate variability is a much more recent endeavor. Although research on decadal climate variability through the use of observations and models is not new [21–26], a community-wide effort in this area is new. The motivation to predict decadal climate variability has arisen in part from a desire to use the climate change projections that appear in the Working Group 1 report of the Intergovernmental Panel on Climate Change [27] to inform sectoral decision making, [28] as well as plans and investments toward climate change adaptation. For these societal needs, climate information for the next 5–20 years becomes more relevant than that for the next 100 years. The other side to the motivation behind experimental decadal predictions is the realization that there are processes inherent in the natural climate system evolving at decadal-to-multidecadal timescales, and the mounting evidence that dynamical models have some ability to simulate some aspects of the observed variability [29, 30].

As the successes and failures in climate prediction are considered, it must be borne in mind that climate predictions are necessarily probabilistic. They indicate the likelihood of a range of possible outcomes. The magnitude of this range of outcomes, often referred to as the uncertainty or probability distribution, is sensitive to uncertainties in the initial conditions from which the predictions evolve, to uncertainties in external forcings, and to errors in prediction models. The value about which the uncertainty is centered is sensitive to the external forcings and to information in the initial conditions that may lead to specific, robust evolution of the climate system. Particularly in the case of decadal prediction, which is still in the experimental phase, success refers to relative performance, or agreement between prediction and observations, compared to the state of predictions beyond the seasonal timescale, namely, climate change projections. In other words, much of the judgment of decadal prediction in these early experiments focuses on the added forecast quality from the initial conditions relative to that from the boundary conditions, or external forcing. Therefore, success in the eyes of the climate community may not constitute information that is accurate enough or specific enough to be actionable.

Although both seasonal prediction and decadal prediction experiments, and climate change projections for that matter, use the same type of dynamical models, substantial differences exist between their application for these different timescales of prediction. The following sections contain discussions of Drivers of Variability, Model Fidelity, Prediction Systems, and Internationally Coordinated Efforts, first for the seasonal-to-interannual timescale, followed by a similar analysis for the decadal prediction problem. The only difference in the structures is that the section on seasonal-to-interannual prediction also contains a discussion of Forecast Skill. The echoed structure is intentional in that many of the issues and approaches will be similar for both timescales. However, there are important differences in what is known about the drivers of climate variability at these different timescales as well as differences in the maturity of the prediction systems.

Seasonal-to-Interannual Prediction

Seasonal-to-Interannual Prediction: Drivers of Interannual Variability

Seasonal-to-interannual prediction derives from initial conditions of the climate system. Unlike weather forecasts, where the relevant initial condition is the atmospheric state and the sea surface temperatures are approximately constant, seasonal forecasts depend more on the initial condition of the ocean. The evolution of the ocean state, particularly the density structure and the currents, leads to changes in the pattern of sea surface temperatures that can then influence the atmospheric circulation.

The dominant pattern of surface temperature variability, after accounting for global warming, is that of the El Niño-Southern Oscillation (ENSO) (e.g., [31]). For this reason ENSO has received a great deal of attention in studies of climate prediction on seasonal-to-interannual timescales. Changes in winds and precipitation are associated with these global temperature pattern changes.

El Niño events recur about every 3–7 years on average, and are somewhat locked to the annual cycle in that they tend to develop and grow through the middle of the year and tend to peak near the end of the year. During an El Niño event when much of the warm water in the western equatorial Pacific moves eastward, the region of deep convection also moves eastward into the central Pacific (Fig. 11.2), and in some cases reaches as far as the coast of South America. Since the equatorial Pacific spans nearly half the circumference of the Earth, a shift of the largest region of deep convection from the far western Pacific to the central equatorial Pacific represents a huge spatial shift in where the tropical atmosphere is heated.

In the tropics, where the effect of Earth's rotation is weaker, the atmospheric response to the pattern of sea surface temperatures is thermally direct. The low-level winds converge toward the warmest water, or equivalently, to the region of lower pressure. This is true of the mean conditions as well as the anomalous conditions. Since the lower atmosphere is very humid in the tropics, the regions of converging low-level winds produce an upward flow of very moist air and heavy precipitation with a very large latent heat release to the atmosphere associated with water vapor condensation. Near the top of the troposphere, relatively dry air is expelled from these regions of strong deep convection, and that air then sinks. The sinking dry air suppresses convection. The regions of warmest sea surface temperatures and associated strong deep convection are located typically over the western Pacific warm pool and the western hemisphere warm pool, which encompasses the northeastern tropical Pacific extending to the northwestern tropical Atlantic. Variations in these warm pool regions have direct impacts on the climate in the neighboring regions, but changes in the strength and location of those convective centers can also impact regional climate remotely though changes in atmospheric circulation.

The resulting changes in the atmospheric circulation can lead to warmer conditions in the other tropical oceans [32, 33], which carry additional regional climate impacts. For prediction of the regional climate due to tropical sea surface temperature changes outside the Pacific, it is important to be able to predict those sea surface temperatures. For example, the tendency for northeastern Brazil to be drier than normal during an El Niño event (Fig. 11.3) is due in part to the anomalous subsidence from the shift in deep convection over the central Pacific, but it is also due to associated warming of the sea surface temperatures over the north tropical Atlantic [34]. Similarly, wetter conditions in eastern Africa associated statistically with El Niño events are now known to result from the warming of SSTs in the western Indian Ocean that are also associated with El Niño events [35]; an El Niño event that is not accompanied by warm SST anomalies in the western Indian Ocean leads to drier conditions over East Africa due to anomalous subsidence resulting from El Niño's enhanced convection in the central Pacific.

El Niño can affect weather and seasonal climate outside the tropics through changes in the position and strength of the storm tracks. When the warm water that normally resides in the western Pacific extends across the Pacific, it changes the large-scale temperature differences between the tropical and the midlatitudes. This allows the storm track associated with the subtropical jet stream to strengthen over the central and eastern subtropical Pacific where it is usually weaker and more variable. Additionally, the warming of the equatorial Pacific region as a whole allows the amount water vapor in the lower atmosphere to increase. The combination brings more frequent and stronger storms into the southern tier of the United States during El Niño events. This impact on extratropical climate is seen in the winter hemisphere because this is when the jet stream is strongest. So although a similar influence can be discerned for storm track headed toward South America, the impact is less robust, since El Niño events are typically growing during southern hemisphere winter in the middle of the year. During the northern hemisphere winter is closer to the time when El Niño events are mature.

It is the large-scale changes in the patterns of low-level heat and moisture that drive changes in the atmospheric circulation. El Niño happens to be the dominant phenomenon influencing that and the focus of those changes is primarily over the tropical oceans. However, changes in land-surface conditions, such as soil moisture or ice, can also influence regional climate. Soil moisture influences the overlying atmosphere primarily through evaporation, which can then influence precipitation as well as near-surface air temperature during certain times of the year [36]. Dry soil conditions, and thus a reduced ability of the surface to cool itself through evaporation, are likely to have contributed to the 2003 European heat wave [37]. Changes in patterns, extent, and timing of snow cover can also impact the atmospheric circulation through changes in land atmosphere energy exchange and may impart predictability to northern hemisphere wintertime temperatures [38] and also the strength of the East Asian monsoon [39].

Seasonal-to-Interannual Prediction: Model Fidelity

Once the main drivers of seasonal-to-interannual climate variability are identified, it is then necessary to ascertain whether the model to be used for seasonal-to-interannual prediction can replicate the drivers with sufficient realism. Change in patterns of SSTs is the dominant driver of seasonal-to-interannual climate variability worldwide. However, the regional terrestrial climate will only be predictable if the relevant SSTs are predictable [40]. Given that the El Niño phenomenon represents the majority of year-to-year variance in SSTs, including influencing the global ocean outside the tropical Pacific [41], most studies of the suitability of a model to predict seasonal-to-interannual climate will focus on the model's ability to predict El Niño. Of course, such studies of model fidelity help further elucidate the processes behind such phenomena.

The first attempt to predict El Niño employed a very simple model of the tropical Pacific Ocean that consisted of a warm, lighter, upper ocean overlying a cold, heavier deep ocean [17]. The depth of the upper layer determined the temperature at the surface in the eastern and central equatorial Pacific where upward currents are known to bring cold water from the deeper ocean into the upper layer and cool the surface; the more shallow the upper layer, the easier for the upwelling currents to bring cold water to the surface. The surface temperature anomalies in the east influence the east–west temperature gradient, which affects the strength of the trade winds, which affect the slope of the interface between the upper and lower ocean layers, and thus affects the eastern equatorial surface temperature. This describes the classic Bjerknes feedback mechanism [15] that maintains the mean state as well as the coupled air-sea feedbacks that can evolve an El Niño or La Niña event. Off the equator in the western Pacific the anomalous winds create depth anomalies of the opposite sign to those in the eastern equatorial Pacific, which can then adjust via equatorial wave dynamics, eventually causing the decline of the current event (e.g., El Niño) and potentially initiating an event of the opposite sign (e.g., La Niña). The positive feedback growth together with the delayed negative feedback that can cause

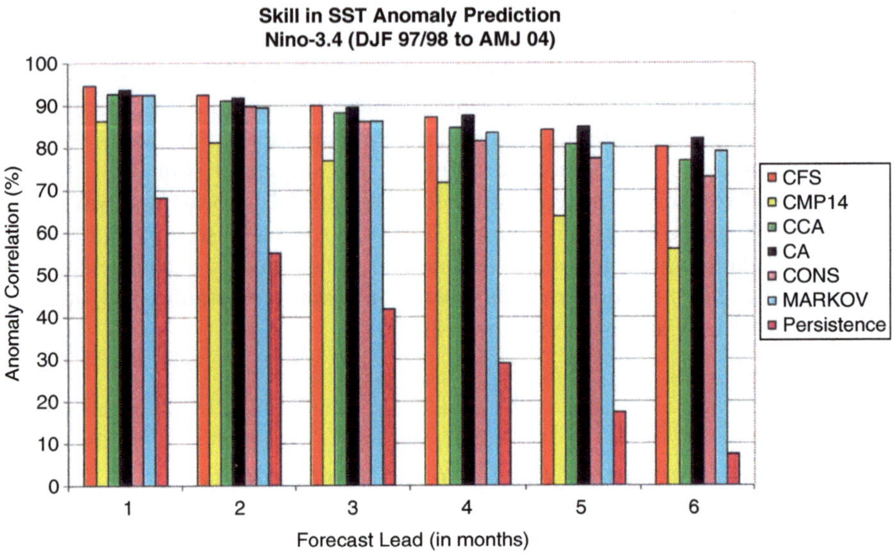

Fig. 11.4 Anomaly correlation (%) by various methods of the seasonal mean Niño-3.4 SST as a function of lead (horizontal; in months). The results are accumulated for all seasons in the (target) period DJF 1997/1998 to DJF 2003/2004. Except for CFS (the Climate Forecast System) coupled ocean–atmosphere model of the National Weather Service's Climate Prediction Center (CPC), all forecasts were archived in real time at CPC from 1996 onward. CMP14 is the previous coupled model, *CCA* is canonical correlation analysis, *CA* is constructed analog, *CONS* is a consolidation (a weighted mean), and MARKOV is an autoregressive method (From [44])

the turnabout from one phase to the next was named the Delayed Oscillator mechanism [19]. Variants on this central idea, such as the Recharge Oscillator [42] have since been formulated as observations of the tropical Pacific became available [43] and as the tropical Pacific air-sea variability was studied in more models.

Although the first models to successfully predict El Niño in the late 1980s were very simplified compared to the complexity of the real ocean–atmosphere system, they still remain viable prediction tools. It is very difficult to represent all the physical processes in the tropical ocean–atmosphere system precisely, and because of the strong interconnectedness of these processes, small errors in the representation of one process leads to associate errors in others. Thus it was not until the early twenty-first century that coupled ocean–atmosphere models of full complexity clearly demonstrated parity with simpler prediction models (Fig. 11.4) [44]. The metric most commonly presented to represent a model's ability to predict El Niño is the NINO3.4 index of sea surface temperature, which is the average of the temperature anomaly over the central equatorial Pacific from 5S–5N and 170W to 120W, as this is the region that exhibits the highest correlation with terrestrial climate anomalies worldwide [45].

However, this simple index does not capture all of the characteristics of El Niño. The timing and spatial structure of El Niño-related sea surface temperature anomalies can also influence the resulting teleconnections [46]. Additionally, single metrics such as correlation or mean error can mask the conditions under which El Niño is predictable. Most dynamical systems, particularly those with a chaotic component, exhibit conditional predictability meaning that there are times when the system is more predictable than others [47]. Thus it is also common to present the prediction history of models, to show how observed sea surface temperatures along the equator vary compared to the predictions as a function of lead time [48]. A common finding from such qualitative examination is that although models may do well in predicting the occurrence of an El Niño event, they have difficulty predicting the magnitude of large events or locating the variability far enough east during strong events [49, 50]. Such biases have repercussions for predicting the associated climate anomalies.

Predicting the driver of the climate anomalies is the first step. Next is to predict the associated climate anomalies. Biases in prediction of the drivers, such as El Niño events that do not exhibit the strength or structure of observed events, lead to biases in regional climate prediction. One way to circumvent some of the error in predicted SSTs is to statistically correct them before providing this information as boundary conditions to the atmospheric model. This approach is known as two-tier forecasting because the SSTs are predicted first and the climate is predicted second using an atmospheric GCM. Changes in the atmospheric circulation do not feed back onto the SST anomalies. Because El Niño is the largest driver of climate anomalies, and El Niño teleconnections are driven by the ocean variability, this is a viable approach. However, outside the tropical Pacific a notable fraction of the ocean variability is driven by the atmosphere, and thus in those regions heat and momentum fluxes will not be properly represented by two-tier forecasts.

One-tier forecasts, where the ocean and atmosphere evolve together, allow for a more physically consistent evolution of the ocean–atmosphere system. Coupled ocean–atmosphere models are increasingly the prediction tools of choice at operational forecast centers around the world [1]. However, due to model biases over some parts of the tropical ocean, regional climate prediction remains problematic with coupled models. In particular coupled models have great difficulty in representing the mean state of the tropical Atlantic, with the warmer water occurring in the western instead of the eastern equatorial Atlantic [51]. As a result, the tropical Atlantic SST variability is not predicted with any skill for most seasons by the current generation of coupled models, and the potential predictability of climate variability over western Africa and northeastern Brazil is substantially degraded compared to what it would be with skillful SST predictions [40].

Other biases that have been known for decades still persist in coupled ocean–atmosphere models and limit the quality of climate predictions. Such systematic biases include a double intertropical convergence zone over the Pacific, poor representation of regions of stratus clouds over the eastern subtropical and extratropical oceans, and vertical temperature gradients that are too diffuse in the equatorial Pacific where the warm upper ocean transitions to the cold deep ocean. The processes responsible for these features in Nature and how they are represented in models are active areas of research. Recent modeling experiments using models with a spatial resolution of tens of kilometers rather than hundreds of kilometers does reduce some of these biases by better resolving certain climate processes.

Seasonal-to-Interannual Prediction: Prediction Systems

Prediction systems are based on observations, models, and their connection through data assimilation systems. The three together form the three-legged chair of prediction systems [9]. Any weak leg compromises the system, and improvements in one leg often lead to improvements in the other legs.

Predictability of seasonal-to-interannual climate variability arises from the initial conditions of the ocean, particularly those conditions in the tropical Pacific Ocean that carry some signal of future El Niño conditions. Observations of upper ocean heat content anomalies in the other tropical oceans are also important for prediction as they can influence the persistence of local sea surface temperature anomalies as well as moderate the impacts of El Niño-related teleconnections in the region. Therefore, it is important to adequately observe the tropical ocean state. However, since models have errors in their representation of the real world, using the observations too faithfully to describe the initial conditions for model forecasts can cause problems. This is where data assimilation is essential to prediction systems.

Data assimilation is the process used to produce initial conditions for a dynamical model by combining observations with other information from a previous simulation of the model. If this is not done carefully, the introduction of the observations into the models can lead to initialization shock when the prediction is started. Initialization shock is a term used to identify the rapid development of model errors when a simulation is started. One approach to minimizing this problem is called anomaly initialization in which observed anomalies of the ocean state rather than the full state are added to the model's mean state to arrive at initial conditions. Other data assimilation methods address the mismatch between the spatial and temporal characteristics of the variability between Nature and the model. Currently, the atmosphere, ocean, and land components of prediction models are initialized separately. The data assimilation efforts are separate, and thus consistency in the initial states and tendencies of these components is not ensured. Methods to assimilate observational data into the coupled model as a whole are being investigated starting with the coupled ocean–atmosphere system [52].

Recent advances in El Niño prediction skill at the European Centre for Medium-Range Weather Forecasts in the United Kingdom were accomplished by both improvements to their model and improvements to the ocean data assimilation system [53]. Additionally, retrospective forecasts, also called "hindcasts," of the NINO3.4 El Niño index from 1960 to present from that forecast system have demonstrated the value of the observations provided by the Tropical Atmosphere–Ocean array of data buoys that measure temperatures of the upper 500 m of the tropical Pacific Ocean, some at the equator also measuring ocean currents, as well as temperature, winds, and humidity at the surface. At the time when the array of buoys was completed in the early 1990s, the forecast error of the NINO3.4 index dropped dramatically [54]. This result is most clearly demonstrated in forecasts that are initiated in February, when the biases in the model are at a minimum. This echoes the connected nature of these three elements of forecast systems; observations and their assimilation into models are crucial for prediction, but better models better elucidate the value of the observing network.

Even at that point when models, the observing network, and the use of those observations for forecast initial conditions becomes essentially perfect, climate forecasts will still contain uncertainties. Small, almost imperceptible, uncertainties in the initial state or the detailed evolution of some small-scale processes will lead to some divergence in the future state. This is the chaotic element of the climate system, sometimes referred to as the "butterfly effect." Where uncertainty is due to errors, there is the potential to reduce it. However, it is not necessarily the goal of forecasters to eliminate uncertainty, as this would be unrealistic, but to quantify it to the extent possible. Better models that can capture the random nature of processes, such as turbulence or convection, would improve process-related contributions to uncertainty. Better representation of such processes may actually increase the uncertainty in forecasts, relative to what models now indicate. Better observations, more complete observational networks, and improved data assimilation techniques can better indicate the uncertainties that arise from initial conditions [55, 56].

The uncertainty in climate forecasts should thus be considered as a range of possible outcomes. Typically the range of possible outcomes, or probabilities, are presented relative to the past climate history of the last several decades. A common format used by many operational forecast centers is tercile classes. For example, the precipitation for a given location over the last 30 years is used to quantify the above-normal category as the wettest 10 years, the below-normal category as the driest 10 years, and the near-normal category as those in between. In this case, the climatological probabilities are 33.3% for any category without any further knowledge. This should be the forecast probability for each category if there is no signal in the current prediction or if the prediction tools have no skill in that region and/or season. If skill and signal exists, then the forecast probabilities will differ from the climatological probabilities (Fig. 11.5). If the signal in the forecast indicates likelihood for wetter conditions, then the probability for above-normal precipitation will be higher than 33.3% and the probability for below-normal precipitation will be less. Alternatively, the forecast can be represented as the probability for exceeding or not exceeding some quantitative value.

One of the most important qualities of probabilistic forecasts is that the probabilities are reliable, or representative of the frequency of occurrence. The other important quality is that they are sharp, or differ substantially from the climatological probabilities. Diagnostics of these forecast characteristics can be visualized through reliability and attributes diagrams (Fig. 11.6) and quantified

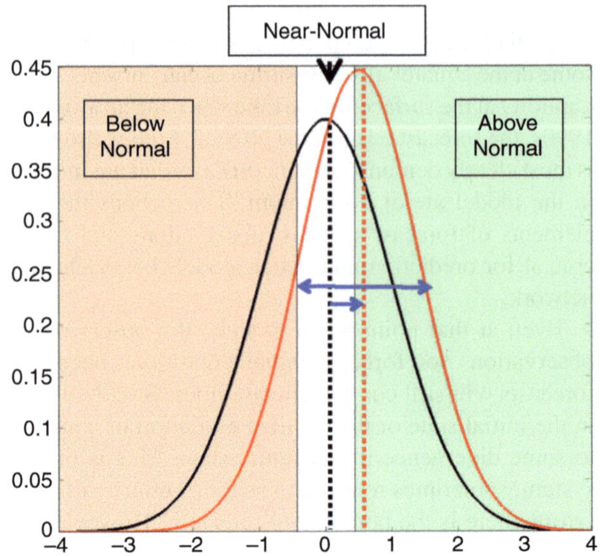

Fig. 11.5 Schematic of a probabilistic forecast distribution relative to a distribution of the historical climate observations. The values (*horizontal axis*) have been normalized with a mean of 0. The *shaded* regions represent the above-normal tercile (*green*) and the below-normal tercile (*brown*) of the historical distribution

through reliability and resolution skill scores, respectively [57]. A reliability diagram shows the complete joint distribution of forecasts and observations for a probabilistic forecast of an event or forecast category (such as the above-normal tercile). In a reliable forecast system, the probability assigned to a particular outcome should be the frequency with which – given the same forecast – that outcome should be observed. The information supplied by reliability diagrams includes calibration, or what is observed given a specific forecast (e.g., under and overforecasting), as well as resolution and refinement which is the frequency distribution of each of the possible forecasts giving information on the degree of aggregate forecaster confidence (small inset graph in Fig. 11.6). Reliability diagrams can further indicate whether there are systematic biases in the forecasts, such as not predicting enough occurrences of above-normal temperatures. Such probabilistic verification, such as reliability diagrams also can be useful for estimating event-specific prediction skill, for example if El Niño events were better predicted than La Niña events or drought conditions were better predicted than very wet seasons. A distinction in prediction skill between the cases of high and low variability calls for further examination of the physical causes of the discrepancy and whether it is inherent to the climate system dynamics or a shortcoming of the model(s).

It is a common feature of dynamical model predictions to be overconfident, indicated by a reliability curve that is more horizontal than the 45° angle that would indicate a reliable prediction system. For example, an overconfident forecast would be one in which a forecast that indicates above-normal rainfall is 80% likely in a given season, but overtime that forecast is followed by observations of above-normal rainfall 40% of the time. Such overconfidence can arise from errors in both the forecast signal and the forecast uncertainty.

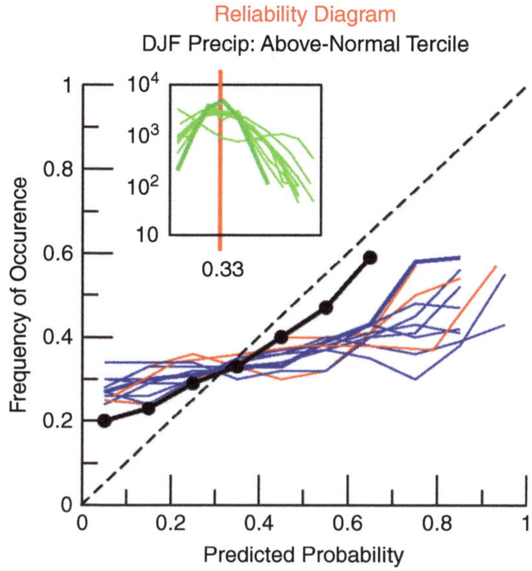

Fig. 11.6 An example of a reliability diagram, which indicates the skill of probabilistic forecasts. The diagram compares the forecasted probability of an event (in this case, above-normal winter rainfall in North America) to its observed frequency. A perfect forecast is represented by the *dashed line*, a *horizontal line* represents a forecast identical to climatology, and *sloped lines* are potentially skillful. The *blue* and *red lines* correspond to individual CGCMs and AGCMs, respectively, and are more horizontal than the *black line*, which represents the mean of these models. While the mean of the models is more reliable than any of the individual models, it tends to be underconfident for rare events (the *black line* lies above the perfect forecast line for low-probability events). Typically, a histogram accompanies a reliability diagram (*inset*), indicating the number of times that forecasts of various confidence levels were issued (Source: Adapted from [58])

Recalibration of predictions and multi-model ensembling are two approaches used to improve forecast reliability. Multi-model ensembling, which combines the prediction of several dynamical models, can improve the reliability and overall skill of predictions in two ways. First, although all models have errors, they do not necessarily have the same errors, thus combining the models reduces the systematic errors that would exist in the prediction from a single model. This can lead to reduced error and thus increased correlation skill in El Niño predictions, for example (Fig. 11.7) [59]. Similarly, it can increase the spatial coverage for where there is skill in capturing the predictable signals in the climate. The second advantage is the improvement in uncertainty estimation by considering the random errors and different parameterizations of random processes that give rise to the range of possible outcomes.

Multi-model ensembling can lead to overall better information on the climate signal and its uncertainty [60], and thus on forecast reliability (Fig. 11.8). Different approaches exist to combine models. The most straightforward is to treat all models equally. Particularly for prediction systems with short retrospective forecast

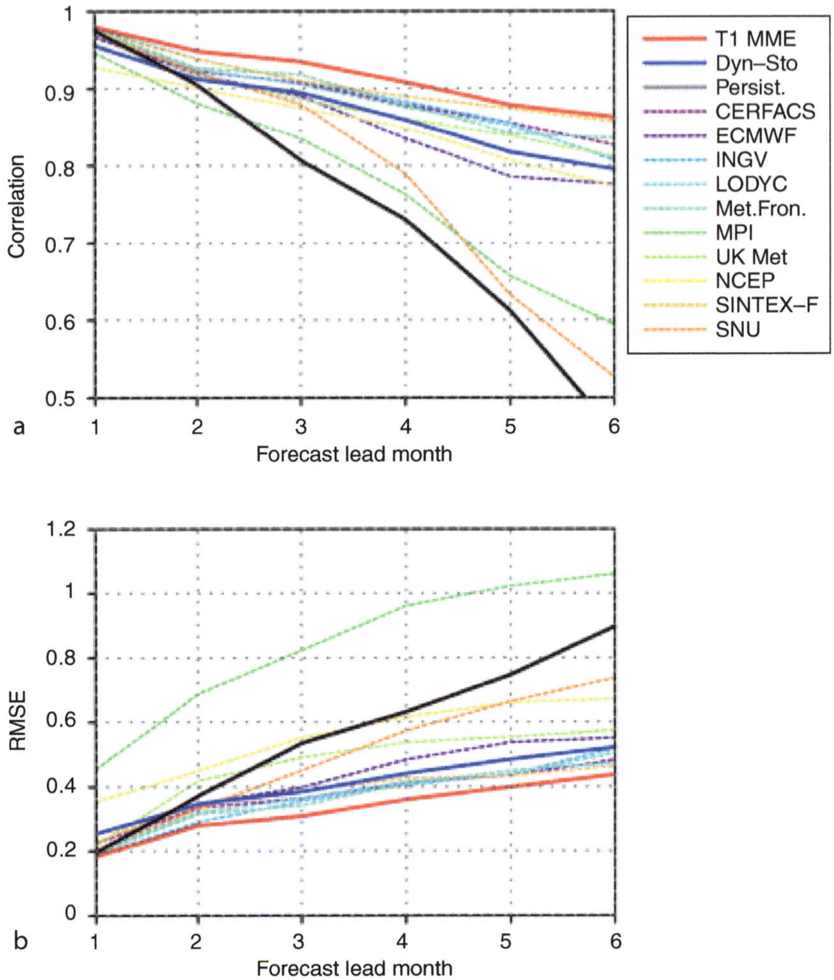

Fig. 11.7 (**a**) Anomaly correlation coefficients and (**b**) RMSE of NINO3.4 index during 1980–2001 with respect to lead time after removing the mean bias. The mean skill for all four cases including February, May, August, and November initial conditions is shown. *Black* for observation, *red* for 10 CGCM multi-model ensemble, *blue* for the Stat-Dyn forecast, and *colored dots* for individual coupled models as shown in the legend, respectively (From [59])

histories of about 25 years or less, it will be difficult to discern differences in forecast quality between comparable models. This is the typically situation with one-tier prediction systems that use coupled ocean–atmosphere models, because the ocean observations used in the forecast initialization is only available since the late 1980s. For two-tiered forecast systems that use atmosphere-only models the ocean temperatures can be predicted statistically, which allows for longer histories of retrospective forecasts. In these systems, it becomes possible to discern differences

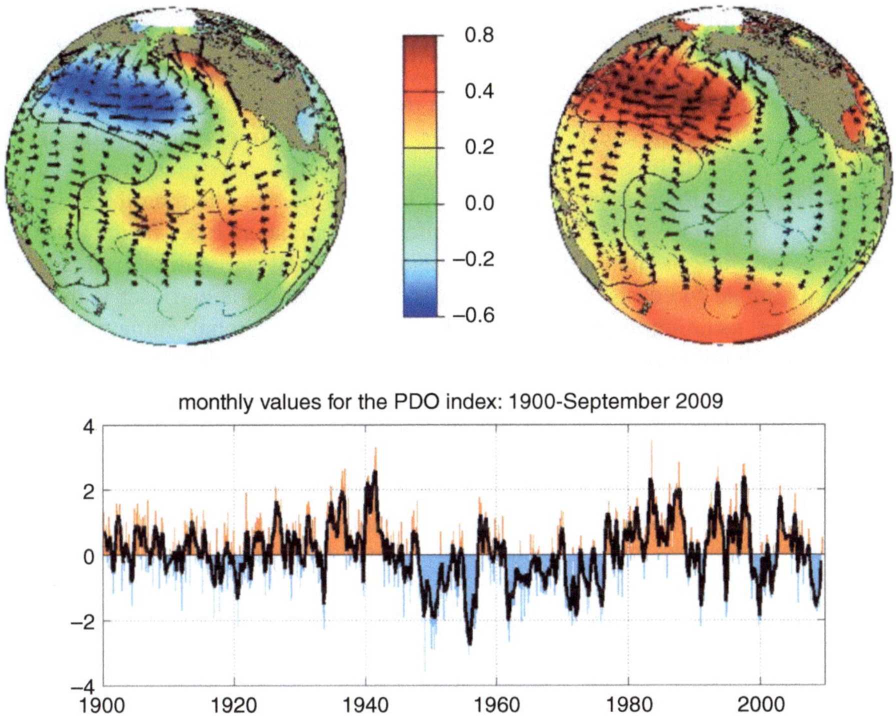

Fig. 11.8 (**a**) The positive (*left*) and negative (*right*) phases of the "Pacific Decadal Oscillation" (PDO), a long-lived El Niño-like pattern of Pacific climate variability, shown in terms of (**a**, *left*) the positive phase in sea surface temperatures (*colors*), sea level pressure anomalies (*contours*), and anomalous low-level winds (*arrows*), and (**b**) the time series of the PDO (Taken from http://jisao.washington.edu/pdo/)

in regional forecast performance and to use that information to give more weight to better performing models, which can lead to further improvements in forecast reliability [61, 62].

An alternative approach to performance-weighting models is to recalibrate the models prior to combination [63]. Recalibration has the advantage of improving forecast quality of individual models. It is also more viable for prediction systems with limited retrospective forecast histories, although a minimum of about 25 years is still required to identify systematic biases in seasonal-to-interannual variability. The recalibration of predictions is an attempt to account for systematic biases in both the signal and uncertainty in the predictions at a given location. Recalibration can also be used to account for spatial biases in the forecasts by comparing observed and predicted seasonal climate over several decades [9].

Reliable forecast information may still not provide enough specificity for those who wish to include seasonal-to-interannual climate forecasts in their decision models, such as those in the agricultural or water sectors. The spatial mismatch of the information, the fact that decision makers in sectors such as agriculture and water require information at much higher resolution, even if it means greater uncertainty, is a commonly cited reason for not using the operational forecasts [64]. There may also exist the desire for greater temporal resolution, such as the characterization of the weather within the climate that might predict the likely number of dry spells of a given duration. In some cases, certain weather characteristics of the seasonal climate may be more predictable than the seasonal totals (e.g., [65]). One way to address the information mismatch between the coarse spatial resolution, or the quality of the higher temporal variability, from global seasonal climate forecasts and the more detailed needs of the end user is through downscaling techniques. In statistical downscaling, the global climate forecast provides the input parameters for an empirical model with high spatial resolution. Statistical techniques can also be used to infer the signal in the weather characteristics relative to the seasonal mean, based on changes in the large-scale background climate, such as those empirically related to ENSO [66, 67], or to changes in atmospheric circulation [65]. In the dynamical downscaling, the global forecast is used to provide lateral boundary conditions to a high-resolution nested regional atmospheric model. While it may provide greater detail of the mean climate by better resolving terrain and coastlines, it has not been robustly demonstrated that dynamical downscaling improves prediction of the climate variability relative to the global model. Dynamical downscaling cannot overcome large-scale errors in the global model driving the nested model, and in many cases will exacerbate those errors. With increases in computing power, global climate models are starting to close the gap by providing fine spatial resolution, and attempting to provide better representation of weather transients that may be of interest to the end user. However, for the next decade or so downscaling techniques, particularly statistical downscaling will continue to add value to seasonal-to-interannual forecasts.

Seasonal-to-Interannual Prediction: Forecast Skill

Forecast skill is a measure of how accurately the prediction system can predict the observed climate variability or how well the probabilities describe the frequency of occurrence of particular outcomes. Measures of accuracy between the best guess, or most likely outcome, of the forecasts and what was observed are often referred to as deterministic measures, meaning they are concerned with verifying the prediction for a single specific outcome, such as a prediction for an above-normal temperature or 2°C warmer than average in the coming season. The metrics for deterministic quantitative forecasts include the Brier skill score and its decomposition, which includes anomaly correlations or root-mean squared errors. The quality, or skill, of deterministic categorical forecasts can be assessed using a variety of measures. There is no single measure of forecast performance that can indicate all aspects of forecast quality [68]. Additionally, forecast producers may be interested in different aspects of forecast performance than users of the forecast information. The World Meteorological Organization has compiled a list of recommended deterministic and probabilistic verification measures for seasonal predictions entitled, The Standard Verification System for Long-Range Forecasts [69].

Keeping in mind that there are different, complementary measures of forecast skill, the accuracy of predictions is typically used to estimate the limit of predictability. The limit of predictability is a function of the predictable signal and the unpredictable chaotic, or noise, component in the climate system. With an ensemble of predictions from a single model or a set of models, the signal and noise can be estimated from that set of information. The signal would be the predicted information that the ensemble has in common and the noise is the range of discrepancy about the signal (Fig. 11.5). When the forecast is initialized, the ensemble contains very little noise, but as the prediction proceeds, the chaotic processes in the climate system lead to divergence of the ensemble members. The limit of temporal predictability is reached once the magnitude of the noise becomes comparable to the signal. This in part, determines how far into the future the certain aspects of the climate can be predicted. Similarly, the average signal-to-noise ratio for a given region, season, variable, etc. describes the expected climate predictability in that case. Since Nature has only one realization, it is not possible to estimate the inherent limit of predictability of the climate system [9]. Estimates of the limit of predictability can be determined in a given prediction system as described above, but that will be only an estimate, and will be different for different forecast systems. At best, the most accurate prediction system for a given region, season, variable, etc. represents the limit of predictability for that case, and should be considered the lower limit of predictability, as the prediction accuracy is found to be at least that good and may improve further with improved models and data assimilation systems.

Given that real-time predictions have been in production for more than a decade now [1], several properties of forecast skill have emerged for seasonal-to-interannual predictions. First, predictions of seasonal mean temperature are more predictable than those for seasonal precipitation totals. This is related in part to the larger-scale nature of temperature anomalies and the processes behind them. Even the coarse resolution global climate models can represent fairly accurately the changes in seasonal temperatures. Precipitation processes and patterns have much smaller spatial scales and are more affected by local scale features. While the global models may be able to capture large-scale shifts in regions of convection and storm tracks, they may have difficulty with the characteristics of storms or local convective activity. The potential importance of local scale processes on precipitation variability also means that the noise component of seasonal precipitation variability is larger than that for temperature. As a result, more ensemble members are required to estimate the seasonal signal for precipitation than for temperature. The second robust property of seasonal predictions is that the tropics are much more predictable than are the extratropics. In the tropics, the atmospheric circulation is more explicitly tied to the changes in patterns of surface temperatures, and the noise in the resulting atmospheric circulation is relatively small. A third and notable property of the predictions, which actually applies to predictions at all timescales, is that there is conditional skill in the expected accuracy. There are times when the initial and evolving state of the climate system carries a much larger predictable signal than other times. For seasonal-to-interannual forecasts this coincides with El Niño events. For seasonal predictions over the United States most predictability derives from El Niño or La Niña conditions [70]. Similar results hold on a global scale too; the fraction of land area over which skillful forecasts can be made is up to twice as large during El Niño or La Niña conditions than in their absence [71]. Moreover, since these events have an inherent timescale of 6–12 months, or longer, the time horizon into the future that skillful forecasts can be issued is also expanded.

Seasonal-to-Interannual Prediction: Internationally Coordinated Efforts

Several internationally coordinated efforts have led to the understanding of seasonal-to-interannual climate variability and its prediction using dynamical models. One of the earliest was the Atmospheric Model Intercomparison Project (AMIP) [72]. This project was organized by the Working Group on Numerical Experimentation as a contribution to the World Climate Research Programme. Different atmospheric models were run with the same observed sea surface temperatures as boundary conditions for the period 1979–1988. The goal was to identify systematic errors as well as systematic responses to the boundary conditions across models. Without such a coordinated effort there had been questions whether the results from a single model were particular to that model or a more robust response expected of the climate system. Other coordinated activities followed.

In the late 1990s, experiments were carried out using different atmospheric models to test the predictability of seasonal climate relative to the variability of sea surface temperatures. Two important issues addressed in that collection of research were the relative impact of initial atmospheric conditions predictability of the seasonal climate and a suggestion that prediction skill could be improved through a multi-model approach. In the United States five modeling centers participated in this research under the Dynamical Seasonal Prediction (DSP) project. On the other side of the Atlantic, 11 different partners throughout Europe contributed to the Prediction of Climate Variations on Seasonal to Interannual Timescales (PROVOST) project [73].

Further research on seasonal predictability and the value of multi-model ensembles was conducted with coupled models from seven European modeling centers under the Development of a European Multimodel Ensemble system for seasonal to inTERannual prediction (DEMETER) project [74]. This project also encouraged research to determine the value of seasonal predictions through their use in models that use the climate data to make prediction over a wide range of interests, from agriculture to health. The next generation of DEMETER was ENSEMBLES, which continued to advance methods and application of seasonal predictions from European Earth system models, thus adding complexity to the dynamical prediction models [75]. The ENSEMBLES project also began to extend those predictions to decadal timescales. The Working Group on Seasonal to Interannual Prediction under the World Climate Research Programme is currently coordinating the Climate-system Historical Forecast Project (CHFP), which will provide access to a wide range of hindcasts to evaluate subseasonal-to-decadal predictions of the climate system, which also aims to quantify the predictability added by elements other than sea surface temperatures, for example through initialization and prediction of the land surface, the cryosphere, and the stratosphere [76].

Decadal Prediction (Experiments)

Decadal Prediction: Drivers of Decadal-Scale Climate

Decadal climate predictions sit between the seasonal-to-interannual forecasts of the next months to a year in the future and the climate change projections of 50–100 years in the future. There are many features of the climate system with timescales that vary over decades (decadal variability). The dominant drivers of climate features over decadal timescales are believed to be changing atmospheric composition, mainly increasing greenhouse gases, and slow changes in ocean circulation that lead to slow changes in the pattern of sea surface temperatures. The changing atmospheric composition changes the energy balance of Earth, which leads to warmer temperatures and other associated climate changes that manifest primarily as trends. The temperature trends are not spatially uniform. Ice-albedo feedback in higher latitudes leads to greater rates of warming there than at low latitudes. Land has a lower heat capacity than water, so the continents warm faster than the oceans. Ocean dynamics also play a role in the patterns of climate change warming, particularly in upwelling regions, where the radiative warming is offset by the upward advection of colder ocean water from depth.

What decadal predictions aim to capture that climate change projections do not is the predicted evolution of naturally occurring decadal-scale features. Climate change projections contain these processes and the associated variability, but since the climate system is not initialized with observations, the decadal evolution will not be temporally consistent with the observations. So one first test of a model is to see whether it is capable of simulating the dominant decadal-scale features observed in Nature.

Decadal-scale variability has been identified in Nature in both the Pacific and Atlantic Oceans. In the Pacific Ocean the variability is referred to as the Pacific Decadal Oscillation (PDO), or more correctly Pacific Decadal Variability (PDV). The pattern of PDV (Fig. 11.8a) has its signature in sea surface temperatures with cooler than normal temperatures in the midlatitudes of the North Pacific Ocean and warmer than normal temperatures in the eastern and central equatorial Pacific Ocean during the positive PDV conditions [22]. The time series associated with the projection of sea surface temperature anomalies on this pattern represents the PDO index (Fig. 11.8b). This sea surface temperature pattern is reminiscent of El Niño conditions, except that the magnitude of sea surface temperature anomalies is larger in the midlatitudes than in the tropics, and the tropical sea surface temperatures have a broader meridional extent. This pattern of sea surface temperatures is accompanied by sea level pressure anomalies in the North Pacific. A measure of the time series of changes in North Pacific sea level pressures is known as the North Pacific Pressure Index (NPPI). It was later realized that there is symmetry in the Pacific decadal variability such that a similar pattern of cooler than normal sea surface temperatures and anomalous low sea level pressure is also found in the midlatitudes of the South Pacific Ocean. The full Pacific view of decadal variability has been named the Interdecadal Pacific Oscillation (IPO, [77]). However, the PDO is the more commonly used index outside Australia.

The symmetry of ocean–atmosphere anomalies outside the tropics, and the resemblance to El Niño, suggests a role for El Niño in driving PDV. It is also notable that there is considerable year-to-year fluctuation in the PDO index. It is very difficult to identify in any particular year what phase, positive or negative, the PDV is in because within the protracted periods in which the PDV is preferentially of one sign or the other, there exist excursions of the index of opposite sign that may only last a year or two.

Simple model experiments have shown that El Niño events can affect the positive phase of PDV [41]. Model analysis suggests an atmospheric Rossby wave train emanating from anomalous convective heating in the central Pacific leads to anomalous low sea level pressure in the region of the Aleutian low, thus strengthening the westerly trade winds. The strengthened winds lead to cooling through enhanced evaporation and also drive southward Ekman flow that brings colder water from the north southward. Those changes in the ocean mixed layer can be sequestered from the atmosphere from one winter to the next due to changes in the ocean mixed layer depth and its connection to the surface from winter, when El Niño peaks, the storm track is strongest and the atmosphere can directly affect the upper ocean, to summer when the previous El Niño would have decayed, the storm track is relatively weak, and increased solar radiation stabilizes the upper ocean. The following winter when the westerly winds of the storm track again increase, the sequestered mixed layer temperature anomalies reemerge [78]. This reemergence mechanism is hypothesized to the main way that the year-to-year variability associated with El Niño and La Niña can be rectified into longer timescale variability. However, other processes may also contribute to PDV. Some mechanisms that have been proposed included ocean–atmosphere coupling of a basin gyre mode [25], excitation of midlatitude oceanic Rossby waves [79], and a complementary, possibly independent oscillation driven by the tropics particularly when El Niño events are focused toward the central equatorial Pacific [80].

Associated with the decadal changes in Pacific Ocean conditions, decadal-scale terrestrial climate anomalies have also been identified over the United States [22] and throughout the Pacific sector [81]. Many of these climate anomalies are consistent with El Niño-related teleconnection patterns, such as wetter conditions in the southern tier of the United States and drier conditions over the Pacific Northwest [82]. Although only a few realizations of each phase of PDV exist in the instrumental records, the broad pattern seems to be consistent across these cases. However, because it is likely that El Niño is a dominant driver of PDV, and is associated with similar terrestrial teleconnections, it is difficult to say with confidence that the PDV is somehow independent of the mere existence of extended periods when El Niño events are stronger or more frequent versus when El Niño events are weaker or less frequent.

Decadal-scale variability in the Atlantic is referred to as the Atlantic Multi-decadal Oscillation (AMO), or more correctly Atlantic Multi-decadal Variability (AMV), because there does not seem to be a spectral peak signaling a true oscillation. Because positive AMV conditions are associated with warming throughout the North Atlantic (Fig. 11.9), the index of AMV is simply the sea surface temperature anomaly averaged over the North Atlantic, and it is often detrended [83]. Other more elaborate means of isolating decadal-scale variability over the Atlantic have been used (e.g., [24]), but result in very similar time series, so the simple index is now the one most widely used.

The hypothesized mechanism driving the AMV is associated with changes in the Atlantic Meridional Overturning Circulation (AMOC). The AMOC brings warm and salty water from the tropical Atlantic poleward. At high latitudes, cold salty water becomes denser than the water below it due to heat fluxes from the westerly storm tracks and brine injection from sea ice formation. The heavy surface water then sinks and flows back equatorward as North Atlantic Deep Water. The sinking water is replaced by the surface flow from tropics to high latitudes. If the rate of sinking increases, the poleward flow of surface water increases, bringing more warm tropical water into the midlatitudes. This represents an increase in the strength of the AMOC, and the AMV index becomes positive. If the North Atlantic water gets too warm or if it freshens the rate of sinking water slows down, and the rate of transport of warm tropical water poleward slows down. This represents a decrease in the strength of the AMOC, and the AMV will become negative.

The AMOC is forced on all timescales. Because the Gulf Stream is the western boundary current of the wind-driven ocean gyre as well as contributing to the AMOC, changes in the winds will affect the AMOC as well as ocean temperatures. However, the multi-decadal-scale variability described above is a much slower process related to the inertia of the overturning circulation and the associated impact on the density properties of the Atlantic Ocean. Since observations of the AMOC have become available only since the end of the twentieth century, there is not enough observational evidence to quantitatively link the sea surface temperatures of the AMV with multi-decadal variability of the AMOC. However, the low-frequency variability of AMOC in some models is associated with a pattern of sea surface temperature anomalies that closely resembles the observed AMV pattern [83]. What has not been resolved is what process or collection of processes can influence the AMOC on long timescales. Some studies point to modification of the strength and local of the intertropical convergence zone over the Atlantic as a way to modify the salinity of the water transported from the tropics [84]. Others suggest that the North Atlantic Oscillation (NAO – also called the Arctic Oscillation, AO) plays a dominant role by influencing the strength of the winds, which then influence the rate of convection, or sinking of heavy water, in the Labrador Sea region with an estimated 10-year lag time [85]. Although this would be a fairly white noise process, the suggestion is that the ocean integrates the noise into a longer timescale red noise process, but one that might still carry some predictability due to persistence.

The teleconnections associated with the positive phase of AMV include wetter conditions over the Sahel and India and drier conditions over northeast Brazil, due to the northward shift of the intertropical convergence zone toward the relatively warmer conditions north of the equator [29]. Also the warm tropical North Atlantic provides more fuel for the growth of tropical storms, and empirically it is seen that more tropical storms grow to hurricane intensity during the positive, warm phase of AMV than during the negative or cool phase.

Decadal Prediction: Model Fidelity

The task of judging whether a model captures decadal variability in the Pacific or Atlantic oceans for the right reason is greatly complicated by the limited history of observations compared to the timescale of the variability. Since the measurements of surface temperature and sea level pressure go back to the nineteenth century, most comparisons are made to these fields. In many cases, the question of model fidelity is closely tied to examination of the processes involved in producing the variability in a particular model and to what extent those are observed in Nature. The difficulty with that approach is as described in the previous section: different models may not agree on which process(es) dominates, or is even involved.

Different models also may yield somewhat different spatial patterns or spectra than suggested by the limited observations [86]. A similar situation exists relative to El Niño in coupled dynamical models [87], and this has not prevented the use of those models for El Niño prediction. Thus, the most important factor may be simply whether or not a model is capable of capturing a reasonable representation of decadal variability in sea surface temperatures, as this is how whatever changes are occurring in the ocean will be communicated to the atmosphere.

Exactly how best to validate decadal variability in the models is an area of active research. To date, more work has been focused around Atlantic variability perhaps because of the recognition of the role of the AMOC in the AMV. The idea is that if the AMOC is responsible for the AMV, then it is the AMOC that a model must be able to predict from a given set of initial conditions. The AMOC must then produce a surface temperature of reasonable resemblance to Nature, and the overlying

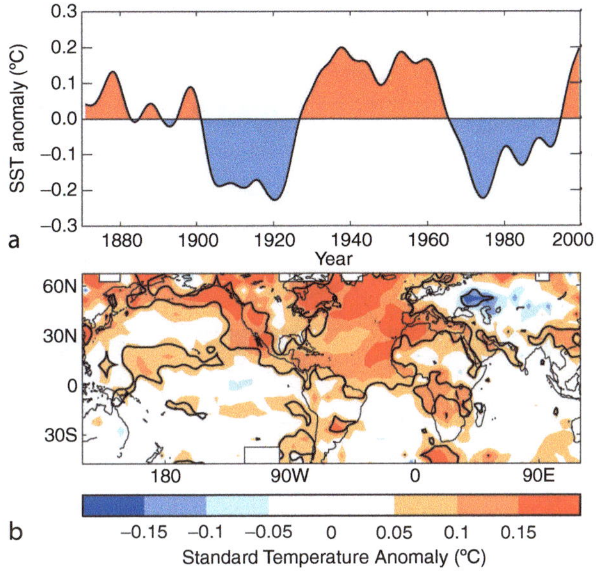

Fig. 11.9 (**a**) AMO index derived from detrended area-weighted mean North Atlantic SST anomalies by using a Chebyshev filter with a half-power period of 13.3 years. (**b**) Surface temperature anomaly associated with one positive standard deviation of the AMO index, calculated by regression of surface temperatures with the index and scaled by its standard deviation. The solid contour bounds regions significant at the 90% limit of a two-sided *t*-test accounting for autocorrelation (Source: From [83])

atmosphere must be able to respond to the changes in surface temperature in a way that captures the observed teleconnections. Again, the difficulty is that very few realizations of the variability exist in the observational record, although paleoclimate reconstructions of past temperature or precipitation suggest that, for example, multi-decadal variability consistent with observed AMV has impacted regional climate 400–500 years back [88]. But the few realizations of the spatial pattern of sea surface temperatures makes it difficult to know which parts of the pattern of anomalies are robust across events and which are variable from one positive phase to another.

For PDV, most models capture the response of the midlatitude Pacific Ocean to El Niño variability. However, they often do not demonstrate the same level of multi-year persistence through a reemergence mechanism. Different models are also influenced to differing degrees by other processes hypothesized to contribute to PDV, including the white noise imposed by variability in the storm track. As a result the patterns of PDV, such as where the sea surface temperature anomalies are focused and the magnitude of that temperature variance, also differ among models.

Some modeling studies have shown that at least the atmospheric models can translate the changes in patterns of sea surface temperature into realistic teleconnections. For example, using observed heat fluxes from the positive, warm phase of AMV in the Atlantic Ocean to drive an atmospheric model leads to decadal-scale changes consistent with observed changes in precipitation over the Sahel and India and also in the wind shear over the tropical North Atlantic relevant to hurricane formation (Fig. 11.10) [30]. Thus as with El Niño, if the sea surface temperatures can be predicted then there may be at least some predictability of the associated terrestrial climate impacts.

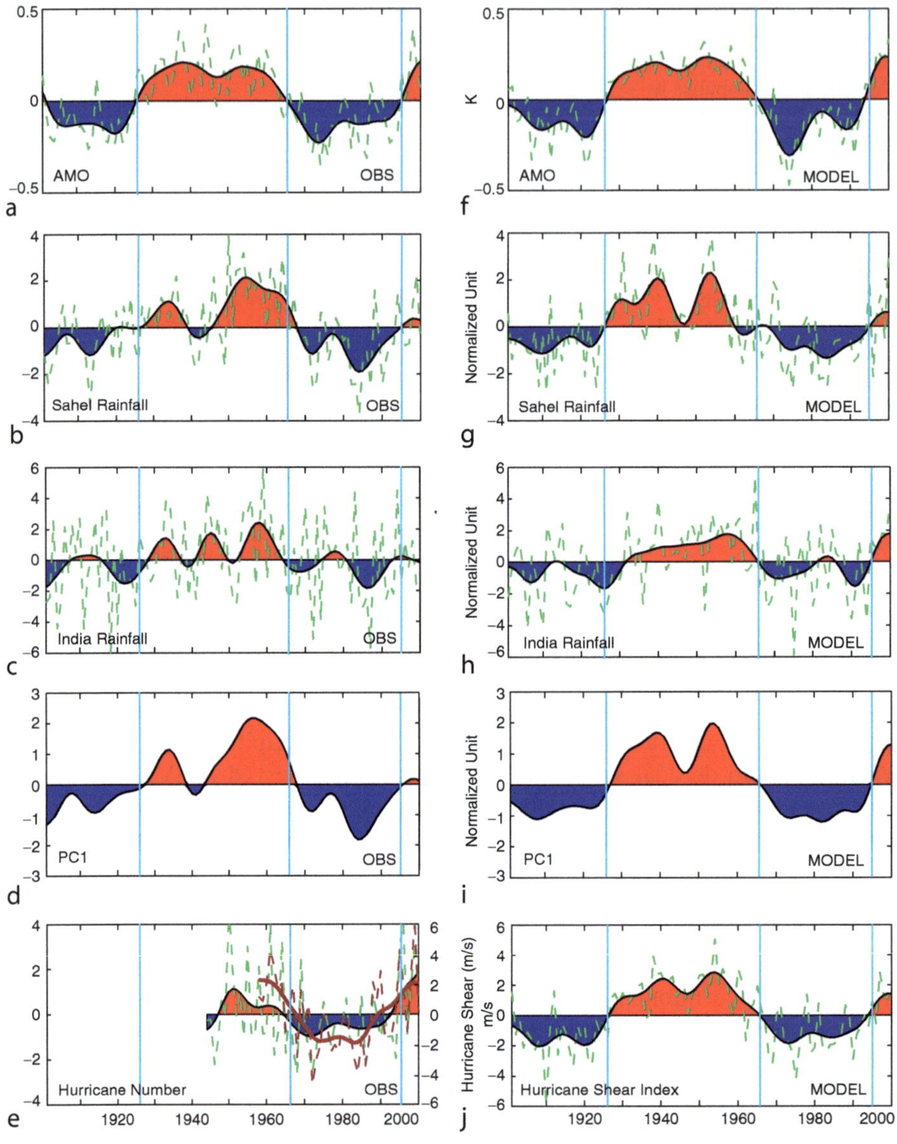

Fig. 11.10 Observed and modeled variability. The *color shading* is the low-pass filtered (*LF*) data and the *green dash line* is the unfiltered data. (**a**) Observed AMO Index (K). (**b**) Observed JJAS Sahel rainfall anomalies (averaged over 20°W–40°E, 10–20°N). (**c**) Observed JJAS west central India rainfall anomalies (averaged over 65–80°E, 15–25°N). (**d**) Observed time series of the dominant pattern (PC 1) of low-frequency JJAS rainfall anomalies. (**e**) Observed anomalous Atlantic major Hurricane number (axis on the *left*, original data from the Atlantic basin hurricane database-HURDAT, with no bias-type corrections from 1944 to 1969, there is no reliable data before 1944) and observed Hurricane Shear Index (1958–2000) (m/s, *brown solid line* for LF data, *brown dash line* for unfiltered data, axis on the *right*). (**f**) Modeled AMO Index (K). (**g**) Modeled JJAS Sahel rainfall anomalies. (**h**) Modeled JJAS west central India rainfall anomalies. (**i**) Modeled PC 1 of LF JJAS rainfall anomalies. (**j**) Modeled Hurricane Shear Index (m/s) (Source: From [30])

Decadal Prediction: Prediction Experiments

Experimental decadal prediction has only recently begun. Decadal prediction differs from climate change projections in the initialization of the climate system, with particular emphasis on initialization of the oceans. The first paper demonstrating actual retrospective, decade-long, initialized forecasts was published in 2007 [3]. They showed improvements in prediction of globally averaged temperatures relative to the un-initialized climate change projections from the same model. However, it was not clear how much predictive information is available at the regional scale from these predictions, and it is not obvious what the main drivers are behind any predictive information they may yield.

The prediction systems for decadal prediction are essentially the same as for seasonal-to-interannual prediction. They require observations, models, and their connection through data assimilation systems. One of the main differences is the requirements on the observations: seasonal-to-interannual predictions mainly need information about the upper several hundred meters of the tropical oceans; decadal predictions require information about the global oceans, including the middle and high latitudes and also to much greater depths to capture information on the lower branch of the AMOC.

Observations needed to produce initial ocean conditions are incomplete. The creation of retrospective forecasts of decadal variability at least several decades into the past requires information on salinity fields that just do not exist. This has tested the limits of ocean state estimation with limited data, and the estimates even for large-scale averages, such as the average salinity anomaly in the upper 700 m of the midlatitudes of the Atlantic Ocean, can vary greatly. The uncertainty among datasets for upper ocean salinity anomalies on basin scales is larger than the variability within a single dataset [89]. Since the beginning of the twenty-first century, however, the Argo program of drifting buoys has provided unprecedented measurements of the upper 2 km of the global ocean. The floats measure temperature and salinity profiles as they descend and ascend the water column about every 10 days. There are currently over 3,000 floats reporting data through satellites (Fig. 11.11). Even with good observational data coverage of the global oceans there will still be challenges in merging those data efficiently with models through data assimilation systems to account for both mean biases and biases in space-time variability.

The first step in exploring decadal prediction has been through perfect model studies. In perfect model studies, a free-running integration of the model is taken as truth; this integration assumes the role of the "observations." Ensemble members are set up to start from a particular point in the free-running integration with small perturbations to the initial state, representing the uncertainty in initial conditions. The ensemble members are then integrated forward to see how well they can track the "truth" of the free-running integration. In this experimental setup, the "observations" are perfect because since they are taken from the model they are known everywhere, and the model is perfect, because it is dynamically consistent

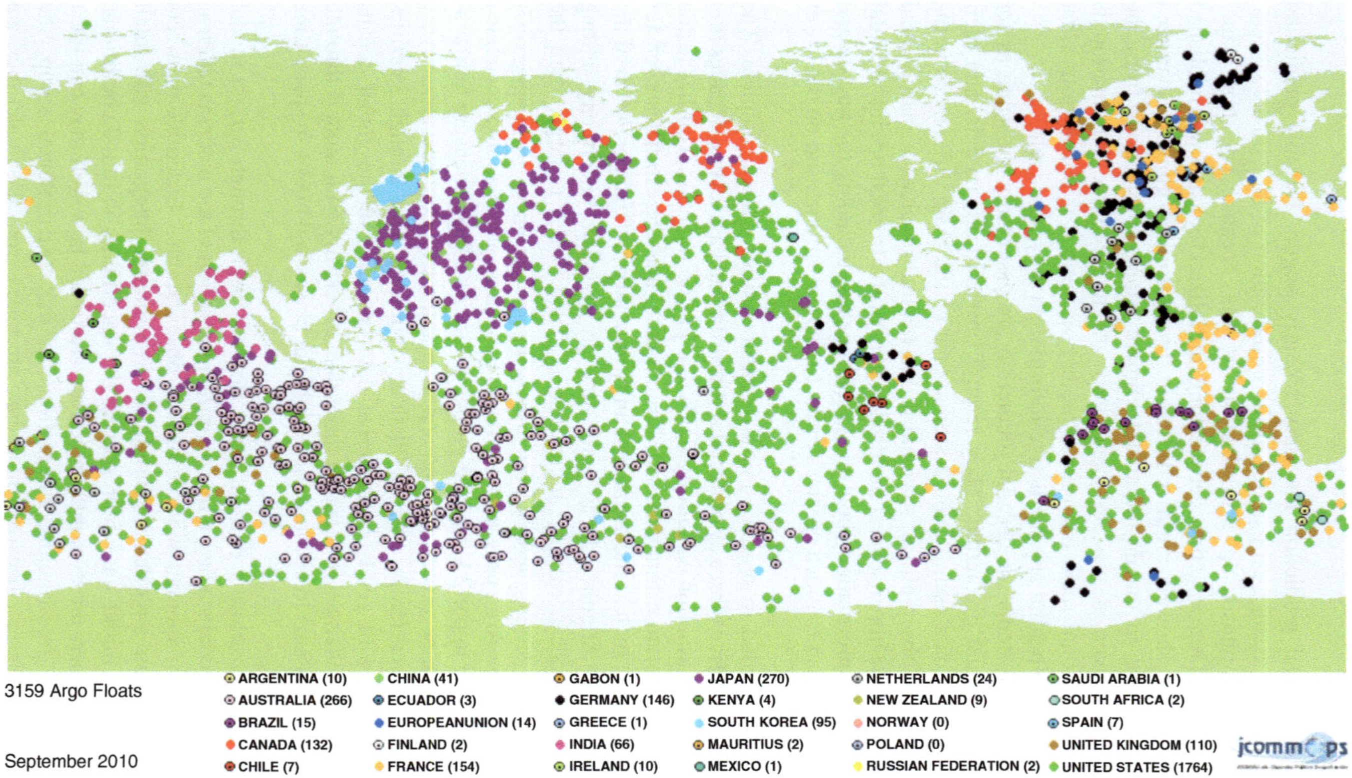

Fig. 11.11 Map of location of Argo floats and the countries that deployed them, as of September 2010. Note that *dot* size is greatly exaggerated relative to size of floats (Source: http://wo.jcommops.org/cgi-bin/WebObjects/Argo.woa)

with the "observations," which are just a snapshot from the model. Similar experiments toward the design of assimilation systems test the insertion of these perfect "observations" but sampled only at locations that the actual observing network could provide data. The idea is to see if in this most idealized of circumstances – perfect observations and perfect model – the model is able to predict the evolution of the "observed" variability taken from the free-running integration. If not, it implies that in the particular model, too much noise exists to extract a predictable signal. The situation will only be worse in a real forecast setting with imperfect observations in the model that is also not perfect.

A number of these perfect model prediction experiments have been carried out since the early part of the twenty-first century. In a coordinated experiment of five European modeling centers, called PREDICATE, two to three experiments were carried out by each group starting their ensemble predictions at different points in time to explore the prediction dependence on the variability of the AMOC. In all cases there was some skill in predicting the evolution of the AMOC (Fig. 11.12). The experiments also demonstrated conditional predictability much like is seen with El Niño predictions. The perfect model predictions started when the AMOC was stronger than average yielded predictability of the AMOC to about 10–15 years into the future; predictions started with a weak AMOC predicted the future evolution of the AMOC only 2–5 years into the future [90]. The PREDICATE experiments were based on model control runs, meaning that atmospheric composition was held fixed. More recent perfect model studies explore the relative predictive signal due to the initial conditions versus due to radiative forcing from increasing greenhouse gases [91]. Several common lessons are beginning to emerge from these studies. One is that the predictable time horizon, when the signal in the ensemble of predictions is larger than the uncertainty across ensemble members, is longer for midlatitudes than for the tropics due to the dominance of year-to-year variability in the tropical oceans. Another lesson is that upper ocean heat content is more predictable than sea surface temperature due to the impact of weather noise on surface temperatures, while the upper ocean temperatures are more reflective of the slow changes in the atmospheric circulation. Thus even if the AMOC is predictable, the surface temperatures connected with that feature will be less so, but it is this surface expression that is necessary for predicting the terrestrial climate impacts. Finally, it appears that the external forcing due to increasing greenhouse gases becomes comparable to the information from ocean initial conditions by 10 years out for the midlatitudes and less in the tropics. Again, these are perfect model results. However, such results only indicate the upper limit of predictability for a particular model, and even though the similar results have been found across several of the current models, it is not to say that different results might be possible from better models.

To date, only a few pioneering attempts have been documented of "retrospective forecasts," which are decadal predictions initialized with real observed initial conditions from some time ago. These prediction experiments not only used different models, they also used very different methods to obtain the initial conditions: one initialized with only sea surface temperatures [4], one initialized the

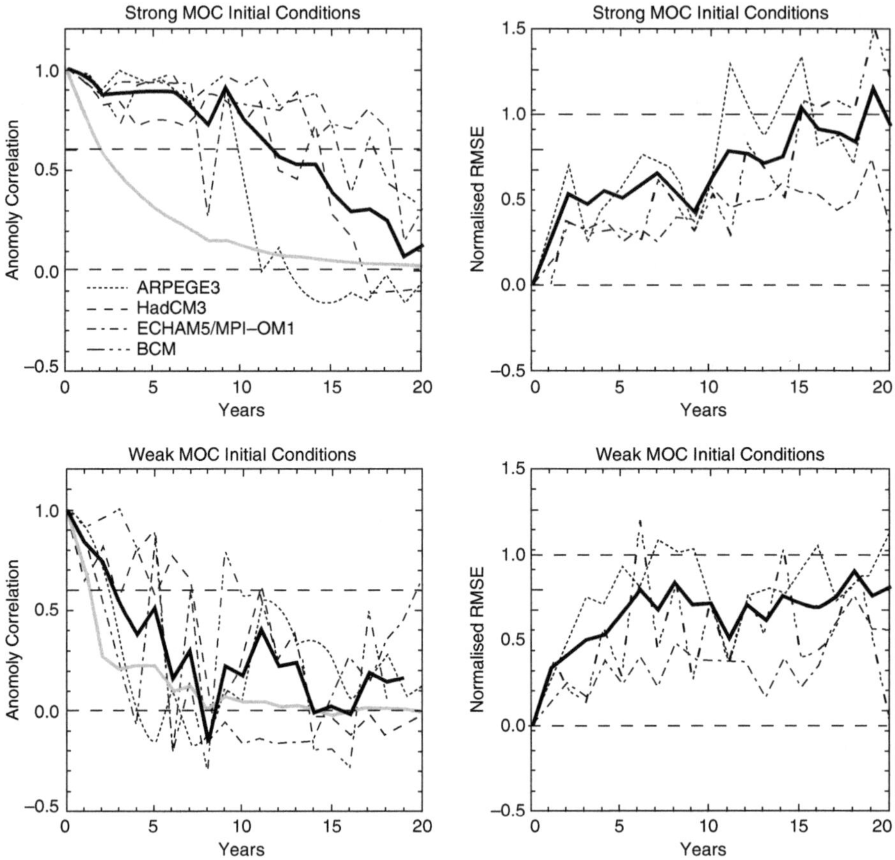

Fig. 11.12 Measures of the potential predictability of variations in the strength of the Atlantic Meridional Overturning Circulation from four of the five coupled models (see legend). (*left*) The anomaly correlation coefficient (ACC: unity for perfect potential predictability; zero for no potential predictability) for (*top*) strong and (*bottom*) weak MOC initial conditions (*right*). The normalized root mean squared error (rmse: zero for perfect potential predictability; unity for no potential predictability) in the same order. Also shown in the figures are the multi-model average ACC and rmse (*thick black line*) and the multi-model average ACC for a simple damped persistence (*thick gray line*) (Source: From [90])

observed data for ocean temperature and salinity anomalies as well as atmospheric anomalies [3], the third nudges their model toward the observational analysis using a different procedure [5].

The results are mixed. All claim to gain benefit from initialization of the climate system compared to the climate change projections that consider only changes in the atmospheric composition. Two of the studies [3, 5] show improvement in global mean temperatures compared to the un-initialized climate change projections; the other study [4] slightly degrades their prediction of global mean temperatures with initialization. All claim, or at least imply, that much of the decadal variability that is captured is due to initialization of the AMOC. Two of the studies [4, 5] demonstrate improved temperature predictions over the eastern North Atlantic region, but for the study that also quantifies the impact of initialization on prediction errors [4] shows larger errors for North Atlantic sea surface temperature in the initialized predictions. The same two studies that show improved correlations for North Atlantic temperature predictions claim that it is due to improved prediction of the AMOC. Since there are no observations of this circulation feature in the twentieth century, the conclusions are based on comparison with the analysis responsible for the initial conditions, which constitutes something like a semi-perfect model result rather than a verified prediction.

Two of the early prediction experiments [3, 4] do show some improvement in regional temperature predictions over land, but how much improvement is not easily discerned, and is difficult to compare across the experiments. There are also some regions where the temperature predictions are less skillful. Maps of skill, or differences in skill, are not provided for precipitation in these studies. These papers are only the beginning of assessment decadal prediction skill.

What these results do or do not show must be viewed with caution though. Several difficulties stand in the way of more conclusive estimates of predictability and prediction skill for decadal climate variability. One difficulty is that the current sets of experiments, and even those that will soon be available (see Decadal Prediction: Internationally Coordinated Efforts) have very few ensemble members. Small ensemble size leads to uncertainty in the predicted signal, and provides very little information about the uncertainty due to uncertain initial conditions in a particular model. Multi-model ensembles will likely be more problematic for decadal predictions given the wide range of approaches to initial conditions; the prediction skill can also be compromised by the data assimilation component, even if the models are of equally high quality. Data assimilation and the development of initialization techniques for current and retrospective decadal predictions require considerable research investment. What experiments do exist and are likely to exist in the next several years will have limited realizations of decadal-scale variability, complicated with the evolution of that variability against a changing background climate due to increasing greenhouse gases. That combined with limited observations, not only of the subsurface ocean but also of terrestrial climate for much of the world, makes verification of retrospective forecasts extremely challenging.

Decadal Prediction: Internationally Coordinated Efforts

Several international efforts have been organized since the beginning of the twenty-first century to ascertain the predictability and prediction skill of decadal climate variability by systematizing the investigation across many models. The PREDICATE project, which was referred to above, provided a systematic comparison of the "perfect model" predictability in five European coupled models. They found potential predictability in the AMOC (Fig. 11.12), and also to some extent in surface air temperatures, that exceeded damped persistence [90]. The PREDICATE project examined the potential predictability of the response of atmospheric models to prescribed sea surface temperatures, such as those associated with the AMV, and found good consistency across the models suggesting potential predictability if the pattern of SST was itself predictable [92].

A more recent activity, also drawing on the European modeling and prediction community is ENSEMBLES ([75]; http://ensembles-eu.metoffice.com/). This was a 5-year climate change research project begun in 2004, and involving 66 research partners across Europe. The project generated retrospective climate forecasts from seasonal to multi-decadal scales, provided local interpolation and/or downscaling, and sought to apply that information to sectoral outlooks, such as agriculture, health, and energy, across Europe.

The most extensive collaboration on decadal prediction experiments is the coordinated experiments designed and being run for the IPCC Fifth Assessment Report [93]. Together with the climate change projections for the next IPCC report, these decadal prediction experiments will be part of the Coupled Model Intercomparison Project-5 (CMIP5). There is a minimal set of runs at the core of the experimental design that requires hindcasts initialized for near the end of 1960 and every 5 years after that to 2005, in each case predicting 10 years past the initialization. Of those 10 sets of experimental start dates, a subset – those initialized at 1960, 1980, and 2005 – will be run out for 30 years. These experiments are to be run with a nominal ensemble size of 3. Of the dozen or more international modeling and prediction centers that will participate in the decadal prediction experiments of CMIP5, several will run with larger ensemble sizes and more start dates.

What is not being coordinated for the CMIP5 decadal prediction experiments is the data assimilation or initialization strategy. The guidelines only require that the predictions begin with a state of the climate system representative of the observations at that time. Thus, although there will likely be prediction systems that perform better than others, considerable analyses and further research will be required to assess which part of various prediction systems are responsible for their relative success or failure.

Future Directions

Although seasonal prediction is a relatively mature activity, considerable room for further improvement exists in the production, provision, and application of seasonal climate forecasts [9]. Dynamical models have many recognized biases in their tropical climate, such as tropical upper ocean structure and mixing, a tendency to produce a double intertropical convergence zone in the Pacific, and poor simulation of the stratus clouds that sit near the coasts along the eastern subtropical oceans.

These problems are probably not unrelated, but they have proved difficult to solve. These tropical biases impact the realism of predicted El Niño events, which introduces biases into the associated teleconnections. Although much of the discussion in this chapter has focused on the climate predictability that arises from tropical SSTs, and especially El Niño, other factors in the climate system that are not well represented of initialized in models may carry additional prediction skill. Such processes include land characteristics [36] such as soil moisture, snow, and vegetation, as well as sea ice, variability in the stratosphere [94], and intra-seasonal variability such as the Madden–Julian Oscillation [95]. The provision of seasonal forecasts has improved since the 1990s; it has become common practice for operational centers to provide probabilistic information. However, that information is not sufficient for many decision makers if it is not accompanied by information on how the forecast is constructed, the past skill of the system, and more flexible or varied information that would allow sophisticated users to incorporate the data into quantitative decision systems. These types of best practices are much easier to address than model biases. Also, if addressed they would allow for broader use of past forecasts for research to underpin the use of current forecasts for decisions.

Decadal prediction is still in the phase of research and experimentation. Thus, decadal prediction itself should be considered a future direction of climate prediction. Although there have been some pioneering studies that present results from decadal prediction systems, there is no community-wide agreement on how decadal prediction systems should be constructed, what information can be provided, with what accuracy, and even how best to verify the information that is predicted [2]. The internationally coordinated set of experiments under CMIP5 should contribute to a better understanding of these prediction systems and their potential. These experimental predictions will build on the current limited understanding to illuminate the relative information provided by initial conditions, in a real forecast setting, compared to the radiative forcing from current and future atmospheric greenhouse gas increases. The prediction experiments taken together should also help identify model biases that are of particular concern to decadal variability and set priorities on the future development and maintenance of the ocean observing system. Also, the added complexities of data assimilation for decadal prediction that will encompass longer timescales, greater depths in the ocean, and more need for salinity information, which was not available for most of the ocean prior to the twenty-first century, will lead to innovations in data assimilation systems (e.g., [96]).

The seasonal and decadal prediction systems share many common elements. In particular, they use the same type of dynamical models, and they both rely heavily on ocean initial conditions interpreted through data assimilation systems. They are both potentially impacted by external forcings such as solar variability and volcanoes. The same model biases that affect seasonal prediction skill will impact decadal predictions also. As the research community develops improved dynamical models, that better represent the Earth system in all its complexity, it will benefit climate predictions at all timescales. The additional observational data and more sophisticated data assimilation systems that are required for initialization of decadal predictions will provide more information of the ocean state that could be relevant for seasonal predictions as well. Already, some efforts to create retrospective seasonal predictions are being run farther into the future to investigate the ability of those systems to predict interannual-to-decadal climate variability. On the other side, retrospective decadal predictions that will contribute to the CMIP5 database already predict through the seasonal timescale.

The larger vision for the future direction of seasonal and decadal prediction is the union of the two efforts. This has been called "seamless prediction" [97, 98], which seeks the seasonal predictions to both the longer-term decadal predictions as well as the shorter-term weather forecasts. Initial steps in bridging the weather and seasonal timescales have been made (e.g., [97]), and since the observational and data assimilation systems are in place and have been well tested for these timescales, it is a sensible starting point. The joining of seasonal and decadal prediction scales would appear to be developing naturally as part of the evolving research into climate variability, predictability, and prediction on these timescales.

Bibliography

Primary Literature

1. Stockdale TN, Alves O, Boer G, Deque M, Ding Y, Kumar A, Kumar K, Landman W, Mason S, Nobre P (2010) Understanding and predicting seasonal-to-interannual climate variability – the producer perspective. Procedia Environ Sci 1:55–80
2. Meehl GA, Goddard L, Murphy J, Stouffer RJ, Boer G, Danabasoglu G, Dixon K, Giorgetta MA, Greene AM, Hawkins E, Hegerl G, Karoly D, Keenlyside N, Kimoto M, Kirtman B, Navarra A, Pulwarty RS, Smith D, Stammer D, Stockdale T (2009) Decadal prediction: can it be skillful? Bull Am Meteorol Soc 90:1467–1485
3. Smith D, Cusack S, Colman A, Folland C, Harris G, Murphy J (2007) Improved surface temperature prediction for the coming decade from a global circulation model. Science 317:796–799
4. Keenlyside NS, Latif M, Jungclaus J, Kornblueh L, Roeckner E (2008) Advancing decadal-scale climate prediction in the North Atlantic sector. Nature 453:84–88
5. Pohlmann H, Jungclaus JH, Köhl A, Stammer D, Marotzke J (2009) Initializing decadal climate predictions with the GECCO oceanic synthesis: effects on the North Atlantic. J Clim 22:3926–3938
6. Seager R, Ting M, Held I, Kushnir Y, Lu J, Vecchi G, Huang H-P, Harnik N, Leetmaa A, Lau N-C, Li C, Velez J, Naik N (2007) Model projections of an imminent transition to a more arid climate in southwestern North America. Science 316:1181–1184
7. Stott PA, Stone DA, Allen MR (2004) Human contribution to the European heatwave of 2003. Nature 432:610–614
8. Trenberth KE, Shea DJ (2006) Atlantic hurricanes and natural variability in 2005. Geophys Res Lett 33:L12704
9. NRC (2010) Assessment of intraseasonal to interannual climate prediction and predictability. The National Academies, Washington, DC
10. Orlove B, Chiang J, Cane M (2000) Forecasting Andean rainfall and crop yield from the influence of El Nino on Pleiades visibility. Nature 403:68–71
11. Walker GS (1923) Correlations in seasonal variations of weather, VIII: a preliminary study of world weather. Mem India Meteorol Dep 24:75–131

12. Ropelewski CF, Halpert MS (1987) Global and regional scale precipitation patterns associated with the El Niño/Southern Oscillation. Mon Weather Rev 115:1606–1626
13. Ropelewski CF, Halpert MS (1989) Precipitation patterns associated with the high index phase of the southern oscillation. J Clim 2:268–284
14. Mason SJ, Goddard L (2001) Probabilistic precipitation anomalies associated with ENSO. Bull Am Meteorol Soc 82:619–638
15. Bjerknes J (1969) Atmospheric teleconnections from the equatorial Pacific. Mon Weather Rev 97:163–172
16. Wyrtki K (1975) El Niño – the dynamic response of the equatorial Pacific Ocean to atmospheric forcing. J Phys Oceanogr 5:572–584
17. Zebiak S, Cane MA (1987) A model of El Nino-Southern Oscillation. Mon Weather Rev 115:2262–2278
18. Battisti DS (1988) Dynamics and thermodynamics of a warming event in a coupled tropical atmosphere–ocean model. J Atmos Sci 45:2889–2919
19. Suarez Max J, Schopf Paul S (1988) A delayed action oscillator for ENSO. J Atmos Sci 45:3283–3287
20. Cane MA, Zebiak SE, Dolan SC (1986) Experimental forecasts of El Niño. Nature 321:827–832
21. Graham NE (1994) Decadal-scale climate variability in the tropical and North Pacific during the 1970s and 1980s: observations and model results. Clim Dyn 10:135–162
22. Mantua NJ, Hare SR, Zhang Y, Wallace JM, Francis RC (1997) A Pacific interdecadal climate oscillation with impacts on Salmon production. Bull Am Meteorol Soc 78:1069–1079
23. Hurrell JW, Van Loon H (1997) Decadal variations in climate associated with the North Atlantic Oscillation. Clim Change 36:301–326
24. Goldenberg SB, Landsea CW, Mestas-Nunez AM, Gray WM (2001) The recent increase in Atlantic hurricane activity: causes and implications. Science 293:474–479
25. Latif M, Barnett TP (1994) Causes of decadal climate variability over the North Pacific and North America. Science 266:634–637
26. Griffies SM, Bryan K (1997) Predictability of North Atlantic multidecadal climate variability. Science 275:181–184
27. IPCC (2007) Contribution of working group 1 to the fourth assessment report of the intergovernmental panel on climate change; Solomon S, Qin D, Manning M, Chen Z, Marquis M, Averyt KB, Tignor M, Miller HL (eds). Cambridge University Press, Cambridge
28. Vera C, Barange M, Dube OP, Goddard L, Griggs D, Kobysheva N, Odada E, Parey S, Polovina J, Poveda G (2010) Needs assessment for climate information on decadal timescales and longer. Procedia Environ Sci 1:275–286
29. Knight JR, Folland CK, Scaife AA (2006) Climate impacts of the atlantic multidecadal oscillation. Geophys Res Lett 33:L17706
30. Zhang R, Delworth TL (2006) Impact of Atlantic multidecadal oscillations on India/Sahel rainfall and Atlantic hurricanes. Geophys Res Lett 33:L17712
31. Schubert S, Gutzler D, Wang H, Dai A, Delworth T, Deser C, Findell K, Fu R, Higgins W, Hoerling M, Kirtman B, Koster R, Kumar A, Legler D, Lettenmaier D, Lyon B, Magana V, Mo K, Nigam S, Pegion P, Phillips A, Pulwarty R, Rind D, Ruiz-Barradas A, Schemm J, Seager R, Stewart R, Suarez M, Syktus J, Ting M, Wang C, Weaver S, Zeng N (2009) A U.S. CLIVAR project to assess and compare the responses of global climate models to drought-related sst forcing patterns: overview and results. J Clim 22:5251–5272
32. Lau N-C, Nath MJ (2003) Atmosphere–ocean variations in the Indo-Pacific sector during ENSO episodes. J Clim 16:3–20
33. Xie S-P, Carton JA (2004) Tropical Atlantic variability: patterns, mechanisms, and impacts. Geophys Monogr 147:121–142
34. Giannini A, Chiang JCH, Cane MA, Kushnir Y, Seager R (2001) The ENSO teleconnection to the Tropical Atlantic Ocean: contributions of the remote and local SSTs to rainfall variability in the Tropical Americas. J Clim 14:4530–4544

35. Goddard L, Graham NE (1999) Importance of the Indian Ocean for simulating rainfall anomalies over eastern and southern Africa. J Geophys Res 104:19099–19116
36. Koster RD, Mahanama SPP, Yamada TJ, Balsamo G, Berg AA, Boisserie M, Dirmeyer PA, Doblas-Reyes FJ, Drewitt G, Gordon CT, Guo Z, Jeong J-H, Lawrence DM, Lee W-S, Li Z, Luo L, Malyshev S, Merryfield WJ, Seneviratne SI, Stanelle T, van den Hurk BJJM, Vitart F, Wood EF (2010) Contribution of land surface initialization to subseasonal forecast skill: first results from a multi-model experiment. Geophys Res Lett 37:L02402
37. Fischer EM, Seneviratne SI, Lüthi D, Schär C (2007) Contribution of land-atmosphere coupling to recent European summer heat waves. Geophys Res Lett 34:L06707
38. Cohen J, Fletcher C (2007) Improved skill of northern hemisphere winter surface temperature predictions based on land–atmosphere fall anomalies. J Clim 20:4118–4132
39. Jhun J-G, Lee E-J (2004) A New East Asian Winter Monsoon Index and associated characteristics of the winter monsoon. J Clim 17:711–726
40. Goddard L, Mason S (2002) Sensitivity of seasonal climate forecasts to persisted SST anomalies. Clim Dyn 19:619–632
41. Alexander MA, Bladé I, Newman M, Lanzante JR, Lau N-C, Scott JD (2002) The atmospheric bridge: the influence of ENSO teleconnections on air–sea interaction over the global oceans. J Clim 15:2205–2231
42. Jin F-F (1997) An equatorial ocean recharge paradigm for ENSO. Part I: conceptual model. J Atmos Sci 54:811–829
43. McPhaden MJ, Busalacchi AJ, Cheney R, Donguy J-R, Gage KS, Halpern D, Ji M, Julian P, Meyers G, Mitchum GT, Niiler PP, Picaut J, Reynolds RW, Smith N, Takeuchi K (1998) The tropical ocean-global atmosphere observing system: a decade of progress. J Geophys Res 103:14169–14240
44. Saha S, Nadiga S, Thiaw C, Wang J, Wang W, Zhang Q, Van den Dool HM, Pan H-L, Moorthi S, Behringer D, Stokes D, Peña M, Lord S, White G, Ebisuzaki W, Peng P, Xie P (2006) The NCEP climate forecast system. J Clim 19:3483–3517
45. Barnston AG, Chelliah M, Goldenberg SB (1997) Documentation of a highly ENSO-related SST region in the equatorial pacific. Atmos–Ocean 35:367–383
46. Larkin NK, Harrison DE (2005) On the definition of El Niño and associated seasonal average U.S. weather anomalies. Geophys Res Lett 32:L13705
47. Chen D, Cane MA, Kaplan A, Zebiak SE, Huang D (2004) Predictability of El Niño over the past 148 years. Nature 428:733–736
48. Kirtman Ben P, Shukla J, Huang B, Zhu Z, Schneider Edwin K (1997) Multiseasonal predictions with a coupled tropical ocean global atmosphere system. Mon Weather Rev 125:789–808
49. Barnston AG, He Y, Glantz MH (1999) Predictive skill of statistical and dynamical climate models in SST forecasts during the 1997–98 El Niño Episode and the 1998 La Niña Onset. Bull Am Meteorol Soc 80:217–243
50. Goddard L, DeWitt DG (2005) Seeking progress. In: El Niño Prediction, vol 3. US CLIVAR Variations
51. Davey M, Huddleston M, Sperber K, Braconnot P, Bryan F, Chen D, Colman R, Cooper C, Cubasch U, Delecluse P, DeWitt D, Fairhead L, Flato G, Gordon C, Hogan T, Ji M, Kimoto M, Kitoh A, Knutson T, Latif M, Le Treut H, Li T, Manabe S, Mechoso C, Power S, Roeckner E, Terray L, Vintzileos A, Voss R, Wang B, Washington W, Yoshikawa I, Yu J, Yukimoto S, Zebiak S, Meehl G (2002) STOIC: a study of coupled model climatology and variability in tropical ocean regions. Clim Dyn 18:403–420
52. Zhang S, Harrison MJ, Rosati A, Wittenberg A (2007) System design and evaluation of coupled ensemble data assimilation for global oceanic climate studies. Mon Weather Rev 135:3541–3564
53. Balmaseda MA, Fujii Y, Alves O, Awaji T, Behringer D, Ferry N, Lee T, Rienecker M, Rosati T, Stammer D, Smith D, Molteni F (2009) Initialization for seasonal and decadal forecasts. http://www.oceanobs09.net/blog/?p=57

54. Stockdale T, Anderson DLT, Balmaseda M, Doblas-Reyes FJ, Ferranti L, Mogensen K, Palmer TN, Molteni F, Vitart F (2011) ECMWF System 3 and its prediction of sea surface temperature. Clim Dyn. doi:s00382-010-0947-3/s00382-010-0947-3
55. Toth Z, Kalnay E (1993) Ensemble forecasting at NMC: the generation of perturbations. Bull Am Meteorol Soc 74:2317–2330
56. Molteni F, Buizza R, Palmer TN, Petroliagis T (1996) The ECMWF ensemble prediction system: methodology and validation. Q J R Meteorol Soc 122:73–119
57. Wilks DS (2006) Statistical methods in the atmospheric sciences, vol 59, International geophysics series. Academic, San Diego
58. Goddard L, Hoerling MP (2006) Practices for seasonal-to-interannual climate prediction. US CLIVAR variations, vol 4
59. Jin EK, Kinter JL, Wang B, Park C-K, Kang I-S, Kirtman BP, Kug J-S, Kumar A, Luo J-J, Schemm J, Shukla J, Yamagata T (2008) Current status of ENSO prediction skill in coupled ocean–atmosphere models. Clim Dyn 31:647–664
60. Hagedorn R, Doblas-Reyes FJ, Palmer TN (2005) The rationale behind the success of multi-model ensembles in seasonal forecasting – I. Basic concept. Tellus A 57:219–233
61. Rajagopalan B, Lall U, Zebiak SE (2002) Categorical climate forecasts through regularization and optimal combination of multiple GCM ensembles*. Mon Weather Rev 130:1792–1811
62. Robertson AW, Lall U, Zebiak SE, Goddard L (2004) Improved combination of multiple atmospheric GCM ensembles for seasonal prediction. Mon Weather Rev 132:2732–2744
63. Weigel AP, Liniger MA, Appenzeller C (2009) Seasonal ensemble forecasts: are recalibrated single models better than multimodels? Mon Weather Rev 137:1460–1479
64. NRC (ed) (2008) Review of CCSP draft synthesis and assessment product 5.3: decision-support experiments and evaluations using seasonal to interannual forecasts and observational data. The National Academies, Washington, DC
65. Robertson AW, Moron V, Swarinoto Y (2009) Seasonal predictability of daily rainfall statistics over Indramayu district, Indonesia. Int J Climatol 29:1449–1462
66. Gershunov A, Barnett TP (1998) ENSO influence on intraseasonal extreme rainfall and temperature frequencies in the contiguous United States: observations and model results. J Clim 11:1575–1586
67. Higgins RW, Leetmaa A, Kousky VE (2002) Relationships between climate variability and winter temperature extremes in the United States. J Clim 15:1555–1572
68. Joliffe IT, Stephenson DB (2003) Forecast verification: a practitioner's guide in atmospheric sciences. Wiley, Chicester
69. WMO (2002) World Meteorological Organization (WMO) Standardized verification system (SVS) for long-range forecasts (LRF) In: New attachment II-9 to the manual on the GDPS (WMO-N. 485), vol 1, Geneva
70. Livezey RE, Timofeyeva MM (2008) The first decade of long-lead U.S. seasonal forecasts. Bull Am Meteorol Soc 89:843–854
71. Goddard L, Dilley M (2005) El Niño: catastrophe or opportunity. J Clim 18:651–665
72. Gates WL (1992) AMIP: the atmospheric model intercomparison project. Bull Am Meteorol Soc 73:1962–1970
73. Palmer TN, Shukla J (2000) DSP/PROVOST. Q J R Meteorol Soc 126:1989–2350, Special Issue
74. Palmer TN, Doblas-Reyes FJ, Hagedorn R, Alessandri A, Gualdi S, Andersen U, Feddersen H, Cantelaube P, Terres J-M, Davey M, Graham R, Délécluse P, Lazar A, Déqué M, Guérémy J-F, Díez E, Orfila B, Hoshen M, Morse AP, Keenlyside N, Latif M, Maisonnave E, Rogel P, Marletto V, Thomson MC (2004) Development of a European multimodel ensemble prediction system for seasonal-to-interannual prediction (DEMETER). Bull Am Meteorol Soc 85:853–872
75. Hewitt CD, Griggs DJ (2004) Ensembles-based predictions of climate changes and their impacts. EOS 85:566

76. Kirtman B, Pirani A (2009) The state of the art of seasonal prediction: outcomes and recommendations from the First World Climate Research Program Workshop on seasonal prediction. Bull Am Meteorol Soc 90:455–458
77. Power S, Casey T, Folland C, Colman A, Mehta V (1999) Inter-decadal modulation of the impact of ENSO on Australia. Clim Dyn 15:319–324
78. Alexander MA, Deser C, Timlin MS (1999) The reemergence of SST anomalies in the North Pacific Ocean. J Clim 12:2419–2433
79. Schneider N, Miller AJ (2001) Predicting Western North Pacific Ocean Climate. J Clim 14:3997–4002
80. Di Lorenzo E, Schneider N, Cobb KM, Franks PJS, Chhak K, Miller AJ, McWilliams JC, Bograd SJ, Arango H, Curchitser E, Powell TM, Rivière P (2008) North Pacific Gyre oscillation links ocean climate and ecosystem change. Geophys Res Lett 35:L08607
81. Deser C, Phillips AS, Hurrell JW (2004) Pacific Interdecadal climate variability: linkages between the Tropics and the North Pacific during Boreal Winter since 1900. J Clim 17:3109–3124
82. Meehl GA, Hu A (2006) Megadroughts in the Indian Monsoon Region and Southwest North America and a mechanism for associated multidecadal Pacific Sea surface temperature anomalies. J Clim 19:1605–1623
83. Knight JR, Allan RJ, Folland CK, Vellinga M, Mann ME (2005) A signature of persistent natural thermohaline circulation cycles in observed climate. Geophys Res Lett 32:L20708
84. Vellinga M, Wu P (2004) Low-latitude freshwater influence on centennial variability of the Atlantic thermohaline circulation. J Clim 17:4498–4511
85. Latif M, Roeckner E, Botzet M, Esch M, Haak H, Hagemann S, Jungclaus J, Legutke S, Marsland S, Mikolajewicz U, Mitchell J (2004) Reconstructing, monitoring, and predicting multidecadal-scale changes in the North Atlantic thermohaline circulation with sea surface temperature. J Clim 17:1605–1614
86. Stoner AMK, Hayhoe K, Wuebbles DJ (2009) Assessing general circulation model simulations of atmospheric teleconnection patterns. J Clim 22:4348–4372
87. AchutaRao K, Sperber KR (2006) ENSO simulation in coupled ocean–atmosphere models: are the current models better? Clim Dyn 27:1–15
88. Gray ST, Graumlich LJ, Betancourt JL, Pederson GT (2004) A tree-ring based reconstruction of the Atlantic multidecadal oscillation since 1567 A.D. Geophys Res Lett 31:L12205
89. Davey M, and co-authors, 2006. Multi-model multi-method multi-decadal ocean analyses from the ENACT project. CLIVAR Exchanges, Vol 11, No. 3, pgs 22–25. http://eprints.soton.ac.uk/41286/1/Exchanges38.pdf
90. Collins M, Botzet M, Carril AF, Drange H, Jouzeau A, Latif M, Masina S, Otteraa OH, Pohlmann H, Sorteberg A, Sutton R, Terray L (2006) Interannual to decadal climate predictability in the North Atlantic: a multimodel-ensemble study. J Clim 19:1195–1203
91. Branstator G, Teng H (2010) Two limits of initial-value decadal predictability in a CGCM. J Clim 23:6292–6311. doi:10.1175/2010JCLI3678.1
92. Rodwell MJ, Drévillon M, Frankignoul C, Hurrell JW, Pohlmann H, Stendel M, Sutton RT (2004) North Atlantic forcing of climate and its uncertainty from a multi-model experiment. Q J R Meteorol Soc 130:2013–2032
93. Taylor KE, Stouffer RJ, Meehl GA (2011) Bulletin of the American Meteorological Society. http://dx.doi.org/10.1175/BAMS-D-11-00094.1
94. Baldwin MP, Dunkerton TJ (2001) Stratospheric harbingers of anomalous weather regimes. Science 294:581–584
95. Gottschalck J, Wheeler M, Weickmann K, Vitart F, Savage N, Lin H, Hendon H, Waliser D, Sperber K, Nakagawa M, Prestrelo C, Flatau M, Higgins W (2010) A framework for assessing operational Madden–Julian oscillation forecasts: a CLIVAR MJO Working Group Project. Bull Am Meteorol Soc 91:1247–1258
96. Zhang S, Rosati A (2010) An inflated ensemble filter for ocean data assimilation with a biased coupled GCM. Mon Weather Rev 138:3905–3931

97. Vitart F, Buizza R, Balmaseda M, Balsamo G, Bidlot J-R, Bonet A, Fuentes M, Hofstadler A, Molteni F, Palmer TN (2008) The new VarEPS-monthly forecasting system: a first step towards seamless prediction. Q J R Meteorol Soc 134:1789–1799
98. Hazeleger W, Severijns C, Semmler T, Ştefănescu S, Yang S, Wang X, Wyser K, Dutra E, Baldasano J, Bintanja R, Bougeault P, Caballero R, Ekman AML, Christensen JH, van den Hurk B, Jimenez P, Jones C, Kållberg P, Koenigk T, McGrath R, Miranda P, van Noije T, Palmer T, Parodi J, Schmith T, Selten F, Storelvmo T, Sterl A, Tapamo H, Vancoppenolle M, Viterbo P, Willén U (2010) EC-Earth: a seamless earth system prediction approach in action. Bull Am Meteorol Soc 91:1357–1363

Books and Reviews

Hurrell JW, Delworth T, Danabasoglu G, Drange H, Griffies S, Holbrook N, Kirtman B, Keenlyside N, Latif M, Marotzke J, Meehl G A, Palmer T, Pohlmann H, Rosati T, Seager, R, Smith D, Sutton R, Timmermann A, Trenberth KE, and Tribbia J (2010) Decadal climate prediction: opportunities and challenges. In: Hall J, Harrison DE, Stammer D (eds) Proceedings of OceanObs'09: sustained ocean observations and information for society, vol 2. ESA Publication WPP-306: Venice, 21–25 Sep 2009. Available at http://www.oceanobs09.net/blog/?p=97

Latif M, Delworth T, Dommenget D, Drange H, Hazeleger W, Hurrell J, Keenlyside N, Meehl GA, and Sutton R (2010) Dynamics of decadal climate variability and Implications for its prediction. In: Hall J, Harrison DE, Stammer D (eds) Proceedings of OceanObs'09: sustained ocean observations and information for society, vol 2. ESA Publication WPP-306: Venice, 21–25 Sep 2009. Available at http://www.oceanobs09.net/blog/?p=104

Mehta V, Meehl GA, Goddard L, Knight J, Kumar A, Latif M, Lee T, Rosati A, Stammer D (2010) The eighth workshop on decadal climate variability decadal climate predictability and prediction: where are we? Bull Am Meteorol Soc. doi:10.1175/2010BAMS3025.1

Murphy J, Kattsov V, Keenlyside N, Kimoto M, Meehl G, Mehta V, Pohlmann H, Scaife A, Smith D (2010) Towards prediction of decadal climate variability and change. Procedia Environ Sci 1:287–304. doi:10.1016/j.proenv.2010.09.018

Chapter 12
Monsoon Systems, Modeling of

Chien Wang and William K.M. Lau

Glossary

Aerosol	Small particles suspended in the atmosphere in solid or liquid phase.
El Niño and Southern Oscillation (ENSO)	Two intimately linked phenomena in tropical regions; El Niño ("the Christ Child" in Spanish) refers to the significant increase in sea surface temperature that irregularly occurs during Christmas time over eastern and central Pacific Ocean; Southern Oscillation refers to the low-latitude oscillation of sea level pressure centered respectively in the eastern Pacific and the western Pacific to Indian Ocean.
General circulation model (GCM)	A computer program that solves numerically the time-dependent governing equations describing the evolution of atmospheric or oceanic circulation.
Intertropical convergence zone (ITCZ)	A longitudinally extended zone near the equator that separates the northeast wind in the Northern Hemisphere from the southeast wind in the Southern Hemisphere near the Earth's surface.

This chapter was originally published as part of the Encyclopedia of Sustainability Science and Technology edited by Robert A. Meyers. DOI:10.1007/978-1-4419-0851-3

C. Wang (✉)
Department of Earth, Atmospheric, and Planetary Sciences,
Massachusetts Institute of Technology, Cambridge, MA 02139, USA
e-mail: wangc@MIT.EDU

W.K.M. Lau (✉)
Laboratory for Atmospheres, NASA Goddard Space Flight Center,
Greenbelt, MD 20771, USA
e-mail: william.k.lau@nasa.gov

P.J. Rasch (ed.), *Climate Change Modeling Methodology: Selected Entries from the Encyclopedia of Sustainability Science and Technology*,
DOI 10.1007/978-1-4614-5767-1_12, © Springer Science+Business Media New York 2012

Madden–Julian oscillation (MJO)	An oscillation of zonal wind in both the boundary layer and upper troposphere propagating eastward with an average speed of 5 m/s across equatorial Indian and western and central Pacific Ocean.
Moist static energy (MSE)	An atmospheric thermodynamic variable defined as: $$\mathrm{MSE} = C_pT + gz + L_vq$$ Here C_p is the specific heat of air, T is air temperature, g is gravity, z is height above surface or a given reference level, L_v is the latent heat of water vaporization, and q is the ratio of water vapor to total air in mass.
Tropical biennial oscillation (TBO)	A zonal wind oscillation in the equatorial stratosphere.

Definition of the Subject

The word monsoon derives from the Arabic word "mausim," referring to the seasonal reversal of prevailing low-level winds blowing from relatively cold and moist ocean to warm land during the wet season (summer), and from cold and dry land to ocean during the dry season (winter). Monsoon systems are found in tropical regions from Africa, India, East Asia, Australia, and the Americas. Deep convection along with heavy rainfall occurs during the wet monsoon season over land as well as ocean. Typically, in monsoon regions the rainfall in wet season accounts for more than half of the annual surface precipitation. Monsoon evolution heavily influences human activities including agricultural practice and societal habits of billions of people living in monsoon regions. Knowledge and improving prediction of the onset, maintenance, variability, and key drivers are critical to the livelihood of these people. Because the monsoon is an integral part of the global climate system, better understanding of the monsoon is pivotal to predict future climate change and also the response of monsoon systems to such change.

The onset and strength of monsoons are determined by dynamical and thermodynamical processes not only locally over monsoon regions but also remotely over other regions. Computer models combining related dynamical, physical, and chemical processes in various scales are hence important tools to examine the current understanding of monsoon dynamics. These models can be used to test various hypotheses, and to actually simulate and forecast monsoon evolution. Simple models used in the earlier days of monsoon research mostly described the monsoon system from an energy budget perspective. These models could capture very rudimentary features of monsoon energy conversion. However, they lack the capability to go further in revealing the details of rainfall intensity and distribution, and particularly the timing of monsoon onset. Sophisticated three-dimensional regional and global climate models have been used in recent years to simulate monsoon systems, and to study the sensitivity of monsoon circulation and precipitation to various factors, including identification of the anthropogenic impact on monsoon system. These models have also been used to project monsoon evolution under different scenarios of possible future climate change.

Introduction

Over 60% of the world's population lives in monsoon regimes with a clear annual cycle of wind and precipitation. Such a cycle consists of a wet and a dry season. Wind in the lower atmosphere blows poleward from a relatively cold ocean to warm land during the wet season, and goes in the opposite direction during the dry season (Fig. 12.1). The monsoon regions include a large part of tropical and subtropical Asia and Africa as well as Australia, where some of the most populous nations in the world are located.

Agricultural activities, water resources, and many societal events in regions with a monsoon climate are strongly influenced by the wind reversal and the uneven distribution of rainfall. Forecasting monsoon rainfall and onset, the sudden transition from dry condition into a heavy downpour, has practical meanings to human activities in these places. In addition, whether future climate change would alter the behavior and strength of monsoon is a critical issue in making climate related strategies. The achievement of an adequate skill to forecast future monsoon evolution relies on a good understanding of fundamental monsoon dynamics along with its variability. This requires knowledge about the formation mechanism and the

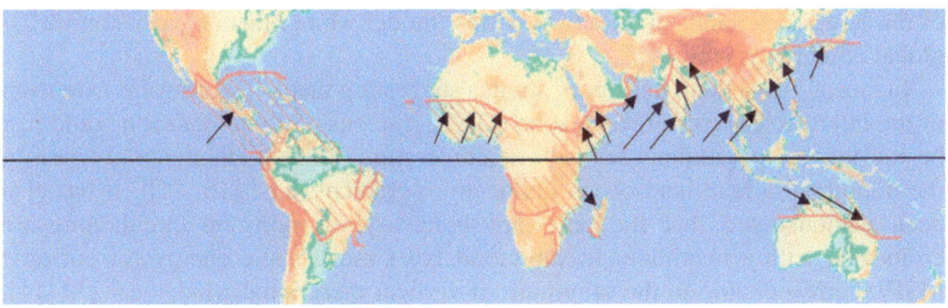

Fig. 12.1 Land areas that have the majority of their rainfall in summer, associated with the poleward motion of deep convection. Where appropriate, low-level wind directions that carry moist warm air are indicated. In areas where there are no arrows, winds are relatively dry, or are weak (as over South America). *Shaded areas* show the normal maximum extent of deep convection (From [1] by J.F.P. Galvin with permissions from the author and Wiley)

major driving factors, including both natural and anthropogenic ones, of the monsoon system.

Monsoon Dynamics Fundamentals

Research to identify the driving forces of monsoon onset and strength has been conducted by correlating various diagnostic quantities with monsoon system characteristics. These characteristics include moisture, clouds, precipitation, and the large-scale circulation. With continued advancement of observational technology, from rain gauge stations to satellite monitoring, such effort has gained momentum, leading to improved knowledge. However, improved knowledge often reveals additional complexity of monsoon systems. This will in turn require even better understanding to further improve theory and modeling. This requirement has led to the building of a hierarchy of regional through global climate-system models for monsoon research.

It has long been held that differential heating on land and ocean following solar insolation cycle to be the major formation mechanism of the large-scale monsoon circulation [2, 3]. Such a heating contrast would force wind blow toward warm region, although because of the geostrophic constraint the actual wind direction is altered. Various rather simple models were developed to simulate monsoon systems based on the differential heating concept. These include zonally symmetric and other types of two-dimensional models that describe the zonal circulation and precipitation from ocean to land over monsoon regions, and often include a description of the planetary boundary layer. These simple models along with their limitations have been discussed extensively in literature [4, 5].

Attempts to forecast monsoons have linked monsoon strength with other phenomena or processes, ranging from the snowfall on the hills of Himalaya in the previous winter [6], mountains [7, 8], to the El Niño and Southern Oscillation (ENSO) ([9] and many others). Some of these factors are still within the framework of the large-scale land-sea thermal contrast model while others clearly connect to global climate dynamics.

In recent years, there have been studies suggesting that as a northward extension of the Intertropical convergence zone (ITCZ), the onset of the monsoon could just be a result of a longitudinal sea surface temperature (SST) gradient, not necessarily the traditionally held land-ocean temperature gradient (e.g., [10, 11]). It has also been demonstrated that the poleward boundary of monsoon circulations are co-located with a maximum in sub-cloud layer moist static energy (or entropy; MSE), corresponding to the minimum of vertical meridional wind shear [12–14]. Such a location and extent of the monsoon would also be influenced by the position of subtropical thermodynamic forcing as well as the advection of MSE.

Due to a meteorological phenomenon called the "thermal wind balance," the heating over land to the north and the cooler ocean to the south during the summer monsoon will produce easterly winds aloft and westerly winds below. When the upper troposphere easterly wind is strong, in the case of a strong monsoon, instability of easterly jet may stimulate the formation of eddies. The formation of these eddies might not always amplify the monsoonal circulation because the northeastward flow from ocean could bring low MSE air to land [5, 12, 13]. A recent proposal [15] actually suggested viewing monsoons as eddy-mediated transitions in the tropical overturning circulation between regimes that are distinct in the degree to which eddy momentum fluxes control the strength of the circulation. In this study, the idealized general circulation model (GCM) simulation on an aqua-planet demonstrated that the role of land in monsoon onset is to just provide a media of low thermal inertia. Whenever such surface differences in heat capacity exist monsoon onset would happen regardless of other surface inhomogeneities. Therefore, interactions between extratropical eddies and the tropical meridional overturning circulation could be essential for monsoons. In addressing the interaction of monsoon and other dynamical system, it was indicated that the feedback of atmosphere to SST forcing might have played a critical role in monsoon evolution [16].

One specific implication of these new hypotheses is on the predictability of the monsoon system. It has been argued that because the dominant forcing of monsoon system are the rather slow processes that control the tropical sea surface temperatures (SST), therefore, the predictability of monsoon rainfall at least in monthly or seasonal scale may be promising [17]. However, should the extratropical eddies and atmosphere to ocean feedback be critical in monsoon onset and evolution, the monsoon predictability issue even on relatively long time scales would be much more complicated. The predictability of the monsoon is further confounded by the ubiquitous presence of monsoon-intraseasonal oscillations (MISO) in both the summer and the winter seasons. These are intrinsic oscillations in the monsoon region, with characteristic timescales of 20–70 days, arising from the organization of tropical convection over the ocean associated, (e.g., planetary scale Madden–Julian Oscillation (MJO); see [18] and many others). MISO are mediated by SST changes as well as the monsoon regional topography, and influence the onset, break, and maintenance of the monsoon as well through interactions with ENSO. Realistic simulations of MISO and MJO have been a challenge even for the state-of-the-art climate models.

Modeling the Monsoon

Because of the complex, multi-scale characteristics of monsoon systems, efforts to understand and to examine various hypotheses about monsoon onset, evolution, and strength have to rely largely on computer models in combination with available data. In order to understand the interaction between this large-scale moist circulation and many other complicated but critical processes, ranging from ocean–atmosphere interaction, extratropical–tropical interaction, MISO, to potential "teleconnection" through synoptic waves between tropical systems in distance, one needs to use a three-dimensional global climate model or a regional climate model interacting with a global climate model.

Simulation of monsoon systems using three-dimensional global models started from the very early stage of atmospheric general circulation models. Typical model used in these early simulations had coarse horizontal resolution (270–540 km) and 11 vertical layers, forced by prescribed seasonal variations of insolation and often sea surface temperature [19]. These simulations were mostly used for exploratory purposes due to their short integration time (often shorter than 3 years), coarse model resolution, and the prescription of some fields of potential importance in modeling the monsoon. Understanding the onset of the monsoon was clearly a far-reach at that stage.

The availability of multi-decadal sea surface temperature data allowed three-dimensional atmospheric general circulation models (AGCM) to simulate monsoon system evolution driven by observed SST time series. This type of simulations follows the procedure of the Atmospheric Modeling Intercomparison Project (AMIP, and AMIP II later; [20]), forced by "real-time" SST data and thus ignored the feedback between the atmosphere and ocean. AMIP models, which generally include interactive land surface models, have the advantage of identifying atmospheric feedback mechanisms without dealing with the complexity of the coupled ocean–atmosphere system. They were the models of choice in the 1980s–1990s. Nowadays, long-term climate simulations are generally done with coupled ocean–atmosphere models. However, AMIP models are still useful when run at high resolutions to test sensitivity to model atmospheric processes of physics, chemistry, and aerosols, and interaction of the atmosphere with surface vegetation. Some current AMIP runs are conducted at mesoscale resolution (<25 km) globally, and others are configured even at higher resolution for global hurricanes studies. With such configurations, models could explore the onset of the monsoon and aspects of the MJO and MISO [21]. The very high-resolution AMIP models are extremely computationally demanding, and can only be run using high-performance computers at large institutions.

Studies exploring multi-decadal to centennial timescale issues that need to consider the role of atmosphere-ocean feedback in monsoon dynamics, typically utilize moderate-to-low resolution (100–200 km) AGCMs coupled with either mixed-layer ocean model or full ocean general circulation model to reduce the computational demand. Regional climate models have also been used for this purpose. The influence of future climate change on monsoon evolution has also been studied mostly using the ensemble results of three-dimensional climate model simulations included in the Fourth Assessment Report (AR4) of the Intergovernmental Panel of Climate Change (IPCC). A fast growing effort in recent years is to study the impacts of anthropogenic forcings particularly of aerosols on monsoon circulation and precipitation.

This entry will begin by describing efforts to use climate models to simulate monsoon systems. Recent research on the potential role of aerosols in monsoon systems is then described. The projections of monsoon system changes under possible climate change scenarios will also be discussed, concluding with an overview of the future opportunities. The discussion will be focused on the utilization of general circulation models and regional climate models, of moderate-to-low resolution in simulating monsoon systems and the study of the sensitivity of monsoon to various climate dynamical processes as well as anthropogenic impacts on equilibrium climate, or on climate time scales of a century or less.

Simulating Monsoon Systems Using Climate Models

Before attempting to use a computer model to forecast monsoon evolution, one would ask the very question that how well the model might reproduce the major observed characteristics and variability of a monsoon, if some of the known factors controlling monsoons were included in the model (of course this condition itself is somewhat a unsettled issue). This actually leads to a type of modeling study, so-called retrospective modeling. In modeling monsoon systems, a retrospect simulation would be performed by prescribing the time series of sea surface temperature, assuming that the ocean is such a large heat reservoir comparing to the atmosphere so that the change in SST reflect the long-term state of energy balance. This type of modeling allows modelers to concentrate on issues other than the feedbacks from the atmosphere to ocean. With the assumption that historical SST change might well represent the effect of all the long-term forcings, this type of simulations is expected to capture major features of the monsoon systems in the past.

In simulating monsoon systems, it is essential for the model to capture certain representative features of the system. These would at least include the reverse atmospheric circulation between the upper and lower levels, the onset of monsoon rainfall, and total precipitation during monsoon season. In addition, the climatological rainfall patterns including land-ocean partition during monsoon is also among important system characters. A more subtle and difficult task in modeling is to capture the weak correlation between ENSO and Indian summer monsoon rainfall [9], and the correlation between anomaly of Gulf of Guinea SST and the dipole rainfall pattern of the West African monsoon over Guinean coast and the Sahel [22]. Interannual and decadal variability is another important test for both retrospect modeling and for revealing the dependency of monsoon systems on critical forcings. Simulating the intraseasonal variation of detailed rainfall strength and distribution (e.g., [23]) would be an important task for regional or high-resolution global models.

Much progress has been made through years of efforts in modeling the monsoon system. In early stage of such attempt, models typically had low resolution and prescribed seasonal forcing of SST. Arguably, the physics processes such as clouds and radiation in those models were also poorly treated comparing to models used today. Nevertheless, the early models captured certain basic features of the monsoon system (mostly on Indian summer monsoon due to its rather clearly defined annual cycle and relatively extensive analyses) such as the reverse low-level circulation over northern Indian Ocean (e.g., [19]). Besides, the role of certain hypothesized driving factors of monsoon circulation such as mountains [7] and the anomaly of Arabian Sea surface temperature [24] had been also examined. The simulated onset of monsoon, however, was much delayed and the distribution of rainfall and intensity has large biases compared to observations.

With the availability of decadal-long observed SST dataset, nearly all the major AGCMs in the world joined the effort of Atmospheric Modeling Intercomparison Project (AMIP) in the 1990s. An AMIP configuration is a typical retrospect simulation, where atmospheric general circulation models were driven by a time series of observed SST data to reproduce the past climate. Lau and Yang had used the Goddard Space Flight Center GCM (4×5 degree resolution along latitude and longitude, respectively and 17 vertical layers) in an AMIP 1979–1988 simulation to examine the Asian monsoon system [25]. The model was able to capture many broad-scale structures of Asian monsoon system, including evolution of global and regional circulation, rainfall, moisture flux, and intraseasonal and synoptic variability. Interestingly, the model also successfully simulated multiple onset of East Asian monsoon along with the onset of Indian summer monsoon. These onsets were initiated by a sudden jump of the ITCZ from the equator to 10°N, related to a northward shift of the ascending branch of the local Hadley circulation. Clearly, this was a significant advance from the early simulations. On the other hand, the model did not reproduce observed rainfall distribution and quantity over many precipitation centers. Intraseasonal transition of ITCZ between equatorial region (ocean) and monsoon land was not well captured. The East Asian monsoon trough was also severely underdeveloped in the model. These shortcomings actually existed in most AMIP models.

In AMIP-type simulations, the atmosphere-ocean feedback is set aside. The modeling focus is instead on the atmospheric simulation, presuming that the SST time series realistically reflects the forcing of the past. Based on the results of ensemble AMIP simulations, Wang et al. indicated that a lack of the atmospheric feedback in simulations forced by observed SST could lead to serious biases in modeled monsoon precipitation [16]. They found that the atmospheric feedback to SST forcing is more significant than SST to atmospheric forcing. Therefore, coupled model would be critical in even retrospect modeling of monsoon and the atmospheric feedback to tropical SST forcing needs to be included. Meehl et al. further demonstrated an improvement of modeled monsoon features using a higher resolution (T85) coupled atmosphere-ocean model compared to a lower resolution (T42) AMIP configuration model [26]. There have also been reports of significant improvement in modeling monsoon features made simply by using high-resolution atmospheric GCM or regional climate model driven by observed SST (see [27]).

When using coupled model to simulate monsoon system, drifts of SSTs away from observation could become an issue. With a rather coarse-resolution coupled model (4.5×7.5 degree and nine layer for the atmospheric model; 5×5 degree and four layer for the ocean model), Meehl indicated that without adopting correction terms to force the coupled model to the observed state, model-simulated SSTs in the tropics tend to be too cold. This bias would enhance land-sea temperature contrast in the monsoon region, yet the pattern of mean monsoon seasonal precipitation and the variability of the simulated South Asian monsoon (SAM) were comparable to the observed pattern [28]. One alternative method to the AMIP configuration is to predict SST using a 2½-layer tropical ocean model between 30°S and 30°N and to prescribe SST in other places [16]. The SST-monsoon rainfall correlations indicated by observations (with 1 month lag) were reflected correctly in the simulation conducted by using this method.

Models have also been used to identify certain hypothesized driving factors behind monsoon variability. For instance, it is known that the Tropical Biennial Oscillation (TBO), a variation in precipitation occurring with approximately a 2-year period, affects monsoon strength. Therefore, identifying the relative importance of various potential conditions leading to TBO transitions could help us to understand the factors that affect monsoon variability. It was found that among three conditions hypothesized to contribute to TBO transitions, tropical Indian Ocean SST anomaly and tropical Pacific Ocean SST anomaly are more effective than anomalous meridional temperature gradients over Asia [29]. The two types of tropical SST anomalies were found to dominate the TBO transitions and thus produce large monsoon response in the model sensitivity results. In addition, the location of the SST anomalies over the tropical Indian Ocean is found to be important. Warm SST anomalies throughout the tropical Indian Ocean enhance rainfall over the ocean and South Asian land areas. Warm SST anomalies near equatorial Indian Ocean produce increased rainfall locally with decreased rainfall over South Asian land areas.

Despite significant progresses achieved, there is still much room for improvement regarding the performance of current climate models in simulating various features of monsoon systems from mean state to variability [30]. For instance, among 18 coupled GCMs that participated the effort of the IPCC Fourth Assessment Report (IPCC AR4), only six of them were found to have realistic representation of South Asian monsoon precipitation climatology in the twentieth century [31]. It is noteworthy that these six models all had large pattern correlation and small root-mean-square differences (RMSD) with observations in modeling June–July–August–September (JJAS) rainfall climatology, both over India (7°–30°N, 65°–95°E) and for the larger monsoon domain (25°S–40°N, 40°E–180°). Only four out of these six, though, exhibited a robust ENSO-SAM teleconnection.

Recent results from the West African Monsoon Model Evaluation (WAMME) project showed that AMIP-type models (both regional and global) generally have reasonable skills in simulating the pattern of the spatial distribution of West African monsoon (WAM) in seasonal mean precipitation, surface temperature, averaged zonal wind in latitude-height cross-section, and low-level circulation [32]. However, there are large differences among models in addition to model biases compared to observations in simulating spatial correlation, intensity, and variability of precipitation.

In a well-designed analysis [22], the abilities of above-mentioned 18 models were evaluated based on whether they can correctly simulate the circulation characteristics that support the precipitation climatology and the physical processes of a prominent mode of WAM variability, that is, the "rainfall dipole" variability that is often associated with dry conditions in the Sahel when SSTs in the Gulf of Guinea are anomalously warm. It was found that each model captured the largest-scale rainfall pattern featuring a zonally oriented precipitation maximum, but about one third of them did not generate the West African monsoon, that is, they did not bring the ITCZ and its associated rainfall onto the African continent during boreal summer. Only three further captured the three precipitation maxima over the continent, that is, the maximum on the west coast, over the eastern portion of the Guinean coast, and over the Ethiopian highlands. It was thus concluded that the current generation of coupled GCMs is much more capable of accurately representing the summer precipitation climatology over North America and Europe than over Africa.

In modeling the Sahel drought during 1970s–1980s, the most pronounced climate signal in WAM regions that had been suggested as a consequence of warm anomalous SST surrounding Africa (e.g., [33]), Lau et al. evaluated the performance of 19 coupled general circulation models (also AR4 models) in twentieth-century simulations [34]. They found that only eight of these models produced a reasonable Sahel drought signal, while others either produced excessive rainfall over the Sahel during the observed drought period or showed no significant deviation from normal. Even the model with the highest prediction skill of the Sahel drought could only predict the increasing trend of severe drought events but not the beginning and duration of the events. Based on the analysis, it was recommended that in order to accurately simulate the Sahel drought, models need to have a strong coupling between Sahel rainfall and the SSTs of both Indian and Atlantic Ocean, in addition to a robust land surface feedback with strong sensitivity of precipitation and land evaporation to soil moisture.

The performance of 12 coupled models in the Coupled Model Intercomparison Project phase 3 (CMIP3; the same group of models that participated in IPCC AR4) in simulating present-day East Asian monsoon has been examined as well [35]. Almost all of these models were found to be able to reproduce observed interannual variability of summer rain belt and associated circulation. The models can also reproduce the interannual variation of the western North Pacific subtropical high (WNPSH) in the lower troposphere, a parameter closely related to the interannual variation of summer rainfall. However, the predicted quantities of interannual variation of WNPSH from these models differ significantly.

Modeling the Impacts of Aerosols on Monsoon System

Atmospheric aerosols serve as a critical player in the climate system. All aerosols attenuate solar radiation through either scattering or absorption, both leading to cooling at the Earth's surface. In addition, absorbing aerosols warm the atmosphere, affecting atmospheric profile and thus dynamical processes. Aerosols also dominate the cloud formation in the atmosphere, serving as cloud condensation nuclei (CCN) or ice nuclei (IN) to provide preexisting surfaces and thus a superior mean for cloud particles to form than homogeneous nucleation. Therefore, changes in aerosol properties such as number concentration, size distribution, or chemical composition (hygroscopicity) are expected to affect atmospheric systems from regional to global scales including the monsoon.

Human activities produce aerosols containing inorganic matters such as sulfate and nitrate, and organic matter as well as black carbon. These anthropogenic aerosols are regarded as an addition to the natural aerosols that mainly include dust, biogenic, and sea salt particles, and hence exert a forcing to the climate system. Studies suggested that the reduction in Indian monsoon strength in recent decades could be a result of an increase of anthropogenic aerosols over monsoon regions, mostly coinciding with the fastest growing economies including China and India as well as Southeast Asia [36].

On the other hand, a recent analysis of 1951–2003 daily gridded rainfall data over India revealed a decreasing trend in both early and late monsoon rainfall and number of rainy days, implying a shorter monsoon over India [37]. There is also a sharp decrease in the area that receives a certain amount of rainfall and number of rainy days during the season. An increase in the frequency of heavy precipitation in the Indian summer monsoon was also identified [38].

A great deal of attention has been paid to the influence of anthropogenic aerosols (particularly absorbing aerosols) on tropical precipitation in recent years. Absorbing aerosols influence the climate in distinctly different ways from aerosols that primarily scatter energy back to space. Studies using different general circulation models all indicate that direct radiative forcing (DRF) of absorbing black carbon (BC) aerosols can lead to a northward shift of precipitation in ITCZ over the Pacific Ocean [39–41]. Modeling studies also suggest that DRF of aerosols could have a significant impact on the monsoon systems as well [42]. Correlations between estimated precipitation/circulation changes with increasing trend of aerosols have unquestionably fueled the researches toward this direction.

Studies of aerosol-monsoon impact are rapidly growing not only for the Indian summer monsoon, but also for the East Asian monsoon, the West African monsoon, and the Australian monsoon. Most of these studies are conducted by using three-dimensional atmospheric GCMs or coupled climate models. Paired simulations driven respectively by including and excluding aerosol effects, or by including a reference and an altered aerosol profile along with aerosol effects, provide a comparison in climate response between different aerosol forcing assumptions. The aerosol effects would be isolated barring the assumption that model's artifact in simulating the monsoon system would not be significantly amplified by using different aerosol profiles. The descriptions of aerosol and aerosol–climate interaction vary in studies though.

Impacts of Aerosols on the Asian Monsoon

Earlier works focused on impacts of absorbing aerosols (black carbon and dust) on atmospheric water cycle in the Asian monsoon, by using prescribed aerosol distributions from global chemistry transport models and/or observations. These studies excluded the dynamical feedbacks between winds and precipitation features and the aerosol distribution. An early exploratory study tested the climate response to absorbing aerosols over China and India [43]. In this study, a 12-layer and 4×5 degree resolution three-dimensional AGCM was used to explore the model response to prescribed aerosol optical depth and single scattering albedo over China and India only. The researchers found that the convection would be enhanced along 20°–30°E in longitude from eastern China to Indian subcontinent in responding to the added aerosol forcing. Despite many discrepancies in detailed results between this study and later ones, perhaps attributed to the regional-only aerosol loading and rather coarse resolution of the model in [43], the general response in large-scale dynamics associated with the monsoon systems caused by the direct radiative effects of absorbing aerosols remains consistent with later studies. For instance, Wang noticed an enhancement of the Indian summer monsoon circulation by the direct radiative forcing of black carbon aerosols in a coupled GCM simulation, though the analysis was done on an annual-mean base so that the seasonal features of the circulation were not discussed [39, 44]. A similar effect of BC aerosols was also found in another study, though where the simulation was driven by prescribed SST [45].

Perhaps the most interesting outcome in recent modeling efforts of aerosol-monsoon studies is the proposals of various hypotheses on the mechanisms of aerosol impact specifically on Indian summer monsoon. The discussions are also centered at the role of absorbing aerosols.

The radiative effects of absorbing (primarily dust and anthropogenic carbonaceous) aerosols in cooling the surface (dimming effect) and in heating the atmosphere can play different roles in affecting the monsoon system. The cooling over land from absorbing aerosols would assist lowering the land-ocean temperature gradient. Ramanathan et al. found that an increase in the BC DRF over Indian Subcontinent and surrounding regions in their model leads to a reduction of monsoon precipitation while an enhancement to the pre-monsoon precipitation of March–April–May (MAM) [36]. Using a coupled atmosphere-ocean general circulation model with prescribed black carbon direct radiative forcing, Meehl et al. found similar circulation and precipitation changes due to BC impact in the pre-monsoon (enhancement) and in monsoon season (reduction) [46]. It was also found that although during the monsoon months the effect of BC is likely to reduce the precipitation over India, it might enhance the precipitation over the elevated Tibetan Plateau. Meehl et al. suggested that BC DRF could weaken the surface temperature gradient between the tropical waters and the land of the Indian Subcontinent. This could serve as the forcing mechanism of BC on the monsoon circulation and precipitation, that is, through the dimming effect.

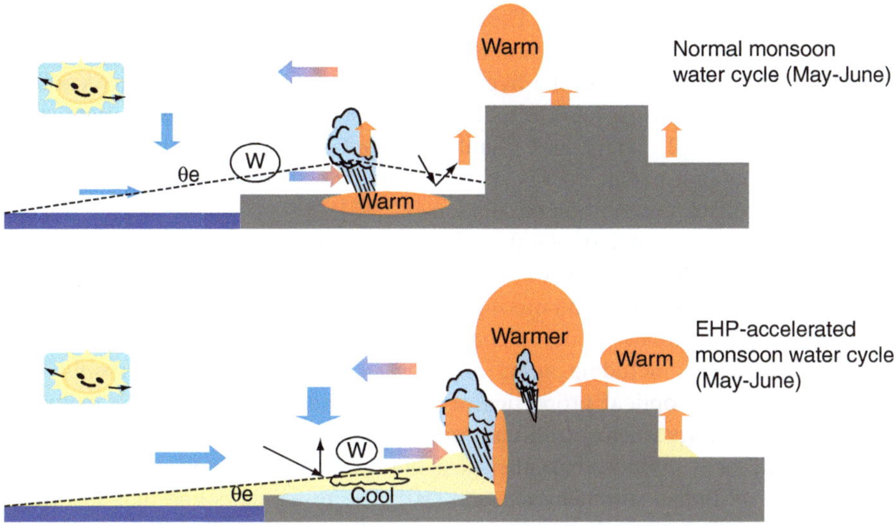

Fig. 12.2 Schematic showing the monsoon water cycle (*top*) with no aerosol forcing and (*bottom*) with aerosol-induced elevated heat pump effect. Low-level monsoon westerlies are denoted by W. The *dashed line* indicates magnitude of the low-level equivalent potential temperature θe. Deep convection is indicated over regions of maximum θe. (See text for further discussions) (Adopted from [42], © American Meteorological Society. Reprinted with permission)

One specific characteristics of absorbing aerosols is its heating to the atmosphere. How would this effect play a role in aerosol-monsoon impact is also discussed. Lau et al. used an atmospheric GCM driven by prescribed global three-dimensional climatology of aerosol optical depth to examine the direct effects of aerosol on the monsoon water cycle variability [34]. The study suggested that, referred to as an "elevated heat pump" effect (EHP) (Fig. 12.2), dust mixed with black carbon aerosols that extend against the foothills of the Himalayas over the Indo-Gangetic Plain (IGP) in April and May could heat the air. This would initiate a positive feedback by drawing water convergence from oceans first and then form condensation and thus further heating over the slope of the Plateau. Based on this hypothesis, monsoon precipitation would be suppressed over central India due to aerosol-induced surface cooling. However, precipitation would come earlier and be enhanced over northern India and the southern slope of the Tibetan Plateau. Monsoon rainfall in July and August over the entire India would also be enhanced. Lately, using satellite aerosol index (AI) data and observed clouds and precipitation data, two studies have demonstrated the existence of anomalous absorbing aerosol loading in late spring over IGP [47, 48]. Both works also suggested a correlation of this aerosol anomaly with variation of monsoon evolution. A widespread warming over the Himalayan-Gangetic region and consequent strengthening of the land-sea thermal gradient was also found recently through satellite microwave sounding data [49]. This trend is most pronounced in the pre-monsoon season, resulting in a warming of 2.7°C in the record. All these observation-based analyses appear to be consistent with the EHP effect. The hypothesis involves both atmospheric-heating and surface-cooling effect of absorbing aerosols as well as induced changes in cloudiness. It is different than the hypothesis that only emphasizes surface cooling. The feedback mechanisms introduced by EHP hypothesis are, however, more complicated. For example, Prive and Plumb indicated that the equatorial oceanic air would have low moist static energy so that the low-level convergence might lead to a negative feedback to monsoon circulation [12, 13].

More recent works include the use of interactive aerosols in the models, that is, the dynamical feedback to aerosol distribution and forcing. For example, the model used in [50] includes a size- and mixing state-dependent aerosol module that is fully coupled with the climate model. Certain sophisticated aerosol microphysical and chemical processes including aging and coating of carbonaceous aerosols with sulfate, along with optical properties of these mixed aerosols are also included. Using this interactive aerosol-climate model coupled with a mixed-layer ocean model, Wang et al. proposed another possible mechanism that absorbing aerosols could affect on Indian summer monsoon [50]. The researchers find that absorbing anthropogenic aerosols, whether coexisting with scattering aerosols or not, can significantly affect the Indian summer monsoon system. This is drawn from a comparison of the results of three simulations. The first two simulations each only included absorbing aerosols (ABS) and scattering aerosols (SCA), respectively; the third one included both types of aerosols (COM). Aerosol-induced climate responses in each of these runs were derived by comparing results to a reference run that excluded the aerosol effect (REF). The similarity in aerosol-induced response between absorbing-aerosol-only case (ABS) and the case with both types of aerosols (COM) was identified. Results of both cases also differ sharply from that of the scattering-aerosol-only case (SCA). The researchers further identified that the influence of absorbing aerosols is reflected in a perturbation to the moist static energy in the sub-cloud layer, initiated as a heating by absorbing aerosols in the planetary boundary layer (Fig. 12.3). The perturbation appears mostly over land, extending from just north of the Arabian Sea to northern India along the southern slope of the Tibetan Plateau. As a result, during the summer monsoon season, modeled convective precipitation experiences a clear northward shift, coincidently in general agreement with observed monsoon precipitation changes in recent decades particularly during the onset season (Fig. 12.4). According to previous works, the northward extent of monsoon convection should collocate with the maximum sub-cloud layer MSE [12–14]. Therefore, a small perturbation in such a location could lead to an observable change in distribution of convection and heavy precipitation. Interestingly, the modeled largest perturbation of absorbing aerosols on sub-cloud layer MSE appeared across this zone. Compared to the forcing required to significantly lower the meridional temperature gradient, the forcing of absorbing aerosols through perturbing MSE to influence monsoon dynamics and precipitation distribution is much more effective. The importance of sub-cloud layer processes, however, does not necessarily preclude the EHP that emphasizes the heating above the boundary layer. It is likely that the heating of the

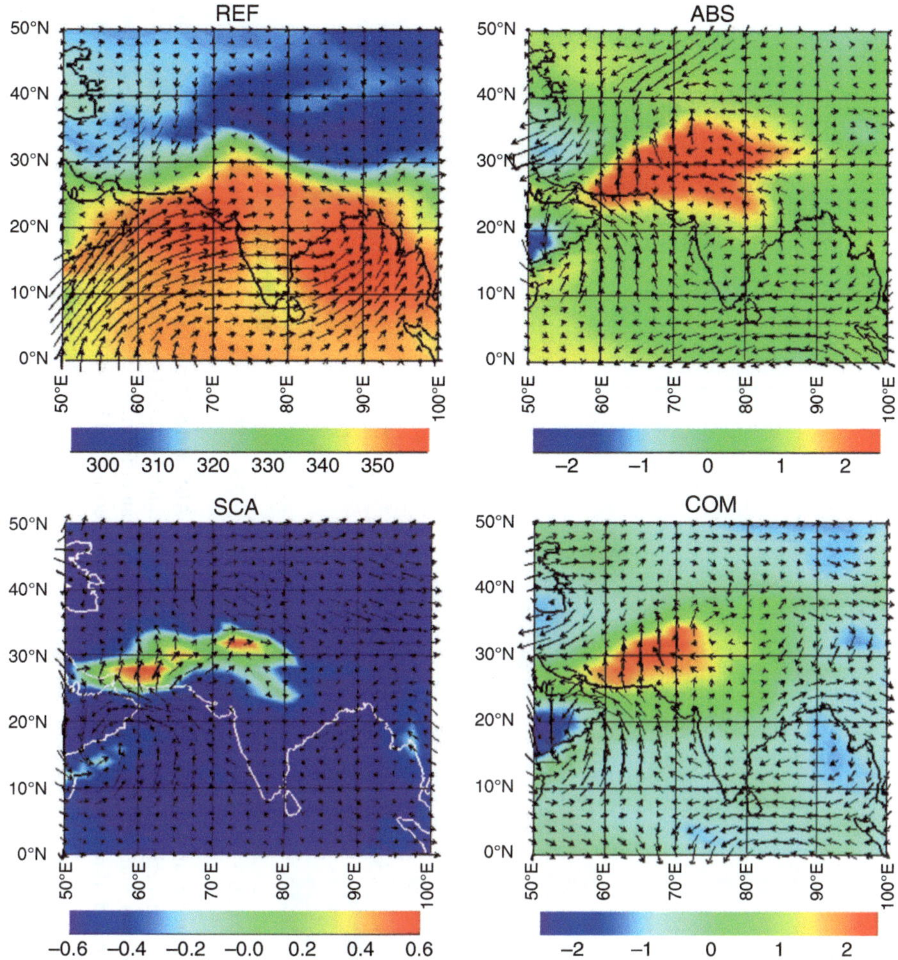

Fig. 12.3 May–June (MJ) mean of wind and moist static energy in the reference run (REF), which excludes the aerosol radiative effect, and anomalies of MJ mean wind and moist static energy derived from three model runs (ABS, SCA, and COM). Data shown are averaged values for the lowermost three atmospheric layers based on year 41–60 means. Unit wind vector = 1 m/s. Moist static energy is in 10^3 J/kg. Note that in the three anomaly plots, a different color scale is used in SCA. A terrain correction has been applied to the REF result (From [50], Copyright 2009 American Geophysical Union. Reproduced/modified by permission of American Geophysical Union)

entire atmospheric column from the boundary layer to the upper troposphere could be important in creating the northward shift of the monsoon rainbelt.

At present, there is still a range of opinions about the reasons behind the impact of aerosols on Indian summer monsoon [42]. Detailed mechanisms of the aforementioned hypothetical impacts still remain to be examined.

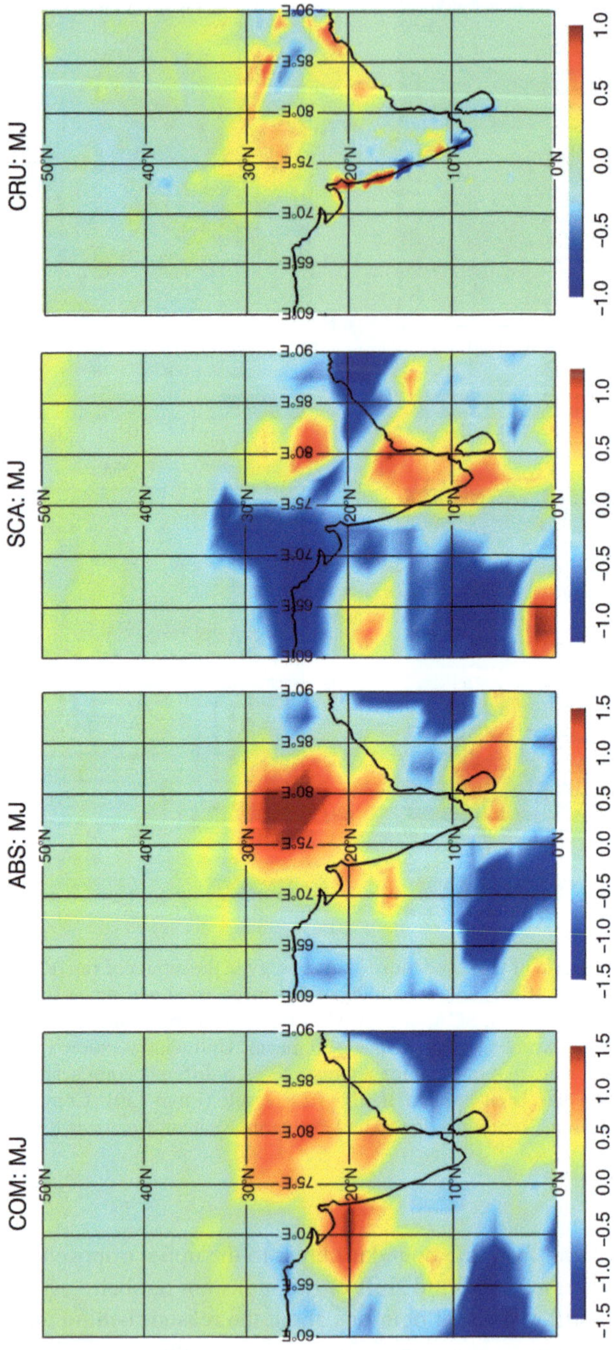

Fig. 12.4 May–June average changes in convective precipitation (dm/season) derived from: COM, ABS, and SCA run for India and surrounding regions, and the observed precipitation change (land-only; dm/season) derived from the data of the Climate Research Unit (CRU) at the University of East Anglia. Model results shown are based on year 41–60 mean differences with REF run. CRU results are derived from differences between 20-year means of 1981–2000 and 1946–1965, and based on the version 2.1 dataset with 0.5° (From [50], Copyright 2009 American Geophysical Union. Reproduced/modified by permission of American Geophysical Union)

Impacts of Aerosols on the West African Monsoon

Aerosol impacts on another monsoon system, the West African Monsoon (WAM) have also been studied. It is known that aerosols over this region are among the most abundant and persistent on the Earth with distinct seasonal variability. The dominant aerosol type is mineral dust from North Africa through May to August and biomass-burning smoke from southern Africa from July to September. The mixture of dust and biomass-burning smoke appear in November–February due to persistent yearlong dust emission from some North African sources and biomass burning in Sahel region [51–56]. Therefore, this region provides an ideal natural test bed for studying aerosol effects on precipitation. Analyses using satellite data have demonstrated that a high concentration of aerosols can induce a significant precipitation reduction in the WAM region along the coast of the Gulf of Guinea, particularly in the boreal late autumn and winter [57, 58]. A recent study [59] further compared the observational results to a global model simulation including only direct radiative forcing of black carbon [39]. It was found from both observations and model simulations that in boreal cold seasons anomalously high African aerosols are associated with significant reductions in cloud amount, cloud top height, and surface precipitation. This result suggests that the observed precipitation reduction in the WAM region is caused by radiative effect of absorbing BC. The mechanism for this reduction, however, remains to be revealed.

In connection to the hypothesis of aerosol-Indian summer monsoon effect proposed by Wang et al. [50], Eltahir and Gong found a correlation of the strength of the West African monsoon to subtropical meridional gradient of sub-cloud MSE [60]. Therefore, similar mechanism could also exist in aerosol-WAM effect.

Recently, Lau et al. showed from GCM experiment that the EHP effect by Saharan dusts and biomass-burning black carbon has a significant impact on the climate and water cycle of the North Atlantic and WAM [61]. They found that during the boreal summer, as a result of large-scale atmospheric feedback triggered by absorbing aerosols, rainfall and cloudiness are enhanced over the West Africa/Eastern Atlantic ITCZ while suppressed over the West Atlantic and Caribbean region. As shown in Fig. 12.5, the elevated dust layer warms the air over West Africa and the eastern Atlantic. As the warm air rises, it spawns an anomalous large-scale onshore flow carrying the moist air from the eastern Atlantic and the Gulf of Guinea. The onshore flow in turn enhances the deep convection over West Africa land, and the eastern Atlantic. The condensation heating associated with the ensuing deep convection drives and maintains an anomalous large-scale east–west overturning circulation, with rising motion over West Africa/eastern Atlantic and sinking motion over the Caribbean region. The response reflects a strengthening of the West African monsoon, manifested in a northward shift of the West Africa precipitation over land, increased low-level westerly flow over West Africa at the southern edge of the dust layer, and a near surface westerly jet underneath the dust layer over the Sahara. The dust radiative forcing also leads to significant changes in surface energy fluxes, resulting in cooling of the West African land and the eastern

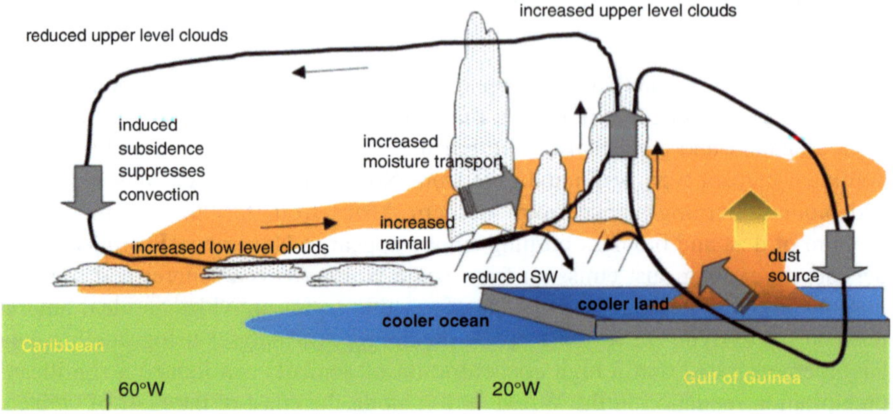

Fig. 12.5 Schematic diagram showing key features in the latitude domain 5–15°N, associated with the "elevated heat pump" mechanism by radiative heating of Saharan dust: anomalous Walker-type and Hadley-type circulations, increased moisture transport from the eastern Atlantic and the Gulf of Guinea to West Africa; enhanced rainfall over the Sahel and the ITCZ off the cost of West Africa; subsidence and suppressed cloudiness in the central Atlantic and Caribbean, and the Gulf of Guinea; cooling of the WAM land, and the upper ocean in the eastern Pacific underneath the dust plume (Adopted from [61] © Author(s) 2009, distributed under the Creative Commons Attribution 3.0 License)

Atlantic, and warming in the West Atlantic and Caribbean. The EHP effect is most effective for moderate to highly absorbing dusts, and becomes minimized for reflecting dust with single scattering albedo at 0.95 or higher.

Additionally, from the same experiments the authors found strong modulation of the diurnal cycle and a northward shift of the African easterly jet, in conjunction with increased cyclonic vorticity to the south of its axis, and increased rainfall in the Sahel [62]. These modeling results are consistent with recent observations [63] showing that during the periods of a strong Sahara dust outbreak, the Atlantic ITCZ tends to be shifted northward of its climatological position, accompanied by a similar shift of the Africa easterly jet.

Impacts of Aerosols on the Australian Monsoon

The anthropogenic aerosol level in the Southern Hemisphere is lower compared to the condition of Northern Hemisphere. Therefore, the impact of local aerosols on Australian monsoon system is expected to be insignificant. However, since monsoon systems are closely associated with large-scale circulation, aerosol effects in the Northern Hemisphere could influence southern hemispheric circulation and thus precipitation by altering the general circulation patterns. This has been suggested in a recent modeling study [64]. Drawn from the results of a pair of ensemble simulations conducted using a coupled atmosphere-ocean climate model respectively including and excluding Asian aerosols, the researchers hypothesized that Asian aerosols could lead to an increase in both rainfall and cloudiness particularly over northwest Australia, which coincides with observed rainfall trend in the region since 1950s. The study suggested that this effect could be implemented through an altered latitudinal gradient of temperature (and thus of pressure) over tropical Indian Ocean by Asian aerosols, which would further enhance monsoonal circulation toward Australia. A recent analysis of the twentieth-century modeling results of 24 CMIP3 models, all including either only direct or both direct and indirect aerosol forcing, however, cannot provide support to the above hypothesis [65]. Despite of the inclusion of aerosol effect in these models, their ensembles did not produce the hypothesized rainfall increase in northwest Australia.

Climate Change and Monsoon System

Analyses using oxygen isotope data from Chinese caves providing information about monsoons over millennia suggest that Asian monsoons are influenced by changes in summer insolation in the Northern Hemisphere [66, 67]. However, variability in shorter terms can also be influenced by other factors. Historically, such variation in Asian monsoon system might have triggered social unrest and thus played a key role in causing demise of several Chinese dynasties [68]. Similarly, the persistent drought in Sahel occurred later last century also led to serious food supply problems and could well be responsible for certain conflicts in Africa. It is thus critical to understand these variabilities of monsoon systems and to identify the natural and anthropogenic influences on such variations.

Several persistent trends have been revealed recently. The reconstructed monsoon winds for the past 1,000 years using fossil *Globigerina bulloides* abundance in box cores from the Arabian Sea suggested an increase in strength during the past four centuries while the Northern Hemisphere has been warming [69]. This implies that the Indian summer monsoon strength could be enhanced during the coming century as greenhouse gas concentrations continue to rise and northern latitudes continue to warm. In the most recent decade, the sea surface winds over the western Arabian Sea have been continually strengthening [70]. Such escalation of summer monsoon winds, accompanied by enhanced upwelling, leads to an increase of more than 350% in average summertime phytoplankton biomass along the coast and over 300% offshore, implying that the current warming trend of the Eurasian landmass is making the Arabian Sea more productive.

Over India, analyses based on rainfall data since early 1950 suggest that in the last half century, the frequencies of moderate and low rain days over the entire country have significantly decreased while the frequency and the magnitude of extreme rain events has significantly increased [38, 71]. Decreasing trends were also found in both early and late monsoon rainfall and number of rainy days, implying a shorter monsoon over India [37]. There is also a sharp decrease in the area that receives a certain amount of rainfall and number of rainy days during the season. One study found that the seasonal mean all-India rainfall did not show a significant trend since 1950s [38]. The researchers argued that this is because that the contribution from increasing heavy events is offset by decreasing moderate events. Apparently, should the trend continue, a substantial increase in hazards related to heavy rain would be expected over central India in the future. In another recent study [72], the authors stressed the need to subdivide Indian rainfall geographically and to distinguish early and peak monsoon seasons for the purpose of rainfall trend detection and attribution. They found fingerprints of absorbing aerosols impact on regional rainfall since 1960s featuring increased rainfall in north and northwestern India in May–June and decreased rainfall in central and southern India in July–August since 1960s. However, whether the observed monsoon systems have strengthened or reduced is still very much an open question, limited by the availability of reliable long-term data record. Based on the dramatic decline in ENSO-monsoon correlation in recent decades, it has been suggested that warming over Eurasia continent might have already led to a favored condition for strong monsoons [9]. Besides the potential cause of global warming behind these trends, anthropogenic aerosols could also be a significant factor based on observation-based and modeling studies discussed in previous sections.

Besides anthropogenic influences, natural variability could also be responsible for decadal to centennial variability of the monsoon. Natural variability in the African monsoon over the past three millennia has been reconstructed using geochemical evidence from the sediments of Lake Bosumtwi, Ghana [73]. It was found that intervals of severe drought lasting for periods ranging from decades to centuries are characteristic of the monsoon and are linked to natural variations in Atlantic sea surface temperatures. The researchers thus believe that the severe drought of recent decades is not anomalous in the context of the past three millennia.

The use of climate models to understand causes and consequences of changes in the past and future monsoon climate system is necessary when complex relationships are under consideration. Patricola and Cook used a regional climate model to study the West Africa monsoon in the African Humid Period (AHP; about 14,800–5,500 years ago) when humidity was increased over Africa based on paleoclimate evidence suggesting that the West African summer monsoon was stronger than today, and the Saharan Desert was green [74]. The model was driven by prescribed changes in insolation, atmospheric CO_2, and vegetation to impose conditions at 6,000 years before present, with SSTs fixed at present-day values. The model simulation produced a precipitation increase across the Sahel and Sahara that is in good agreement with the paleoclimate data. They found the precipitation increase in the Sahel is related to a northward shift of monsoon, the elimination of the African easterly jet, and intensification and deepening of the low-level westerly jet on the west coast. Interestingly, the thermal low-Saharan high system of the present-day climate is replaced by a deep thermal low. Even though solar forcing is the ultimate cause of the AHP, the model responded more strongly to the vegetation forcing, emphasizing the importance of vegetation in maintaining the intensified monsoon system.

Takata et al. carried out a pair of climate model simulations using according land use estimates over China and India in 1700 and 1850 [75]. The comparison between these two runs isolates the climate responses to the two different land use estimations. It was found that land use change over China and India from 1700 to 1850 due to population growth (forest to cropland) led to a reduction of surface roughness and thus a weakening of monsoon circulation and precipitation in India.

Various groups have also studied the influence of projected future climate warming on monsoon evolution. Coupled atmosphere-ocean GCM simulations suggested that the increase of surface temperature due to a doubling of CO_2 concentration could enhance mean precipitation of Indian summer monsoon and, as a partial consequence, interannual variability of area-averaged monsoon rainfall [76]. This is believed to be consistent with the observed large variability associated with warm surface temperature.

The performance of the current GCMs in retrospective modeling of monsoon evolution raises issues on both the capability of these models in projecting monsoon in future climate and certain variabilities used to evaluate the models (e.g., the ENSO-Indian summer monsoon rainfall relation). In a recent study, four out of 18 models participated in IPCC AR4 were selected based on their performance of the twentieth-century modeling of monsoon evolution [31]. An analysis was done then for each of these four models using their results from integrations in which the atmospheric CO_2 concentration doubled over preindustrial values. These selected models in the double CO_2 simulations all projected an increase both in the mean monsoon rainfall over the Indian subcontinent (by 5–25%) and in its interannual variability (5–10%). For each model the ENSO-monsoon correlation in the global warming runs is very similar to that in the twentieth-century runs, suggesting that the ENSO-monsoon connection will not weaken as global climate warms. This result is, however, curiously inconsistent with the finding in [9]. In addition, the diversity as seen in the simulations of ENSO variability of these coupled models suggests that these results should be taken with caution.

In a similar study to explore model performance in simulating twentieth-century WAM climatology, Cook and Vizy selected three best-performing models out of 18 coupled GCMs participated in IPCC AR4 to analyze their twenty-first-century integrations under various assumptions about future greenhouse gas increase [22]. Interestingly, each of these three models behaved differently in the twenty-first-century simulations. Only one model projected wet Guinea coast and more frequent dry year in Sahel that are consistent with predicted warming in the Gulf of Guinea based on known dynamic mechanism of the precipitation dipole. The authors thus concluded that there is no consensus among the models concerning the future of the West African monsoon system under greenhouse gas forcing. In another study based on the results from the CMIP3 models, a similar conclusion was obtained that the outlook for Sahel precipitation in these simulations of the twenty-first century is very uncertain, with different models disagreeing even on the sign of the trends [77]. It is especially surprising because most of these models in the twentieth-century integration reproduced the links of Sahel rainfall anomalies to tropical SST anomalies at interannual time scales as shown in observations. Conversely, such a relationship does not explain the rainfall trend in the twenty-first century in a majority of the models.

Future Directions

A key component that limits modeling of monsoon system is observation. The availability of high-resolution surface precipitation data, based on satellite retrievals and surface rain gauge measurements in recent 2 decades, has generally improved the situation. To analyze the longer-term variability, results derived from such dataset need to be compared more carefully with that from the high-density local meteorological measurements. For certain regions such as India, the latter type of data covers more than half century.

Computational technology is advancing rapidly, providing opportunity to model monsoons at higher resolution. However, the speed or memory gain from hardware or software advancements needs to be harnessed not only to increase model resolution but also for improved treatments of physical, chemical, and biogeophysical processes. For modeling the monsoon systems, previous experience suggests that one might need both. One issue that remains a huge challenge in the field of global climate modeling ever since its earliest day is the parameterization of convection. Because the requirement of a-few-kilometer resolution to resolve convection has been a far stretch for GCMs (and will likely still be the case in the near future), description of convection processes for the monsoon system in these models were empirically formulated using parameters resolved in model grid scale. Though still expensive, today's GCMs are already being run in horizontal resolution much closer to the cloud-resolving scale in exploratory and short tests, and perhaps will reach that scale earlier than one would expect. An immediate issue that would come along with this advancement, however, is the realization that variability of monsoon features will also increase at such a high resolution. Clearly, an accurate characterization of monsoon features at the cloud scale exceeds all the current available measurement networks, providing a challenge in the constraint of the high-resolution global climate models by observations.

Instead of using a global high-resolution climate model, the monsoon system can be also simulated in high resolution by coupling the global model with a regional climate model. The latter model usually has a more realistic description of various physical and chemical processes. When needed, running the regional rather than global climate in cloud-resolving scale would greatly reduce the demand for computation. There are numerous attempts reported in literature of so-called downscaling modeling, mostly done by driving a regional model using the output from a global model with a given increment of time (e.g., 6 h). The two-way coupling of such approach, that is, to include certain feedbacks of regional processes to influence the global model, however, is still rare. For modeling the monsoon system, the two-way coupling would provide a better description in the global model of atmospheric feedback to external forcing such as SST anomalies, presuming that the coupled atmosphere and ocean GCMs will remain dominant type of models to cover the global scale.

Studies in recent years have demonstrated the potential influence of aerosols on monsoon circulation and precipitation. Models need to be improved to include more physical- and chemical-based aerosol descriptions, for example, the mixing of anthropogenic aerosol species with dust by the chemical and physical evolution of those particles. The treatment of aerosol–cloud interactions (the so-called indirect effect) is currently quite crude and dependence of these interactions on cloud dynamics is also very crude. Both sets of processes need to be improved in next generation models in order to explore their effects on monsoons. The role of absorbing aerosols revealed in recent studies and limited by measurements of aerosol absorption strength and the distribution of absorbing aerosols also remains a challenge and needs to be much improved.

The onset and evolution of the monsoon system can be influenced by both natural and anthropogenic factors, which are often intertwined both in space and time. Future modeling study should work toward unraveling these two major impacts through better simulation design and advanced statistical analysis. Separating anthropogenic impact from natural variability is essential in narrowing uncertainties in projecting future changes of global climate including monsoon system.

Bibliography

Primary Literature

1. Galvin JFP (2008) The weather and climate of the tropics: part 6 – monsoons. Weather 63:129–137
2. Halley E (1686) A historical account of the trade winds and the monsoon, observable in the seas between and near the tropicks, with an attempt to assign the physical cause of the said winds. Phil Trans R Soc London 16:153–168
3. Webster PJ (1987) The elementary monsoon. In: Fein JS, Stephens PL (eds) Monsoon. Wiley, New York, pp 3–32
4. Krishnamurti TN (1987) Monsoon models. In: Fein JS, Stephens PL (eds) Monsoon. Wiley, New York, pp 467–522
5. Plumb RA (2007) Dynamical constraints on monsoon circulations. In: Schneider T, Sobel AH (eds) The global circulation of the atmosphere. Princeton University Press, Princeton, pp 252–266
6. Blanford HF (1884) On the connexion of the Himalaya snowfall with dry winds and seasons of drought in India. Proc R Soc London 37:3–22
7. Hahn DG, Manabe S (1975) The role of mountains in the South Asian monsoon circulation. J Atmos Sci 32:1515–1541

8. Boos WR, Kuang Z (2009) Dominant control of the South Asian monsoon by orographic insulation versus plateau heating. Nature 463:218–223
9. Kumar KK, Rajagopalan B, Cane MA (1999) On the weakening relationship between the Indian monsoon and ENSO. Science 284:2156–2159
10. Chao WC (2000) Multiple quasi equilibria of the ITCZ and the origin of monsoon onset. J Atmos Sci 57:641–651
11. Chao WC, Chen B (2001) The origin of monsoons. J Atmos Sci 58:3497–3507
12. Prive NC, Plumb RA (2007a) Monsoon dynamics with interactive forcing. Part I: axisymmetric studies. J Atmos Sci 64:1417–1430
13. Prive NC, Plumb RA (2007b) Monsoon dynamics with interactive forcing. Part II: impact of eddies and asymmetric geometries. J Atmos Sci 64:1431–1442
14. Emanuel KA, Neelin JD, Bretherton CS (1994) On large-scale circulations in convecting atmospheres. Quart J R Meteor Soc 120:1111–1143
15. Bordoni B, Schneider T (2008) Monsoons as eddy-mediated regime transitions of the tropical overturning circulation. Nature Geos 1:515–519
16. Wang B, Ding Q, Fu X, Kang I-S, Jin K, Shukla J, Doblas-Reyes F (2005) Fundamental challenge in simulation and prediction of summer monsoon rainfall. Geophys Res Lett 32: L15711. doi:10.1029/2005GL022734
17. Charney JG, Shukla J (1981) Predictability of monsoons. In: Lighthill J, Pearce RP (eds) Monsoon dynamics. Cambridge University Press, New York, pp 99–109
18. Lau KM, Walsier D (2005) Intraseasonal variability in the atmosphere-ocean climate system. Springer Praxis, Chichester, 289 pp
19. Manabe S, Hahn DG, Holloway JL Jr (1974) The seasonal variation of the tropical circulation as simulated by a global model of the atmosphere. J Atmos Sci 31:43–83
20. Gates WL et al (1999) An overview of the results of the atmospheric model intercomparison project (AMIP I). Bull Am Meteorol Soc 80:29–56
21. Shen B-W, Tao W-K, Lau WK, Atlas R (2010) Improving tropical cyclogenesis prediction with a global mesoscale model: hierarchical multiscale interactions during the formation of tropical cyclone nargis (2008). J Geophys Res 115:D14102. doi:10.1029/2009JD013140
22. Cook KH, Vizy EK (2006) Coupled model simulations of the West African monsoon system: twentieth- and twenty-first-century simulations. J Clim 19:3681–3703
23. Krishnamurthy V, Shukla J (2007) Intraseasonal and seasonally persisting patterns of Indian monsoon rainfall. J Clim 20:3–20
24. Shukla J (1975) Effect of Arabian sea-surface temperature anomaly on Indian summer monsoon: a numerical experiment with the GFDL model. J Atmos Sci 32:503–511
25. Lau K-M, Yang S (1996) Seasonal variation, abrupt transition, and intraseasonal variability associated with the Asian summer monsoon in the GLA GCM. J Clim 9:965–985
26. Meehl GA, Arblaster JM, Lawrence D, Seth A, Schneider EK, Kirtman BP, Min D (2006) Monsoon regimes in the CCSM3. J Clim 19:2482–2495
27. Christensen JH, Hewitson B, Busuioc A, Chen A, Gao X, Held I, Jones R, Kolli RK, Kwon W-T, Laprise R, Magaña Rueda V, Mearns L, Menéndez CG, Räisänen J, Rinke A, Sarr A, Whetton P (2007) Regional climate projections, in climate change 2007: the physical science basis. In: Solomon S, Qin D, Manning M, Chen Z, Marquis M, Averyt KB, Tignor M, Miller HL (eds) Contribution of working group I to the fourth assessment report of the intergovernmental panel on climate change. Cambridge University Press, New York
28. Meehl GA (1994) Coupled land-ocean-atmosphere processes and South Asian monsoon variability. Science 266:263–267
29. Meehl GA, Arblaster JM (2002) Indian monsoon GCM sensitivity experiments testing tropospheric biennial oscillation transition conditions. J Clim 15:923–944
30. Shukla J (2007) Monsoon mysteries. Science 318:204–205
31. Annamalai H, Hamilton K, Sperber KR (2007) The South Asian summer monsoon and its relationship with ENSO in the IPCC AR4 simulations. J Clim 20:1071–1092

32. Xue YK et al. (2010) Intercomparision and analyses of the West African monsoon in the West African monsoon model evaluation (WAMME) project: first model intercomparison experiment. Climate Dynamics 35:3–27. doi:10.1007/s00382-010-0778-2
33. Giannini A, Saravanan R, Chang P (2003) Oceanic forcing of Sahel rainfall on interannual to interdecadal time scales. Science 302:1027–1030
34. Lau KM, Kim MK, Kim KM (2006) Asian monsoon anomalies induced by aerosol direct effects. Clim Dyn 26:855–864. doi:10.1007/s00382-006-0114-z
35. Lu R, Fu Y (2010) Intensification of East Asian summer rainfall interannual variability in the twenty-first century simulated by 12 CMIP3 coupled models. J Clim 23:3316–3331
36. Ramanathan V, Chung C, Kim D, Bettge T, Buja L, Kiehl JT, Washington WM, Fu Q, Sikka DR, Wild M (2005) Atmospheric brown clouds: impact on South Asian climate and hydrologic cycle. Proc Natl Acad Sci USA 102:5326–5333
37. Ramesh KV, Goswami P (2007) Reduction in temporal and spatial extent of the Indian summer monsoon. Geophys Res Lett 34:L23704. doi:10.1029/2007GL031613
38. Goswami BN, Venugopal V, Sengupta D, Madhusoodanan MS, Xavier PK (2006) Increasing trend of extreme rain events over India in a warming environment. Science 314:1442–1445
39. Wang C (2004) A modeling study on the climate impacts of black carbon aerosols. J Geophys Res 109:D03106. doi:10.1029/2003JD004084
40. Roberts DL, Jones A (2004) Climate sensitivity to black carbon aerosol from fossil fuel combustion. J Geophys Res 109:D16202. doi:10.1029/2004JD004676
41. Chung SH, Seinfeld JH (2005) Climate response of direct radiative forcing of anthropogenic black carbon. J Geophys Res 110:D11102. doi:10.1029/2004JD005441
42. Lau K-M, Ramanathan V, Wu G-X, Li Z, Tray SC, Hsu C, Sikka R, Holben B, Lu D, Tartari G, Chin M, Koudelova P, Chen H, Ma Y, Huang J, Taniguchi K, Zhang R (2008) The joint aerosol-monsoon experiment, a new challenge for monsoon climate research. Bull Am Meteorol Soc 89:369–383
43. Menon S, Hansen J, Nazarenko L, Luo Y (2002) Climate effects of black carbon aerosols in China and India. Science 297:2250–2253
44. Wang C (2007) Impact of direct radiative forcing of black carbon aerosols on tropical convective precipitation. Geophys Res Lett 34:L05709. doi:10.1029/2006GL028416
45. Randles CA, Ramaswamy V (2008) Absorbing aerosols over Asia: a geophysical fluid dynamics laboratory general circulation model sensitivity study of model response to aerosol optical depth and aerosol absorption. J Geophys Res 113:D21203. doi:10.1029/2008JD010140
46. Meehl GA, Arblaster JM, Collins WD (2008) Effects of black carbon aerosols on the Indian monsoon. J Clim 21:2869–2882. doi:10.1175/2007JCLI1777.1
47. Lau K-M, Kim K-M (2006) Observational relationships between aerosol and Asian monsoon rainfall, and circulation. Geophys Res Lett 33:L21810. doi:10.1029/2006GL027546
48. Bollasina M, Nigam S, Lau K-M (2008) Absorbing aerosols and summer monsoon evolution over South Asia: an observational portrayal. J Clim 21:3221–3239
49. Gautam R, Hsu NC, Lau K-M, Tsay S-C, Kafatos M (2009) Enhanced pre-monsoon warming over the Himalayan-Gangetic region from 1979 to 2007. Geophys Res Lett 36:L07704. doi:10.1029/2009GL037641
50. Wang C, Kim D, Ekman AML, Barth MC, Rasch PJ (2009) Impact of anthropogenic aerosols on Indian summer monsoon. Geophys Res Lett 36:L21704. doi:10.1029/2009GL040114
51. Carlson TN, Prospero JM (1972) The large-scale movement of Saharan air outbreaks over the Northern equatorial Atlantic. J Appl Meteor 11:283–297
52. Prospero JM, Lamb JP (2003) African droughts and dust transport to the Caribbean: climate change and implications. Science 302:1024–1027
53. Dwyer E, Pinnock S, Gregoire JM, Pereira JMC (2000) Global spatial and temporal distribution of vegetation fire as determined from satellite observations. Int J Rem Sens 21:1289–1302
54. Duncan BN, Martin RV, Staudt AC, Yevich R, Logan JA (2003) Interannual and seasonal variability of biomass burning emissions constrained by satellite observations. J Geophys Res 108:4100. doi:10.1029/2002JD002378

55. Ito A, Penner JE (2005) Historical emissions of carbonaceous aerosols from biomass and fossil fuel burning for the period 1870–2000. Global Biogeochem Cy 19:GB2028. doi:10.1029/2004GB002374
56. Huang J, Zhang C, Prospero JM (2009) African dust outbreaks: a satellite perspective of temporal and spatial variability over the tropical Atlantic ocean. J Geophys Res 115:D05202. doi:10.1029/2009JD012516
57. Huang J, Zhang C, Prospero JM (2009) Large-scale effects of aerosols on precipitation in the West African monsoon region. Q J R Meteorol Soc 135:581–594
58. Huang J, Zhang C, Prospero JM (2009) African aerosol and large-scale precipitation variability over West Africa. Environ Res Lett 4:015006. doi:10.1088/1748-9326/4/1/015006
59. Huang J, Adams A, Wang C, Zhang C (2009) Black carbon and West African monsoon precipitation: observations and simulations. Ann Geophys 27:4171–4181
60. Eltahir E, Gong C (1996) Dynamics of wet and dry years in West Africa. J Climate 9:1030–1042
61. Lau KM, Kim KM, Sud YC, Walker GK (2009) A GCM study of the response of the atmospheric water cycle of West Africa and the Atlantic to Saharan dust radiative forcing. Ann Geophys 27:4023–4037, http://www.ann-geophys.net/27/4023/2009/
62. Kim KM, Lau KM, Sud YC, Walker GK (2010) Influence of aerosol-radiative forcing on the diurnal and seasonal cycles of rainfall over West Africa and the Eastern Atlantic using GCM simulations. Clim Dyn. doi:10.1007/s00382-010-0750-1
63. Wilcox E, Lau WKM, Kim KM (2010) A northward shift of the Inter-tropical convergence zone in response to summertime Saharan dust outbreak. Geophys Res Lett 37:L04804. doi:10.1029/2009GL041774
64. Rotstayn LD, Cai W, Dix MR, Farquhar GD, Feng Y, Ginoux P, Herzog M, Ito A, Penner JE, Roderick ML, Wang M (2007) Have Australian rainfall and cloudiness increased due to the remote effects of Asian anthropogenic aerosols? J Geophys Res 112:D09202. doi:10.1029/2006JD007712
65. Cai W, Cowan T, Sullivan A, Ribbe J, Shi G (2011) Are anthropogenic aerosols responsible for the Northwest Australia summer rainfall increase? A CMIP3 perspective and implications. J Climate 24:2556–2564. doi:10.1175/2010JCLI3832.1
66. Wang Y, Cheng H, Edwards RL, He Y, Kong X, An Z, Wu J, Kelly MJ, Dykoski CA, Li X (2005) The Holocene Asian monsoon: links to solar changes and North Atlantic climate. Science 308:854–857
67. Wang Y, Cheng H, Edwards RL, Kong X, Shao X, Chen S, Wu J, Jiang X, Wang X, An Z (2008) Millennial- and orbital-scale changes in the East Asian monsoon over the past 224,000 years. Nature 451:1090–1093
68. Zhang P, Cheng H, Edwards RL, Chen F, Wang Y, Yang X, Liu J, Tan M, Wang X, Liu J, An C, Dai Z, Zhou J, Zhang D, Jia J, Jin L, Johnson KR (2008) A test of climate, sun, and culture relationships from an 1810-year Chinese cave record. Science 322:940–942
69. Anderson DM, Overpeck JT, Gupta AK (2002) Increase in the Asian Southwest monsoon during the past four centuries. Science 297:596–599
70. Goes JI, Thoppil PG, Gomes HDR, Fasullo JT (2005) Warming of the Eurasian landmass is making the Arabian Sea more productive. Science 308:545–547
71. Dash SK, Kulkarni MA, Mohanty UC, Prasad K (2009) Changes in the characteristics of rain events in India. J Geophys Res 114:D10109. doi:10.1029/2008JD010572
72. Lau KM, Kim K-M (2010) Fingerprinting the impacts of aerosols on long-term trends of the Indian summer monsoon regional rainfall. Geophys Res Lett 37:L16705. doi:10.1029/2010GL043255
73. Shanahan TM, Overpeck JT, Anchukaitis KJ, Beck JW, Cole JE, Dettman DL, Peck JA, Scholz CA, King JW (2009) Atlantic forcing of persistent drought in West Africa. Science 324:377–380
74. Patricola CM, Cook KH (2007) Dynamics of the West African monsoon under mid-Holocene precessional forcing: regional climate model simulations. J Clim 20:694–716

75. Takata K, Saito K, Yasunari T (2009) Changes in the Asian monsoon climate during 1700–1850 induced by preindustrial cultivation. Proc Natl Acad Sci USA 106:9586–9589
76. Meehl GA, Washington WM (1993) South Asian summer monsoon variability in a model with doubled atmospheric carbon dioxide concentration. Science 260:1101–1104
77. Biasutti M, Held IM, Sobel AH, Giannini A (2008) SST forcings and Sahel rainfall variability in simulations of the twentieth and twenty-first centuries. J Clim 21:3471–3486

Books and Reviews

Fein JS, Stephens PL (1987) Monsoon. Wiley, New York

Lighthill J, Pearce RP (eds) (1981) Monsoon dynamics. Cambridge University Press, New York

Wang B (2006) East Asian monsoon. Springer/Praxis, Chichester

Webster PJ, Magana VO, Palme TN, Shukla J, Tomas RA, Yanai M, Yasunari T (1998) Monsoons: processes, predictability, and the prospects for prediction. J Geophys Res 103:14451–14510

Index

P.J. Rasch (ed.), *Climate Change Modeling Methodology: Selected Entries from the Encyclopedia of Sustainability Science and Technology*,
DOI 10.1007/978-1-4614-5767-1, © Springer Science+Business Media, New York 2012

MIX
Papier aus verantwortungsvollen Quellen
Paper from responsible sources
FSC® C105338

If you have any concerns about our products,
you can contact us on
ProductSafety@springernature.com

In case Publisher is established outside the EU,
the EU authorized representative is:
Springer Nature Customer Service Center GmbH
Europaplatz 3, 69115 Heidelberg, Germany

Printed by Libri Plureos GmbH
in Hamburg, Germany